Active Contours

Springer

London
Berlin
Heidelberg
New York
Barcelona
Budapest
Hong Kong
Milan
Paris
Santa Clara
Singapore
Tokyo

Andrew Blake and Michael Isard

Active Contours

The Application of Techniques from Graphics, Vision, Control Theory and Statistics to Visual Tracking of Shapes in Motion

 Springer

Andrew Blake and Michael Isard
Department of Engineering Science
University of Oxford
Parks Road
Oxford OX1 3PJ

Cover image created by Andrew Blake with the assistance of Josephine Sullivan

ISBN-13:978-1-4471-1557-1   e-ISBN-13:978-1-4471-1555-7
DOI: 10.1007/978-1-4471-1555-7
British Library Cataloguing in Publication Data
Blake, Andrew
    Active contours : the application of techniques from
    graphics, vision, control theory and statistics to visual
    tracking of shapes in motion
    1.Computer vision 2.Pattern recognition systems
    I.Title II.Isard, Michael
    006.4

Library of Congress Cataloging-in-Publication Data
A catalog record for this book is available from the Library of Congress

Softcover reprint of the hardcover 1st edition 1998

Printed and bound at Cambridge University Press, Cambridge, England
34/3830-543210 Printed on acid-free paper

Contents

Foreword

The field of computer vision has its sights set on nothing less than enabling computers to see. This monumental challenge has absorbed many creative minds over the course of more than three decades. A basic premise held by the computer vision community is that vision may be understood in precise computational terms, and that doing so raises the possibility of engineering camera-equipped computer systems with human-like perceptual abilities. Once envisioned only in science fiction, powerful machine vision systems are now more than ever poised to become science fact. This is due in part to the advent of increasingly potent microprocessors, as predicted by Moore's law, and in part to the slow but steady unraveling, on multiple scientific fronts, of the mystery that is visual perception in living systems.

Model-based vision is a major trend in the field that approaches computational problems attendant to vision using mathematical models. To see familiar objects as normal people evidently do with ease, computer vision systems must be able to analyze object shape and motion in real time. To this end, in the early 1980s, my colleagues and I introduced a family of mathematical models, known as "deformable models". The motivation was to formulate visual models that unify the representation of shape and motion by combining geometry and physics; in particular, free-form (spline) geometry and the dynamics of elastic curves, surfaces, and solids. We anticipated that deformable models would lead to vision systems capable of interpreting video sequences in terms of rigid and nonrigid objects moving before the camera. Perhaps the simplest deformable model, deformable contours confined to the plane, also known as "active contours" or "snakes", quickly gained popularity following early public demonstrations of these contours actively conforming to the shapes and tracking the motions of object boundaries in video sequences.

I have admired Andrew Blake and his work for many years. His contribution to computer vision is undeniable. A very readable author, Blake's book on *Visual Reconstruction* has become a classic in the field. It gives me great pleasure to see the concept of active contours developed to the remarkable degree evident in this, his latest book, which is authored with his talented student, Michael Isard.

In a characteristically no-nonsense, mathematically solid treatment, Blake and Isard take the subject of active contours to new heights of theoretical sophistication and practical application. The latter addresses the difficult task of visually tracking the motions of a variety of complex objects captured by a video camera feeding a frame-rate video digitizer. The impressive technical achievement of the book is a novel, probabilistic interpretation of active contours built on a geometric substrate that combines B-spline curve bases with shape spaces defined by global deformations. This combination leads to a new class of very fast and highly robust (non-Gaussian) active contour propagation algorithms. Another noteworthy achievement is the ability of these new tracking algorithms to learn the complex motions of specific objects through observation, thereby automatically tuning the tracker with greater selectivity to objects of interest, further enhancing its robustness.

This book defines the state-of-the-art of contour-based object tracking algorithms. It is required reading for anyone interested in computer vision theory and in the design of working computer vision systems.

Demetri Terzopoulos
November, 1997

Preface

In the seventies and eighties, interest in Computer Vision was concentrated on the development of general purpose seeing machines. There was wide agreement on research priorities, developing "bottom-up" computer algorithms that would organise the raw intensity values in images into a more compact form. The purpose of this was not just to compress the data but also to extract its salient features. Salient features could include corners, edges and surface fragments, to be used in identifying objects and deducing their positions. However, experience suggests strongly that general purpose vision is too difficult a goal for the time being.

If general purpose vision is abandoned, what alternative approach could be taken? One answer is that generality can be abated by introducing some "prior" knowledge — knowledge that is specific to the objects that the computer is expected to see. An extreme form of this approach is exemplified by automatic visual inspection machines of the kind used on factory assembly lines. In that context, it is known in advance precisely what objects are to be inspected — it is rare, after all, for potatoes streaming along a conveyor to give way, without notice, to a crop of spanners or chocolate bars. When computer hardware and software are specialised entirely to deal with one object, phenomenal performance can be obtained. A striking example is the "Niagara" machine (Sortex, UK Ltd) for sorting rice grains which "sees" 70,000 grains every second and almost literally spits out the rejects.

It is a commonly held view that it is hard to make progress in research by building such specialised machines because general principles are lost to engineering detail. That is a fair point but by no means, in our view, outlaws the use of prior knowledge about shape in computer vision research. Instead, we would argue, scientific principles for representing prior knowledge need to be developed. Then, when a new problem area is addressed, the principles can be applied to "compile" a new vision system as rapidly as possible. This includes such issues as how to represent classes of shapes that are defined loosely. Potatoes, for instance, might be characterised as roundish but with substantial size variations, with or without knobs. On the other hand, the class of human faces could be represented in terms of a common basic layout, but with considerable variation in the sizes and separations of features. Modelling classes of shapes, their variability and their motion is one of the principal themes of the book. The use of those models to help interpret moving images is the other central theme.

We have tried to present ideas about shape and motion in a way that will be readable not only by specialists, but also by those who are not regularly immersed in the ideas of machine vision. In particular we would hope that those with backgrounds in graphics or signal processing or neural computing would find the book a useful and accessible guide.

Acknowledgements

We have enjoyed and are grateful for discussions at various times and for proof-reading by Benedicte Bascle, Fiona Blake, Mike Brady, Roger Brockett, Roberto Cipolla, Rupert Curwen, Ron Daniel, Colin Davidson, Hugh Durrant-Whyte, Andrew Fitzgibbon, Robert Kaucic, Pete Lindsey, John Kent, John MacCormick, Elisa Martínez Marroquín, David Mumford, David Murray, Alison Noble, Ben North, Ian Reid, David Reynard, Brian Ripley, Jens Rittscher, Simon Rowe, Steve Smith, Lionel Tarassenko, Andrew Wildenberg, Alan Yuille and Andrew Zisserman. We would like to acknowledge support at various stages of the work from the EPSRC and the EU, and experimental assistance from Sarah Blake.

Chapter 1

Introduction

Psychologists of vision have delighted in various demonstrations in which prior knowledge helps with interpreting an image. Sometimes the effects are dramatic, to the point that the viewer can make no sense of the image at all until, when cued with a single word, the object pops out of the image. This idea of "priming" with prior knowledge is illustrated (light-heartedly) in figure 1.1. Priming in that example is

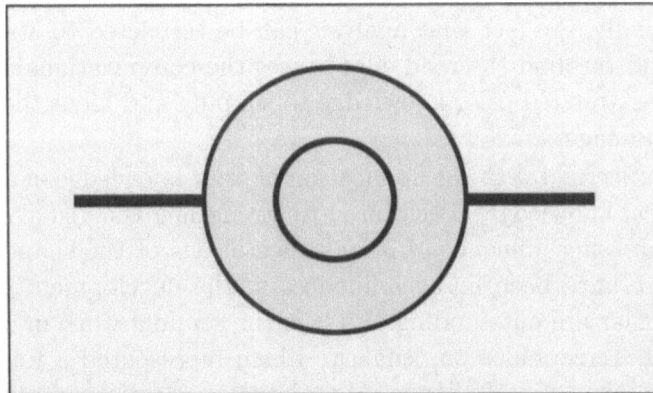

Figure 1.1: Priming with prior knowledge. *If you have never seen it before this figure probably means little at first sight. Now look for a cyclist in a Mexican hat.*

rather "high-level," calling on some intricate and diverse common-sense knowledge, concerning wheels, hats and so on. The aim of this book is to look at how prior knowledge can be applied in machine vision at the lower level of shapes and outlines.

The attraction of using prior knowledge in machine vision is simply that it is so hard to make progress without it, as a decade or more of research around the 1970s showed. There was considerable success in converting images into something like line drawings without resorting to any but the most general prior knowledge about smoothness and continuity. That led to the problem of "grouping" together the lines belonging to each object which is difficult in principle, and very demanding of computing power. One effective escape from this bind has been to design vision processes in a more goal-directed fashion and this is part of the philosophy of the notably successful "Active Vision" paradigm of the 1980s. Consider the task of examining visually the field of view immediately in front of a driverless vehicle, in order to steer automatically along the road. If the nature of the task is taken into account from the outset, it is quite unnecessary to examine an entire image; it is sufficient to focus on the expected appearance and position of the road edge at successive times. Deviations of actual from expected position can be treated as an error signal to control steering. This has two great advantages. First there is no need to organise or group features in the image; the relevant area of the image is simply tested directly against its expected appearance. Secondly, the fact that analysis can be restricted to a relatively narrow "region of interest" (around the road edge) eases the computational load. Active Vision, then, uses task-related prior knowledge to simplify and focus the processing that is applied to each image.

This book is concerned with the application of prior knowledge of a particular kind, namely geometrical knowledge. The aim is to strengthen the visual interpretation of shape via the stabilising influence of prior expectations of the shapes that are likely to be seen. There have been many influences in the development of this approach and two in particular are outstanding. First is the seminal work in 1987 of M. Kass, A. Witkin and D. Terzopoulos on "snakes" which represented a fundamentally new approach to visual analysis of shape. A snake is an elastic contour which is fitted to features detected in an image. The nature of its elastic energy draws it more or less strongly to certain preferred configurations, representing prior information about shape which is to be balanced with evidence from an image. If also inertia is attributed to a snake it acquires dynamic behaviour which can be used to apply prior knowledge of motion, not just of shape. Snakes are described in detail in the next chapter. The second outstanding influence is "Pattern Theory" founded by U. Grenander in the 70s and 80s and a popular basis for image interpretation in the statistical community. It puts the treatment of prior knowledge about shape into a probabilistic context by regarding any shape as the result of applying some distortion to an ideal prototype shape. The nature and extent of the distortion is governed by an appropriate probability distribution which then effectively defines the range of likely shapes.

Defining a prior distribution for shape is only part of the problem. The complete image interpretation task is to modify the prior distribution to take account of image features, arriving at a "posterior" distribution for what shape is actually likely to be present in a particular image. Mechanisms for fusing a prior distribution with "observations" are of crucial importance. Suffice it to say here that a key idea of pattern theory is "recognition by synthesis," in which predictions of likely shapes, based on the prior distribution, are tested against a particular image. Any discrepancy between what is predicted and what is actually observed can be used as an error signal, to correct the estimated shape. Fusion mechanisms of this general type exist in the snake, in the ubiquitous "Kalman filter" described in the next chapter, and in other more general forms described later in the book.

1.1   Organisation of the book

The organisation of material in the book is as follows. This chapter concludes by illustrating a range of applications and the next introduces active contour models. The book is then divided into two parts. Part I deals with the fundamentals of representing curves geometrically using splines, including basic machinery for least-squares approximation of spline functions, an essential topic not normally dealt with in graphics texts. Chapter 4 lays out a design methodology for linear, image-based, parametric models of shape, an important tool in applying shape constraints. Then algorithms for image processing and fitting splines to image features are introduced, leading to practical deformable templates in chapter 6. At this stage, a tool-set has been amassed sufficient for fitting curves to individual images, under a whole spectrum of prior assumptions, ranging from the least constrained snake to a two-dimensional rigid template. The treatment of part I aims to be thorough and complete, accessible by readers who are not necessarily familiar with the techniques of computer vision, given just a reasonable background in computing and vector algebra. (Appendix A reviews the necessary background in vectors and matrices, and gives some additional implementation details on spline curves.)

Part II introduces two new themes: models of motion and deformation, and probabilistic treatment of shape and motion. It begins (chapter 8) by reinterpreting the deformable templates of part I, in probabilistic terms. This is extended to dynamical models in chapter 9, as a preparation for fully probabilistic dynamical contour tracking, by Kalman filter, in chapter 10. By this stage, there are numerous parameters to be chosen to build a competent tracker and clear design guidelines are given on setting those parameters and on their intuitive physical interpretations. The most effective dynamical models derive, however, from learning procedures, as described in chapter 11, in which tracking performance improves automatically with experience. Finally, probabilistic modelling up to this point has been based on Gaussian distributions. Chapter 12 shows that for the hardest tracking problems, involving dense background clutter, non-Gaussian models are essential. They can be applied via random sampling algorithms, at increased computational cost, but to very considerable effect in terms of enhanced robustness.

As far as writing conventions go, references to books and papers have been kept out of the main text, to improve readability, and collected in separate bibliographic notes, appearing at the end of each chapter. These notes give sources for the ideas introduced in the body of the text and pointers to references on related ideas. Again for readability, mathematical derivations are kept from intruding on the main text by the use of two devices. The most important derivations are sandwiched (stealing a convention from Knuth's TEX manual) between

double-bend   and   **all-clear**

road signs in the margins. These are optional reading for those who want the mathematical details. Still more optional are the results and proofs in appendix B which support chapter 9 on dynamical models.

Web page

A web page for the book is at URL `http://www.robots.ox.ac.uk/~contours/` and contains MPEG sequences and additional material for those interested in exploring further the ideas discussed in the book.

1.2 Applications

A decade ago, it seemed unlikely that the research effort invested in Computer Vision would be harvested practically in the foreseeable future. Partly this reflected the lack of computational power of hardware available at the time, a limitation which has been greatly eased by the passing years. Partly though it was the result of an ambitious view of the problems of vision, in which the aim was to build a general purpose vision engine, rather than particular applications. More recently, that view has been rather overtaken by a more focused, algorithmically driven approach. The result is that Computer Vision ideas are working their way into a variety of practical applications, particularly in the areas of robotics, medical imaging and video technology.

The active contour approach is a prime candidate for practical exploitation. This is because active contours make effective use of specific prior information about objects and this makes them inherently efficient algorithms. Furthermore, active contours apply image processing selectively to regions of the image, rather than processing the entire image. This enhances efficiency further, allowing, in many cases, images to be processed at the full video rate of 50/60 Hz. Incidentally, the ability to do vision at real-time rate has an important spin-off in stiffening criteria of acceptability, amounting to a qualitative re-evaluation of standards. As an example, an algorithm that locates the outline of a mouth in a single image nine times out of ten might be considered quite successful. Let loose on a real-time image sequence of a talking mouth, this is re-interpreted as abject failure — the mouth is virtually certain to be "lost" within a second or so, and the loss is usually unrecoverable. The ability to follow the mouth while it speaks an entire paragraph, tracking through perhaps 1000 video frames is an altogether more stringent test.

Ten examples of applications follow. Earlier ones are already promising candidates for commercial application while later ones are more speculative.

Actor-driven facial animation

A deforming face is reliably tracked to relay information about the variation over time of expression and head position to a Computer Graphics animated face. The relayed expression can be reproduced or systematically exaggerated. Tracking can be accomplished in real time, keeping pace with rate of acquisition of video frames so the actor can be furnished with valuable visual feedback. Systems currently available commercially rely on markers affixed to the face. Visual contour tracking allows marker-free monitoring expression, given a modicum of make-up applied to the face, something to which actors are well accustomed. An example of real-time re-animation is illustrated for a cartoon cat in figure 1.2. This was done using two SGI INDY workstations, linked by network, one for visual tracking and one for mapping tracked motion onto the cat animation channels and display.

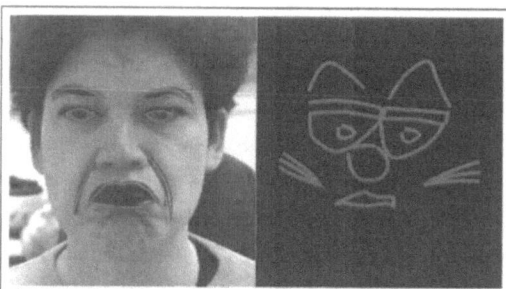

Figure 1.2: Actor-animated cat. *Tracked facial motions drive input channels to a cartoon cat, programmed with some exaggeration of expression. (Figure courtesy of Benedicte Bascle, Ben North and Julian Morris.)*

Traffic monitoring

Roadside video cameras are already familiar in systems for automated speed checks. Overhead cameras, sited on existing poles, can relay information about the state of traffic — its density and speed — and anomalies in traffic patterns. Contour tracking is particularly suited to this task because vehicle outlines form a tightly constrained class of shapes, undergoing predictable patterns of motion. Already the state of California has sponsored research leading to successful prototype systems.

Work in our laboratory, monitoring the motion of traffic along the M40 motorway near Oxford, is illustrated in figure 1.3. Vehicle velocity is estimated by recording

Figure 1.3: Traffic monitoring. *By automatically tracking cars, the emergency services can, for example, obtain rapid warning of an accident or traffic jam. (Illustration taken from (Ferrier et al., 1994).)*

the distance traversed by the base of a tracked vehicle contour over a known elapsed time. The measured distance is in image coordinates and this must be converted to world coordinates to give true distance. The mapping between coordinate systems is determined as a projective mapping between the image plane and the ground plane. The mapping is calibrated in standard fashion from the corners of a rectangle on the ground of known dimensions (known by reference to roadside markers which are standard fittings on British motorways), and the corresponding rectangle in the image plane, as in figure 1.4. Analysis of speeds shows clearly a typical pattern of UK motorway traffic with successively increasing vehicle speeds towards the centre lanes of the carriageway. This is summarised in the table in figure 1.5.

Automatic crop spraying

Agricultural systems for crop spraying suffer from limited ability to control overspray. Excess fertiliser seeps into the water table, a problem that is increasingly becoming a

Figure 1.4: Calibration of the image-ground mapping. *Positions of the four corners of a known rectangle on the ground and its projection onto the image plane are sufficient to determine the mapping, using standard projective methods. (Illustration taken from (Ferrier et al., 1994).)*

target of legislators. It is clearly also highly desirable to ensure that toxic chemicals used to control weeds are directed away from plants intended for human consumption. Segmentation of video images on the basis of colour can be an effective means of visually separating plant from soil but is disrupted by shadows cast by the moving tractor. Contour tracking, as in figure 1.6, offers an alternative means of detecting plants that is somewhat immune to such disruption.

Robot grasping

The use of vision in robotics is commonplace in commercial practice, both for inspection and for coordination of grasp. Figure 1.7 shows an experimental system designed for use with a camera mounted on the robot's wrist, to determine stable two-fingered grasps. A snake is used to capture the outline shape, and geometric calculations along the B-spline curve, using first and second derivatives to calculate orientation and curvature, establish a set of safe grasps.

region	start	exit	distance	speed	av spd
(lane)	(sec)	(sec)	(yards)	(mph)	(mph)
1	269.28	273.96	132	58	
1	275.92	279.72	127	68	
1	297.86	301.56	129	72	
1	303.96	308.40	130	60	68
1	314.12	317.24	133	87	
1	321.76	325.24	126	74	
1	330.20	334.04	132	70	
1	343.16	347.58	123	57	
2	687.38	692.18	158	67	
2	708.46	712.36	164	86	
2	727.26	731.20	155	80	76
2	733.12	737.72	164	73	
2	749.12	753.64	169	77	
3	506.78	510.66	156	83	79
3	513.04	517.04	148	75	

Figure 1.5: Analysis of data from tracked cars. *Vehicle velocities are measured between gates space 150 yards apart. (Data from experiments reported in (Ferrier et al., 1994).)*

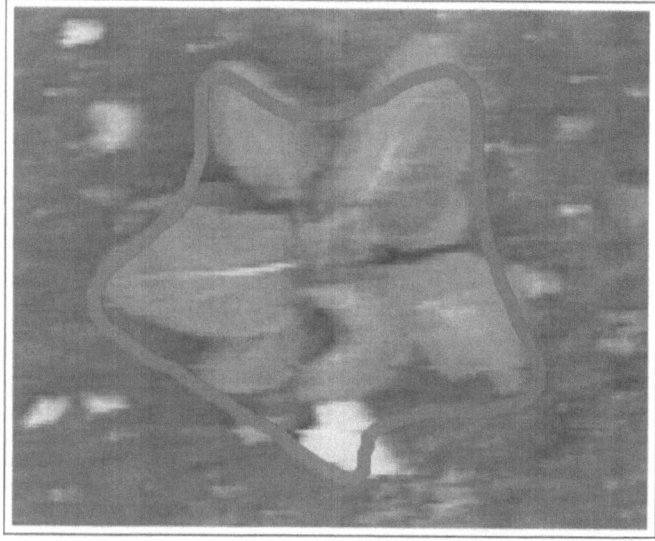

Figure 1.6: Robot tractor. *An autonomous tractor carrying a camera and computer for video analysis has the task of spraying earth and plants automatically, using an array of independently controlled spray nozzles. Plants can be segmented dynamically from the earth and weeds around it, the spraying of fertiliser and weed-killer to be directed onto or away from plants as appropriate. (Figures courtesy of David Reynard, Andrew Wildenberg and John Marchant.)*

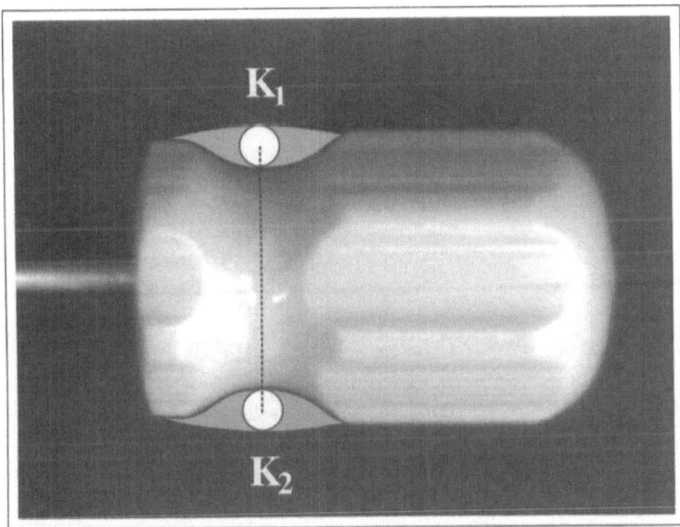

Figure 1.7: Robot hand-eye coordination. *The white circles are placement regions for two thin fingers, computed automatically from the outline of the screwdriver handle. Provided each finger lies within its circle, closing the gripper is bound to capture the screwdriver. (Figure courtesy of Colin Davidson.)*

Surveillance

A combination of visual motion sensing and contour tracking is used to follow an intruder on a security camera in figure 1.8. The camera is mounted on a computer controlled pan-tilt platform driven by visual feedback from the tracked contour.

Biometrics: body motion

This application (figure 1.9) involves the measurement of limb motion for the purposes of analysis of gait as a tool for planning corrective surgery. The tool is also useful for ergonomic studies and anatomical analysis in sport. It is related to the facial animation application above, but more taxing technically. Again, marker based systems exist and are commercially successful as measurement tools both in biology and medicine but it is attractive to replace them with marker-free techniques. There are also increasingly applications in Computer Graphics for whole body animation. Capture of the motion of an entire body from its outline looks feasible but several problems remain to be solved: the relatively large number of degrees of freedom of the articulating body

Figure 1.8: Tracking a potential intruder on security video. *(Figure courtesy of Simon Rowe, David Murray.)*

poses stability problems for trackers; the agility of, say, a dancing figure requires careful treatment of "occlusion" — periods during which some limbs and body parts are obscured by others.

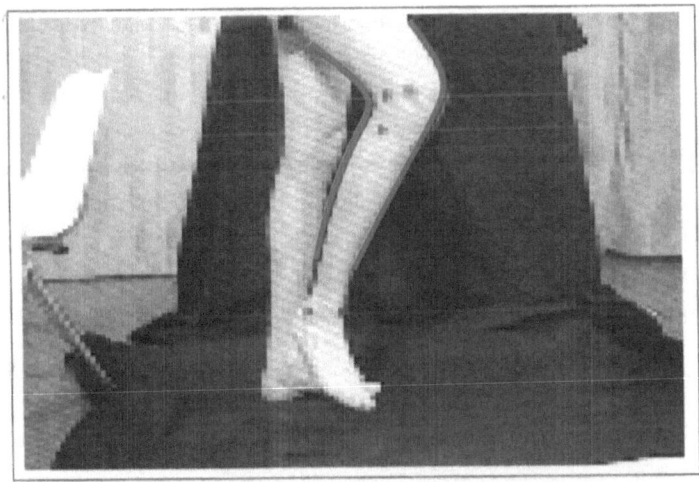

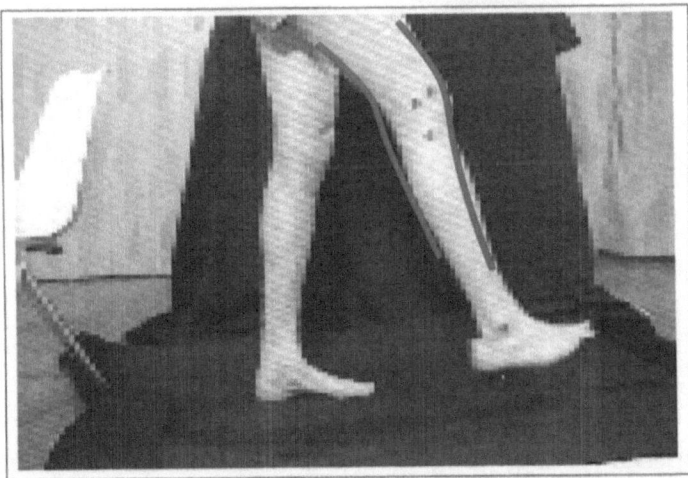

Figure 1.9: Biometrics. *Tracking the articulated motion of a human body is applicable both to biometrics and clinical gait analysis and for actor-driven whole body animation. (Figure courtesy of Rupert Curwen and Julian Morris.)*

Audio-visual speech analysis

Automatic speech-driven dictation systems are now available commercially with large vocabularies though often restricted to separately articulated words. The functioning of such a system is dependent on very reliable recognition of a small set of keywords. In practice, adequately reliable keyword recognition has been realised in low-noise environments but is problematic in the presence of background noise, especially cross-talk from other speakers. Independent experiments in several laboratories have suggested that lip-reading has an important role to play in augmenting the acoustic signal with independent information that is immune to cross-talk. Active contour tracking has been shown to be capable of providing this information (figures 1.10 and 1.11), robustly and in real time, resulting in substantial improvements in recognition-error rates.

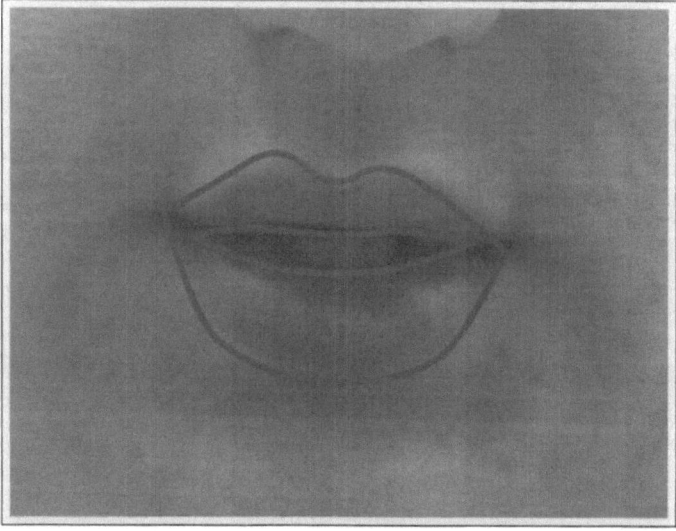

Figure 1.10: Speech-reading. *Performance in automatic speech recognition can be enhanced by lip-reading. This is done by tracking visually the moving outlines of lips to obtain visual signals which are synchronised with the acoustic signal. (Figure courtesy of Robert Kaucic and Barney Dalton.)*

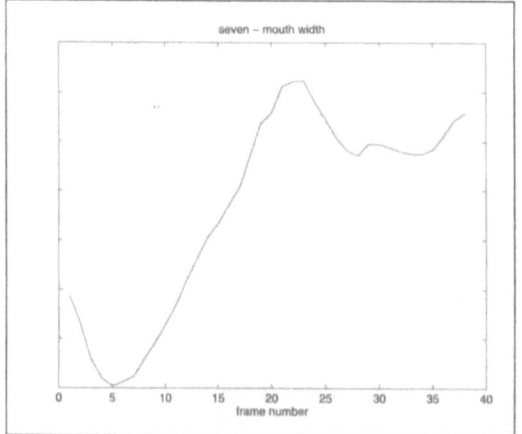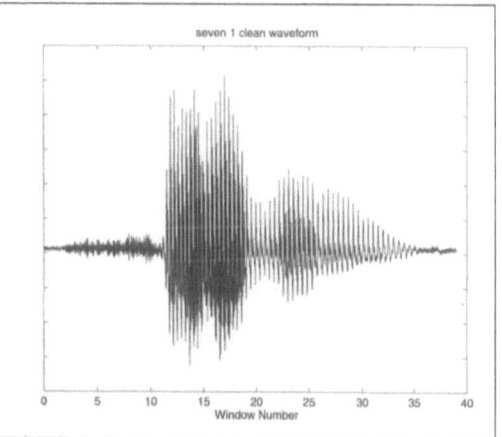

Figure 1.11: Audio and visual speech signals *This figure shows visual (left) and audio (right) signals for the spoken word "seven," over a duration of 0.6 s.*

Medical diagnosis

Ultrasound scanners are medical diagnostic imaging devices that are very widely available owing to their low cost. They are especially suited to dynamic analysis owing to their ability to deliver real-time video sequences. There are numerous potential applications for automated analysis of the real-time image sequences, for example the analysis of abnormalities in cardiac action as in figure 1.12. Noisy artifacts — ultrasound speckle — make these images especially hard to analyse. In this context, active contours are particularly powerful because speckle-induced error tends to be smoothed by the averaging along the contour that is a characteristic of active contour fitting. Broadly tuned, learned models of motion are used in tracking as prior constraints on the moving subject, to aid automated perception. The research issue here is how to learn more finely tuned models to classify normal and aberrant motions.

Another important imaging modality for medical applications is "Magnetic Resonance Imaging" (MRI). It is an expensive technology, but popular because it is as benign as ultrasound, yet as detailed as tomographic X-rays. Applications are pervasive, and one specific example concerning measurements of the cerebral hemispheres of the brain is illustrated in figure 1.13. In each of successive slices of the brain image, two separate snakes lock onto the outlines of the left and the right hemispheres. Geometric coherence in successive slices means that a fitted snake from one slice can

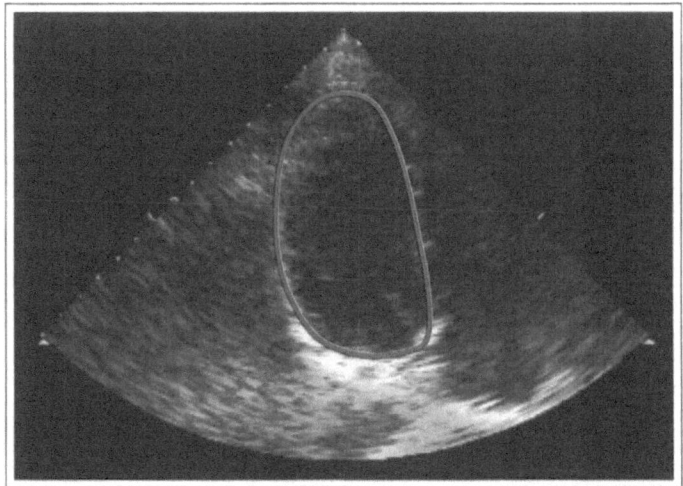

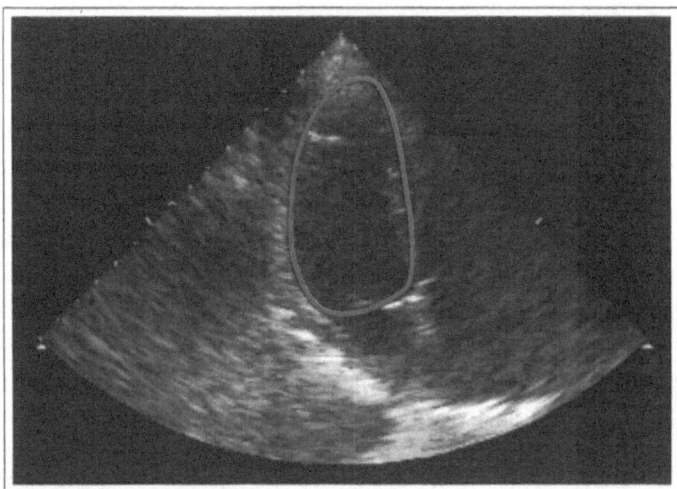

Figure 1.12: Medical echocardiogram analysis. *The left ventricle beating heart is tracked by ultrasound imaging for use in medical diagnosis. (Figure courtesy of Alison Noble and Gary Jacob.)*

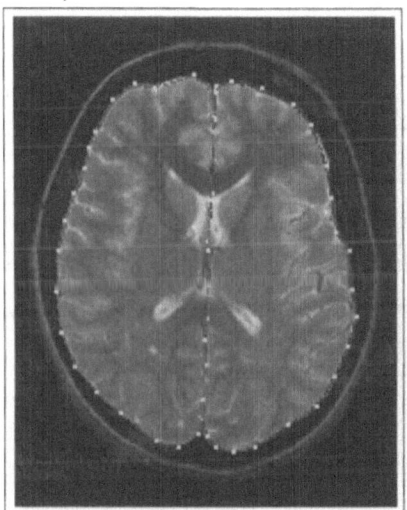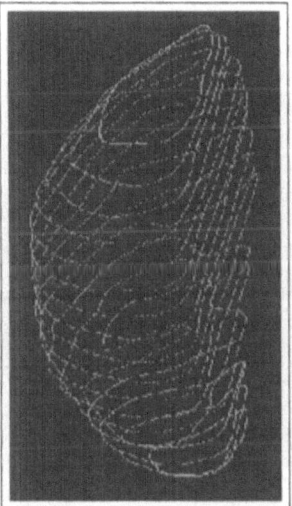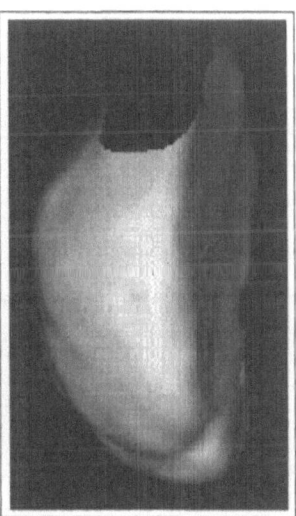

Figure 1.13: MRI imaging of brain hemispheres. *Each MRI scan (left) of the brain images one cross-sectional slice of the brain. Separate snakes trace outlines of the left and right hemispheres. Slices from one hemisphere are stacked (middle), converted to a mesh and finally rendered as a solid (right). (Figures reproduced from (Marais et al., 1996).)*

be used as the initial snake for the next. The entire fitting process can therefore be initialised by hand fitting snakes around outlines in the first slice. The degree of symmetry of the reconstructed hemispheres has been proposed as a possible diagnostic indicator for schizophrenia.

Automated video editing

It is standard practice to generate photo-composites by "blue-screening" in which a foreground object, photographed in motion against a blue background is isolated electronically. It can then be superimposed against a new background to create special effects. Contour tracking raises the possibility of doing this with objects photographed against backgrounds that have not been prepared specially in any way, as in figure 1.14. This increases the versatility of the technique and raises the possibility of extracting moving objects from existing footage for re-incorporation in new video sequences. In a second example (figure 1.15), the motion of a cluster of leaves is not only tracked, but also interpreted as a three-dimensional displacement, so that a computer-generated

Figure 1.14: Automated video editing. *Tracking the outline of a foreground object allows it to be separated automatically from the background, and manipulated as desired, a special effect which can otherwise only be achieved by "blue-screening" from specially prepared footage.*

object can be "hung" from the cluster and added to the animation. This is achieved despite the heavy clutter in the background that makes tracking harder by tending to camouflage the moving leaves.

User interface

The use of body parts as input devices for graphics has of course been thoroughly explored in "Virtual Reality" applications. Current devices such as data-gloves and infra-red helmets are cumbersome and restrictive to the wearer. Visual tracking technology raises the possibility of flexible, non-contact input devices as in figure 1.16. One aim is to use tracking to realise the "digital desk" concept in which a user manipulates a mixture of real and virtual documents on a desk, the virtual ones generated by an overhead video-projector.

Figure 1.15: Automated re-animation. *A cluster of leaves is tracked as it moves (top), its motion interpreted three-dimensionally, and computer-generated pot and flowers are added. This technique is then applied to a sequence with the leaf cluster moving against heavy clutter (bottom).*

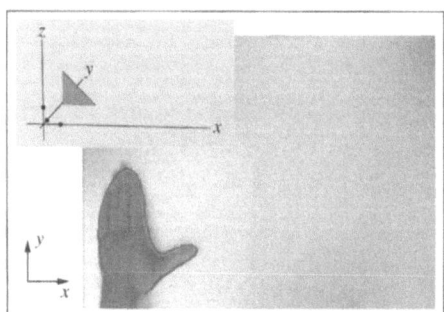 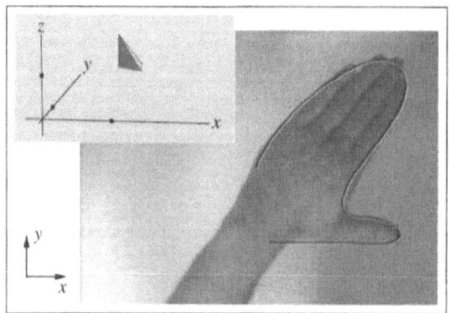

Hand translates in x, y and z directions and rotates; object follows hand's motion.

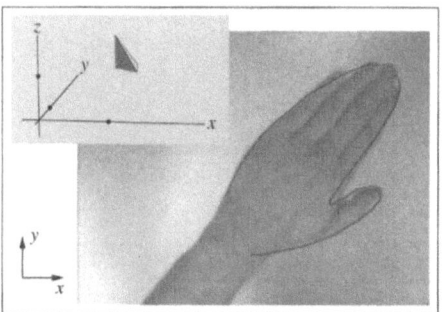 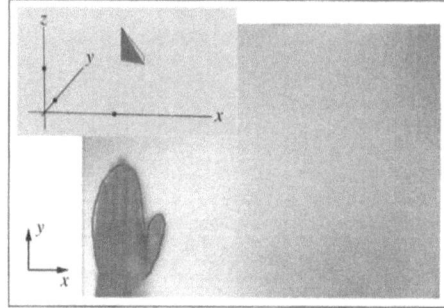

Thumb closed to "lock" object while hand returns to start.

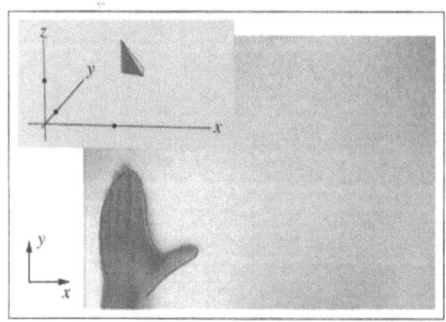 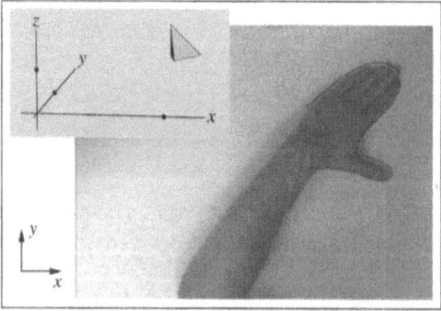

Thumb open: object follows hand translating and rotating.

Figure 1.16: *A hand tracked in real time by a video camera acts as a three-dimensional mouse. Moving the thumb towards the hand acts as an "indexing" gesture, equivalent to lifting a conventional mouse off the desk to reposition it without moving the pointer. (Figure courtesy of Ben North.)*

Bibliographic notes

Despite enormous research effort, the pinnacle of which is represented by (Marr, 1982), the goal of defining general low-level processes for vision has proved obstinate and elusive. Much effort was directed towards finding significant features in images. The theory and practice of image-feature detection is very fully developed — some of the landmarks include (Roberts, 1965; O'Gorman, 1978; Haralick, 1980; Marr and Hildreth, 1980; Canny, 1986; Perona and Malik, 1990) on feature detection and (Montanari, 1971; Ramer, 1975; Zucker et al., 1977) on grouping them into linear structures. See also (Ballard and Brown, 1982) for a broad review. The challenge lies in recovering features undamaged and free of breaks, and in successfully grouping them according to the object to which they belong. In some cases subsequent processes can tolerate errors — gaps in contours and spurious fragments — and this is particularly true of certain approaches to object recognition, for instance (Ballard, 1981; Grimson and Lozano-Perez, 1984; Faugeras and Hebert, 1986; Mundy and Heller, 1990). Another important theme in "low-level" vision has been matching using features, including (Baker and Binford, 1981; Buxton and Buxton, 1984; Grimson, 1985; Ohta and Kanade, 1985; Pollard et al., 1985; Ayache and Faverjon, 1987; Belhumeur, 1993), mostly applied to matching pairs of stereoscopic images.

One notably successful reaction against the tyranny of low-level vision was "active vision" (Aloimonos et al., 1987; Bajcsy, 1988) whose progress and achievements are reviewed in (Blake and Yuille, 1992; Aloimonos, 1993; Brown and Terzopoulos, 1994). Another radical departure was the "snake", for which the original paper is (Kass et al., 1987), and many related papers are given in the bibliography to the following chapter. Pattern theory is a general statistical framework that is important in the study of active contours. It was developed over a number of years by Grenander (Grenander, 1981), and a lucid summary and interpretation can be found in (Mumford, 1996). Again, many related papers following the pattern theory approach are given in the course of the book.

Applications

Actor-driven animation is a classic application for virtual reality systems. Tracking of changing expressions can be done using VR hardware, or visually with reflective markers (Williams, 1990), using active contours (Terzopoulos and Waters, 1990; Terzopoulos and Waters, 1993; Lanitis et al., 1995) or using so-called "optical flow" (Essa and Pentland, 1995; Black and Yacoob, 1995). Underlying muscular motion may be modelled to constrain tracked expressions.

Traffic monitoring is firmly established as a viable application for machine vision, for traffic information systems, non-contact sensors, and autonomous vehicle control (Dreschler and Nagel, 1981; Dickmanns and Graefe, 1988a; Sullivan, 1992; Dickmanns, 1992; Koller et al., 1994; Ferrier et al., 1994). Projective (homogeneous) transformations (Mundy and Zisserman, 1992; Foley et al., 1990) are used for the conversion between image and world coordinates.

Automated crop-handling based on vision has become a realistic possibility in the last decade (Marchant, 1991; Plá et al., 1993), and active contour tracking has a role to play here (Reynard et al., 1996).

A series of theories of determining stable grasps based on an outline have been proposed (Faverjon and Ponce, 1991; Blake, 1992; Rimon and Burdick, 1995a; Rimon and Burdick, 1995b; Rimon and Blake, 1996; Ponce et al., 1995; Davidson and Blake, 1998) and are particularly suited to real-time grasp planning with active contours (Taylor et al., 1994).

A pioneering advance in the visual tracking of human motion was Hogg's "Walker" (Hogg, 1983) which used an articulated model of limb motion to constrain search for body parts. Active contours have been applied with some success to tracking whole bodies and body parts (Waite and Welsh, 1990; Baumberg and Hogg, 1994; Lanitis et al., 1995; Goncalves et al., 1995), though methods based on point features can also be useful for coarse tracking (Rao et al., 1993; Murray et al., 1993).

Audio-visual speech analysis, or speech-reading, has been the subject of psychological study for some time (Dodd and Campbell, 1987). The computational problem has received a good deal of attention recently, using both active contours (Bregler and Konig, 1994; Bregler and Omohundro, 1995; Kaucic et al., 1996) and methods based more directly on image intensities (Petajan et al., 1988), or using artificial facial markers (Finn and Montgomery, 1988; Stork et al., 1992). Generally, as in conventional speech recognition, Hidden Markov Models (HMMs) (Rabiner and Bing-Hwang, 1993) are used for classification of utterances, e.g. (Adjoudani and Benoit, 1995).

Several researchers have investigated the application of active contours to the interpretation of medical images, for example (Amini et al., 1991; Ayache et al., 1992; Cootes et al., 1994).

The technique of rotoscoping allows film-makers to transfer a complex object from one image sequence to another. This can be done automatically using blue-screening (Smith, 1996) if the object can be filmed against a specially prepared background.

Computer-aided techniques for object segmentation are also of great interest for augmented reality systems, which attach computer-generated imagery to real scenes. Traditionally mechanical or magnetic 3D tracking devices have been used (Grimson et al., 1994; Pelizzari et al., 1993; Wloka and Anderson, 1995) to solve this problem, but they are inaccurate and cumbersome. Vision-based tracking has been used instead (Kutulakos and Valliano, 1996; Uenohara and Kanade, 1995; State et al., 1996; Heuring and Murray, 1996), especially for medical applications, mostly restricted to tracking artificial markers. Graphical objects can be made to pass behind real ones (State et al., 1996), by building models of the real-world objects off-line, using scanned range maps.

Effective ways of using a gesturing hand as an interface are yet to be generally established. One very appealing paradigm is the "digital desk" (Wellner, 1993) in which moving hands interact both with real pieces of paper and with virtual (projected) ones, on the surface of a real desk. Other body parts may also be useful for controlling graphics, for instance head (Azarbayejani et al., 1993) and eyes (Gee and Cipolla, 1994). Gestures need not only to be tracked but also interpreted by classifying segments of trajectories, either in configuration space or phase space (Mardia et al., 1993; Campbell and Bobick, 1995; Bobick and Wilson, 1995). This is related both to classification of speech signals (see above) and to classification of signals in other domains, such as electro-encephalograph (EEG) in sleep (Pardey et al., 1995).

Chapter 2

Active shape models

Active shape models encompass a variety of forms, principally *snakes*, *deformable templates* and *dynamic contours*. Snakes are a mechanism for bringing a certain degree of prior knowledge to bear on low-level image interpretation. Rather than expecting desirable properties such as continuity and smoothness to emerge from image data, those properties are imposed from the start. Specifically, an elastic model of a continuous, flexible curve is imposed upon and matched to an image. By varying elastic parameters, the strength of prior assumptions can be controlled. Prior modelling can be made more specific by constructing assemblies of flexible curves in which a set of parameters controls kinematic variables, for instance the sizes of various subparts and the angles of hinges which join them. Such a model is known as a deformable template, and is a powerful mechanism for locating structures in an image.

Things become more difficult when it is necessary to locate moving objects in image sequences — the problem of tracking. This calls for dynamic modelling, for instance invoking inertia, restoring forces and damping, another key component of the original snake conception. We refer to curve trackers that use prior dynamical models as "dynamic contours." Later parts of the book are all about understanding, specifying and learning dynamical prior models of varying strength, and applying them in dynamic contour tracking.

2.1   Snakes

The art of feature detection has been much studied (see bibliographic notes for previous chapter). The principle is that a "mask" or "operator" is designed which produces an output signal which is greatest wherever there is a strong presence in an image of a feature of a particular chosen type. The result is a new image or "feature map" which codes the strength of response for the chosen feature type, at each pixel. Examples of feature maps for three different kinds of feature are illustrated in figure 2.1. Details of the designs of masks and the application to images by digital convolution are given in chapter 5. For now it is sufficient to say that the operator is a sub-image which is scanned over an image using "mathematical correlation" or "convolution" (this is explained in chapter 5). The mask is a prototype image, typically of small size, of the feature being sought: for a valley feature, for instance, the mask would be a V-shaped intensity function. The output of the correlation process is a measure of goodness of fit of the prototype to the image, in each of the image locations evaluated.

However, feature maps are only the beginning. They enhance features of the desired type but do not unambiguously detect them. Detection requires a decision to be made at each pixel, the simplest decision rule being that a feature is marked wherever feature strength exceeds some preset threshold. A constant threshold is rarely adequate except for the simplest of situations such as an opaque object on a back-lit table, as commonly used in machine vision systems. However, the features on a face cannot be back-lit and, if the threshold is set high, gaps appear in edges. If the threshold is low, spurious edges appear, generated by fine texture. Often no happy medium exists. More subtle decision schemes than simple thresholds have been explored but after around two decades of concerted research effort, one cannot expect to do very much better than the example in figure 2.2. The main structure is present but the topology of hand contours is disrupted by gaps and spurious fragments.

The lesson is that "low-level" feature detection processes are effective up to a point but cannot be expected to retrieve entire geometric structures. Snakes constitute a fundamentally new approach to deal with these limitations of low-level processing. The essential idea is to take a feature map $F(\mathbf{r})$ like the ones in figure 2.1, and to treat $(-F(\mathbf{r}))$ as a "landscape" on which the snake, a deformable curve $\mathbf{r}(s)$, $0 \leq s \leq 1$, can slither. For instance, a filter that gives a particularly high output where image contrast is high will tend to attract a snake towards object edges. Equilibrium equations for $\mathbf{r}(s)$ are set up in such a way that $\mathbf{r}(s)$ tends to cling to high responses of F, that is, maximising $F(\mathbf{r}(s))$ over $0 \leq s \leq 1$, in some appropriate sense. This ten-

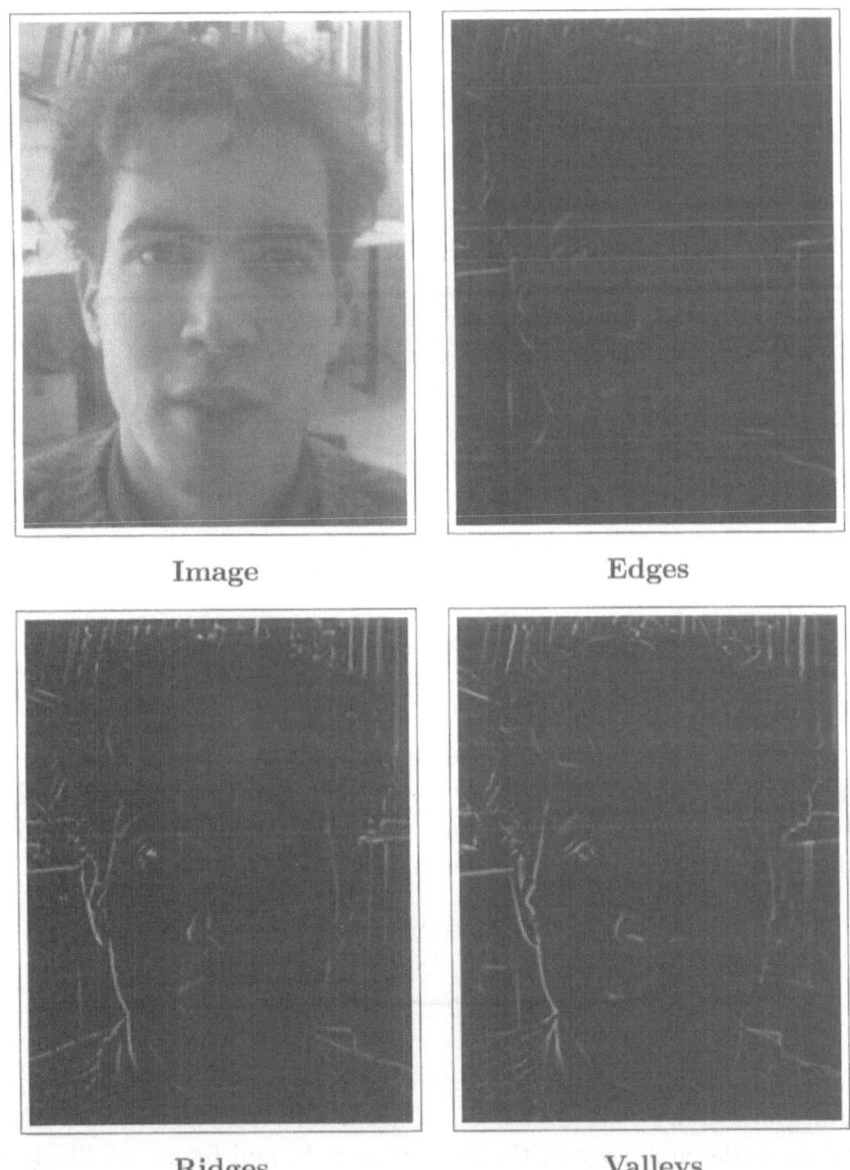

Image   Edges

Ridges   Valleys

Figure 2.1: Image-feature detectors. *Suitably designed image filters can highlight areas of an image in which particular features occur. The examples shown here filter for areas of high contrast ("edges"), peaks of intensity ("ridges") and intensity troughs ("valleys").*

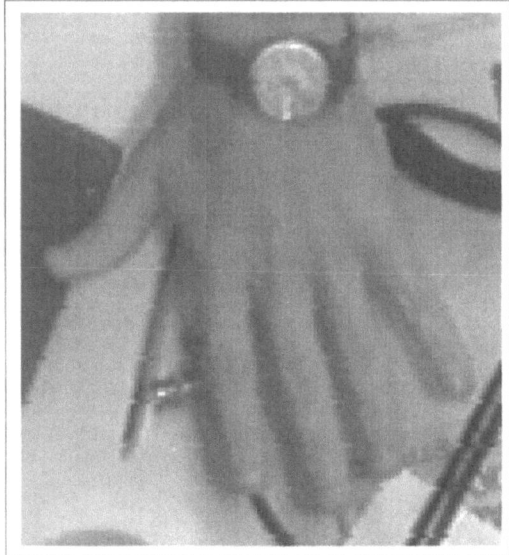

Figure 2.2: Detecting edges. *Edges (right) are generated from the image (left) using horizontally and vertically oriented masks and a decision process (Canny, 1986) that attempts to repair gaps. Nonetheless, there are breaks at critical locations such as corners or junctions, and spurious fragments that disrupt the topology of the hand.*

dency to maximise F is formalised as the "external" potential energy of the dynamical system. It is counterbalanced by "internal" potential energy which tends to preserve smoothness of the curve. The equilibrium equation is:

$$\underbrace{\left(\frac{\partial(w_1\mathbf{r})}{\partial s} - \frac{\partial^2(w_2\mathbf{r})}{\partial s^2}\right)}_{\text{internal forces}} + \underbrace{\nabla F}_{\text{external force}} = 0. \tag{2.1}$$

(Note: s and t subscripts denote differentiation with respect to space and time, and ∇F is the spatial gradient of F.) If (2.1) is solved iteratively, from a suitable configuration, it will tend to settle on a ridge of the feature map F, and figure 2.3 illustrates this. The coefficients w_1 and w_2 in (2.1), which must be positive, govern the restoring forces associated with the elasticity and stiffness of the snake respectively. Either of these coefficients may be allowed to vary with s, along the snake. For example, allowing w_2 to dip to 0 at a certain point $s = s_0$ will allow the snake to kink there, as illustrated

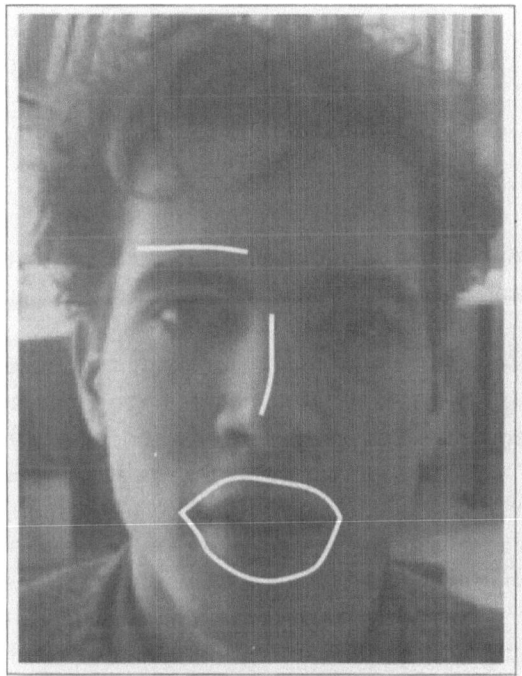 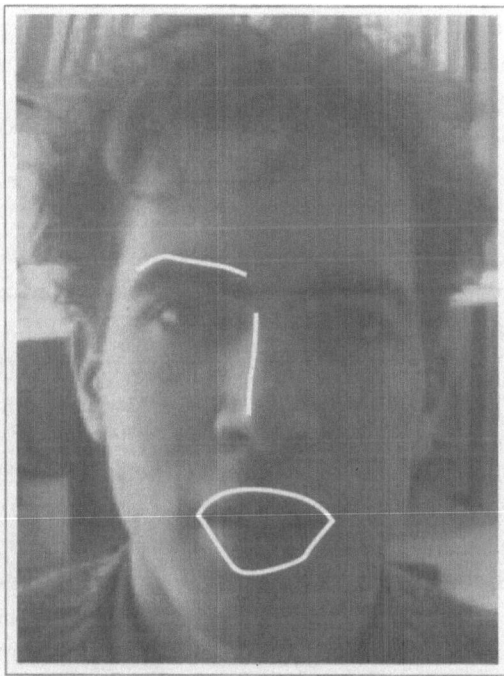

| Initial configuration | Final configuration |

Figure 2.3: Snake equilibrium. *Snakes are shown in initial and final configurations. The eyebrow snake moves over an edge-feature map. The mouth snake is also attracted to edge-features; smoothness constraints are suspended at mouth corners, to allow the snake to kink there. Given that the strongest feature on the nose is a ridge (see figure 2.1), the nose snake is chosen to be attracted to ridges.*

at the mouth corners in figure 2.3. Increasing w_2 encourages the snake to be smooth, like a stiff but flexible rod, but also increases its tendency to regress towards a straight line. Increasing w_1 makes the snake behave like stretched elastic which encourages an even parameterisation of the curve, but increases the tendency to shortness, even collapsing to a point unless counterbalanced by external energy or constraints.

Discrete approximation

Practical computations of $\mathbf{r}(s)$ must occur over discrete time and space, and approximate the continuous trajectories of (2.1) as closely as possible. The original snake represented $\mathbf{r}(s)$ by a sequence of samples at $s = s_i$, $i = 1, \ldots, N$, spaced at intervals of length h, and used "finite differences" to approximate the spatial derivatives $\mathbf{r}_s$ and $\mathbf{r}_{ss}$ by

$$\mathbf{r}_s(s_i) = \frac{\mathbf{r}(s_i) - \mathbf{r}(s_{i-1})}{h} \quad \text{and} \quad \mathbf{r}_{ss}(s_i) = \frac{\mathbf{r}(s_{i+1}) - 2\mathbf{r}(s_i) + \mathbf{r}(s_{i-1})}{h^2}$$

and solve the resulting simultaneous equations in the variables $\mathbf{r}(s_1), \ldots, \mathbf{r}(s_N)$. The system of equations is "sparse," so that it can be solved efficiently, in time $O(N)$ in fact.

In finite difference approximations, the variables $\mathbf{r}(s_i)$ are samples of the curve $\mathbf{r}(s)$, at certain discrete points, conveying no information about curve shape between samples. Modern numerical analysis favours the "finite element" method in which the variables $\mathbf{r}(s_i)$ are regarded as "nodal" variables or parameters from which the continuous curve $\mathbf{r}(s)$ can be completely reconstructed. The simplest form of finite-element representation for $\mathbf{r}(s)$ is as a polygon with the nodal variables as vertices. Smoother approximations can be obtained by modelling $\mathbf{r}(s)$ as a polynomial "spline curve" which passes near but not necessarily through the nodal points. This is particularly efficient because the spline maintains a degree of smoothness, a role which otherwise falls entirely on the spatial derivative terms in (2.1). The practical upshot is that with B-splines the smoothness terms can be omitted, allowing a substantial reduction in the number of nodal variables required, and improving computational efficiency considerably. For this reason, the B-spline representation of curves is used throughout this book. Details are given in chapter 3.

Robustness and stability

Regularising terms in the dynamical equations are helpful to stabilise snakes but are rather restricted in their action. They represent very general constraints on shape, encouraging the snake to be short and smooth. Very often this is simply not enough, and more prior knowledge needs to be compiled into the snake model to achieve stable behaviour. Consider the following example in which a snake is set up with internal constraints reined back to allow the snake to follow the complex outline of the leaf in figure 2.4. In fact it is realised as a B-spline snake with sufficient control points to do justice to the geometric detail of the complex shape. Suppose now the snake is required to follow an image sequence of the leaf in motion, seeking energy minima repeatedly, on successive images in the sequence. If all those control points are allowed to vary

Figure 2.4: The need for shape-spaces. *The white curve is a B-spline with sufficient control points to do justice to the complexity of the leaf's shape. Control point positions vary over time in order to track the leaf outline. However, if the curve momentarily loses lock on the outline it rapidly becomes too tangled to be able to recover. (Figure by courtesy of R. Curwen.)*

somewhat freely over time, the tracked curve can rapidly tie itself into unrecoverable knots, as the figure shows. This is a prime example of the sort of insight that can be gained from real-time experimentation. A regular snake, with suitably chosen internal energy may succeed in tracking several dozen frames off-line. However, once tracking is seen as a *continuous* process, and this is the viewpoint that real-time experiments enforce, the required standards of robustness are altogether more stringent. What was an occasional failure in one computation out of every few, becomes virtually certain eventual failure once the real-time process is allowed to run. It is of paramount importance that recovery from transients — such as a gust of wind causing the leaf to twitch — is robust.

This need for robustness is what drives the account of active contours given in this book. General mechanisms for setting internal shape models are not sufficient. Finely tunable mechanisms are needed, representing specific prior knowledge about classes of objects and their motions. The book aims to give a thorough understanding of the components of such models, initially in geometric terms, and later in terms of probability, as a means of describing *families* of plausible shapes and motions.

2.2   Deformable templates

The prior shape constraints implicit in a snake model are soft, encouraging rather than enforcing a particular class of favoured shapes. What is more, those favoured shapes have rather limited variety. For example, in the case that $w_1 = 0$ in (2.1)), they are solutions of

$$\mathbf{r}_{ss} = 0$$

which are simply straight lines. Models of more specific classes of shapes demand some use of hard constraints, and "default" shapes more interesting than a simple straight line. This can be achieved by using a parametric shape-model $\mathbf{r}(s; \mathbf{X})$, with relatively few degrees of freedom, known as a "deformable template." The template is matched to an image, in a manner similar to the snake, by searching for the value of the parameter vector $\mathbf{X}$ that minimises an external energy $E_{\text{ext}}(\mathbf{X})$. Internal energy $E_{\text{int}}(\mathbf{X})$ may be included as a "regulariser" to favour certain shapes.

As an example of a deformable template, Yuille and Hallinan's eye template is illustrated in figure 2.5, showing how the template is parameterised, and results of fitting to an image of a face. The template $\mathbf{r}(s; \mathbf{X})$ has a total of 11 geometric parameters in the parameter vector $\mathbf{X}$ and it varies non-linearly with $\mathbf{X}$. The non-linearity is evident because, for example, one of the parameters is an angle θ whose sine and cosine appear in the functional form of $\mathbf{r}(s; \mathbf{X})$. The bounding curves of the eye are parabolas which also vary non-linearly, as a function of length parameters a, b and c. The internal energy $E_{\text{int}}(\mathbf{X})$ is a quadratic function of $\mathbf{X}$ that encourages the template to relax back to a default shape. The external energy $E_{\text{ext}}(\mathbf{X})$ comprises a sum of various integrals over the image-feature maps for edges, ridges and valleys. Each integral is taken over one of the two regions delineated by the eye model or along a template curve, which causes E_{ext} to vary with $\mathbf{X}$. Finally the total energy is minimised by iterative, non-linear gradient descent which will tend to find a good minimum, in the

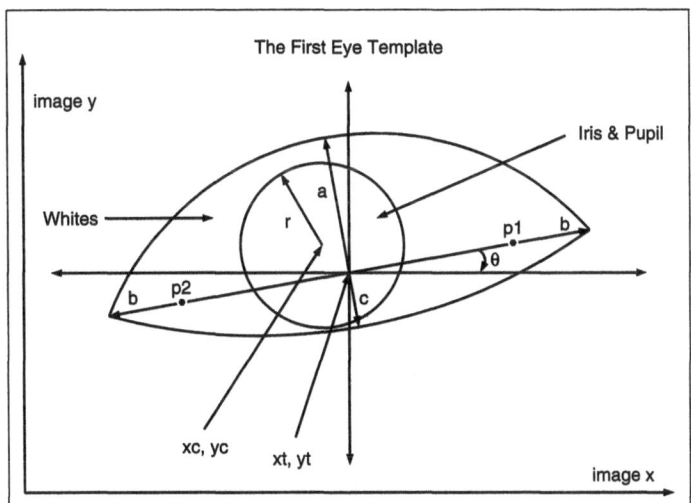

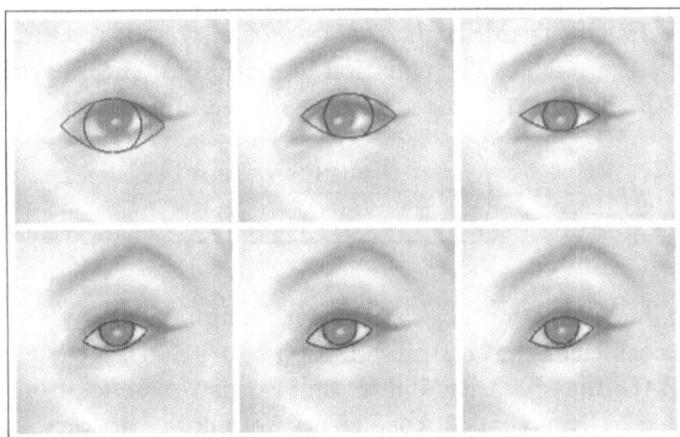

Figure 2.5: Deformable eye template *An eye template is defined (top) in terms of a modest number of variable geometric parameters. In successive iterations of a "gradient descent" algorithm, an equilibrium configuration is reached in which the template fits the eye closely. (Figure reprinted from (Yuille and Hallinan, 1992) which also gives details of external and internal energy functions.)*

sense of giving a good fit to image data, provided the initial configuration is not too far from the desired final fit.

A methodology for setting up linearly parameterised deformable templates — we term them "shape-spaces" — will be described in chapter 4. Restriction to linear parameterisation has certain advantages in simplifying fitting algorithms and avoiding problems with local minima. It is nonetheless surprisingly versatile geometrically. It should be pointed out that some elegant work has been done with three-dimensional parametric models (see bibliographic notes) but this is somewhat outside the scope of this book. Here we deal with three-dimensional motion by modelling directly its effects on image-based contour models using "affine spaces" amongst other devices.

2.3   Dynamic contours

Active contours can be applied either statically, to single images, or dynamically, to temporal image sequences. In dynamic applications, an additional layer of modelling is required to convey any prior knowledge about likely object motions and deformations. Now both the active contour $\mathbf{r}(s,t)$ and the feature map $F(t)$ vary over time. The contour $\mathbf{r}(s,t)$ is drawn towards high responses of $F(t)$ as if it were riding the crest of a wave on the feature map. The equation of motion for such a system extends the snake in (2.1) with additional terms governing inertia and viscosity

$$\underbrace{\rho\,\mathbf{r}_{tt}}_{\text{inertial force}} = -\underbrace{\left(\gamma\mathbf{r}_t - \frac{\partial(w_1\mathbf{r})}{\partial s} + \frac{\partial^2(w_2\mathbf{r})}{\partial s^2}\right)}_{\text{internal forces}} + \underbrace{\nabla F}_{\text{external force}}. \qquad (2.2)$$

This is Newton's law of motion for a snake with mass, driven by internal and external forces. New coefficients in (2.2), in addition to w_1 and w_2 the elastic coefficients from (2.1), are ρ the mass density and γ the viscous resistance from a medium surrounding the snake. Given that all coefficients are allowed to vary spatially, there is clearly considerable scope for setting them to impose different forms of prior knowledge. The spatial variation also introduces a multiplicity of degrees of freedom and potentially complex effects. One of the principal aims of the book is to attain a detailed understanding of those effects, and to harness them in the design of active contours.

Most powerful of all is to combine dynamical modelling as in (2.2) with the rich geometrical structures used in deformable templates, and this is the basis of the dynamic contour. It involves defining parameterised shapes $\mathbf{r}(s; \mathbf{X})$ as for deformable templates and then specifying a dynamical equation for the shape parameter $\mathbf{X}$. In the dynamic contour equation (2.2), prior constraints on shape and motion were implicit, but to facilitate systematic design it is far more attractive that they should be explicit. This can be achieved by separating out a dynamical model for likely motions from the influence of image measurements. The dynamic contour becomes a two-phase process in which a dynamical model is used for *prediction*, to extrapolate motion from one discrete time to the next. Then the predicted position for each new time-step is refined using measured image features, as in figure 2.6. The "Kalman filter" is a

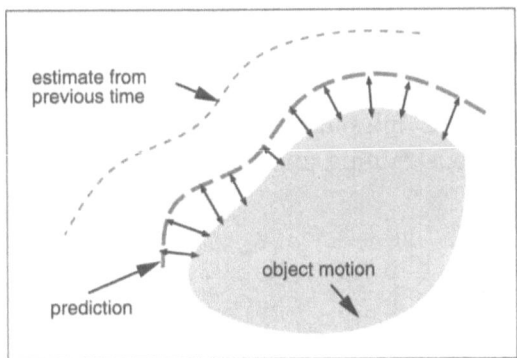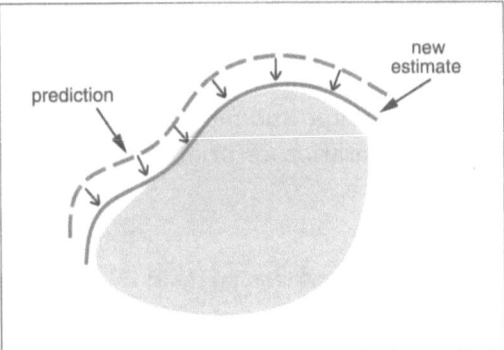

Figure 2.6: Prediction and measurement. *Dynamic contour tracking involves a two-phase process at each successive time. Past motion history and prior knowledge of motion arc extrapolated to predict the displacement between successive times, then predicted position is refined using image features.*

ready made engine for applying the two-phase cycle, and for this reason has been a very popular and successful paradigm for tracking (see bibliographic notes). It is a probabilistic mechanism and this is one reason that probabilistic modelling pervades the treatment of the second part of this book.

Intuitively, predictive models demand probabilistic treatment in order to avoid being too strong. The two-phase cycle fuses a prediction with some measurements. If the prediction were deterministic with no allowance for uncertainty, it would dominate the measurements, which would therefore be ignored. As an example, consider the task of tracking a pendulum in motion. If the pendulum is believed to be executing perfect harmonic motion, free of external disturbances, then provided initial conditions are known, the future motion of the pendulum is entirely determined. Knowing initial conditions, any subsequent observation of the pendulum is redundant. Realistic visual tracking problems are more like observing a pendulum oscillating in a turbulent airflow. The mean behaviour of the pendulum may be explained as deterministic simple harmonic motion, but the airflow drives the motion with random external forces. In terms of the shape parameter $\mathbf{X}$, this implies a dynamical equation of the form

$$\ddot{\mathbf{X}} = f(\dot{\mathbf{X}}, \mathbf{X}, \mathbf{w}), \qquad (2.3)$$

where $\dot{\mathbf{X}}$ and $\ddot{\mathbf{X}}$ are the first and second temporal derivatives of $\mathbf{X}$ and $\mathbf{w}$ is a random disturbance. Thus the value of initial conditions weaken over time, as the motion of the pendulum is progressively perturbed away from the ideal deterministic motion. This increasing uncertainty generates a "gap" in information which sensory observations can fill. A primary aim of the book is to define design principles for probabilistic models of shape and motion and explain those principles in terms of their effects both on representation of prior knowledge and in constraining and conditioning tracking performance.

Bibliographic notes

The seminal paper on snakes is (Kass et al., 1987). This spawned many variations and extensions including the use of Fourier parameterisation (Scott, 1987), incorporation of hard constraints (Amini et al., 1988) and incorporation of explicit dynamics (Terzopoulos and Waters, 1990; Terzopoulos and Szeliski, 1992). Realisation of snakes using B-splines was developed by (Cipolla and Blake, 1990; Menet et al., 1990; Hinton et al., 1992) and combined with Lagrangian dynamics in (Curwen et al., 1991). B-splines used in this way are a form of "finite element," a standard technique of numerical analysis for solving differential equations by computer (Strang and Fix, 1973; Zinkiewicz and Morgan, 1983).

The idea of deformable templates predates the development of snakes (Fischler and Elschlager, 1973; Burr, 1981; Bookstein, 1988) but has enjoyed a revival inspired by the snake. Variations on the deformable template theme rapidly emerged (Yuille et al., 1989; Yuille, 1990; Bennett and Craw, 1991; Yuille and Hallinan, 1992; Hinton et al., 1992; Cootes and Taylor, 1992; Cootes et al., 1993; Cootes et al., 1995). A good deal of research has been done on matching with three-dimensional models, both rigid (Thompson and Mundy, 1987; Lowe, 1991; Sullivan, 1992; Lowe, 1992; Harris, 1992b; Gennery, 1992) and deformable (Terzopoulos et al., 1988; Terzopoulos and Fleischer,

1988; Cohen, 1991; Terzopoulos and Metaxas, 1991; Rehg and Kanade, 1994) but is somewhat outside the scope of this book. As models become more detailed, and search becomes more exhaustive, the three-dimensional approach merges into visual object recognition (Grimson, 1990).

The Kalman filter (Gelb, 1974; Bar-Shalom and Fortmann, 1988) is very widely used in control theory and for target tracking (Rao et al., 1993) and sensor fusion (Hallam, 1983; Durrant-Whyte, 1988; Hager, 1990) and has become a standard tool of computer vision (Ayache and Faugeras, 1987; Dickmanns and Graefe, 1988b; Dickmanns and Graefe, 1988a; Matthies et al., 1989; Deriche and Faugeras, 1990; Harris, 1992b; Terzopoulos and Szeliski, 1992; Faugeras, 1993).

Finally, it seems appropriate at least to give some pointers to approaches to visual tracking that are rather outside the active contour paradigm.

- (Black and Yacoob, 1995) uses the visual motion field over a region to track and identify movement

- (Bray, 1990) tracks using a mixture of polyhedral, model-based vision to initialise and optic-flow vectors along contours for incremental displacement

- (Fischler and Bolles, 1981; Gee and Cipolla, 1996) are very elegant uses of random generation and testing of point-correspondence hypotheses, respectively for static and dynamic image matching problems

- (Huttenlocher et al., 1993) used the "Hausdorff metric" to match successive views in a sequence; the beauty of the approach is that it requires almost no prior model of shape or motion

- (Allen et al., 1991; Papanikolopoulos et al., 1991; Mayhew et al., 1992; Brown et al., 1992; Murray et al., 1992; Heuring and Murray, 1996) are control theoretic approaches to visual-servoing, real-time tracking with robot hands and heads

Part I

Geometrical Fundamentals

Chapter 3

Spline curves

Throughout this book, visual curves are represented in terms of parametric spline curves, as is common in computer graphics. These are curves $(x(s), y(s))$ in which s is a parameter that increases as the curve is traversed, and x and y are particular functions of s, known as splines. A spline of order d is a piecewise polynomial function, consisting of concatenated polynomial segments or *spans*, each of some polynomial order d, joined together at *breakpoints*. Parametric spline curves are attractive because they are capable of representing efficiently sets of boundary curves in an image (figure 3.1). Simple shapes can be represented by a curve with just a few spans. More complex shapes could be accommodated by raising the polynomial order d but it is preferable to increase the number of spans used. Usually the polynomial order is fixed at quadratic $(d = 3)$ or cubic $(d = 4)$[1]. Maintaining a fixed, low polynomial degree, even in the face of geometric complexity, makes for computational stability and simplicity.

The chapter begins by explaining spline functions and their construction. Later sections explain how parametric curves are constructed from spline functions and introduce methods for matching one curve to another. This forms the basis for the algorithms developed in subsequent chapters for active contour matching.

---

[1]The *order* of a polynomial is the number of its coefficients. Hence a quadratic function $a + bx + cx^2$ has order $d = 3$. Its *degree* — the highest power of x — is 2.

Figure 3.1: Image edges represented as parametric spline curves.

3.1   B-spline functions

B-splines are a particular, computationally convenient representation for spline functions. In the B-spline form, a spline function $x(s)$ is constructed as a weighted sum of N_B *basis functions* (hence 'B'-splines) $B_n(s)$, $n = 0, \ldots, N_B - 1$. In the simplest ("regular") case, each basis function consists of d polynomials each defined over a *span* of the s-axis. We take each span to have unit length. The spans are joined at *knots* as in figure 3.2. It shows the simplest case in which the knots are evenly spaced and the joins between polynomials are *regular* — that is, as smooth as possible, having $d - 2$ continuous derivatives. The quadratic spline, for instance, has continuous gradient in the regular case. The constructed spline function is

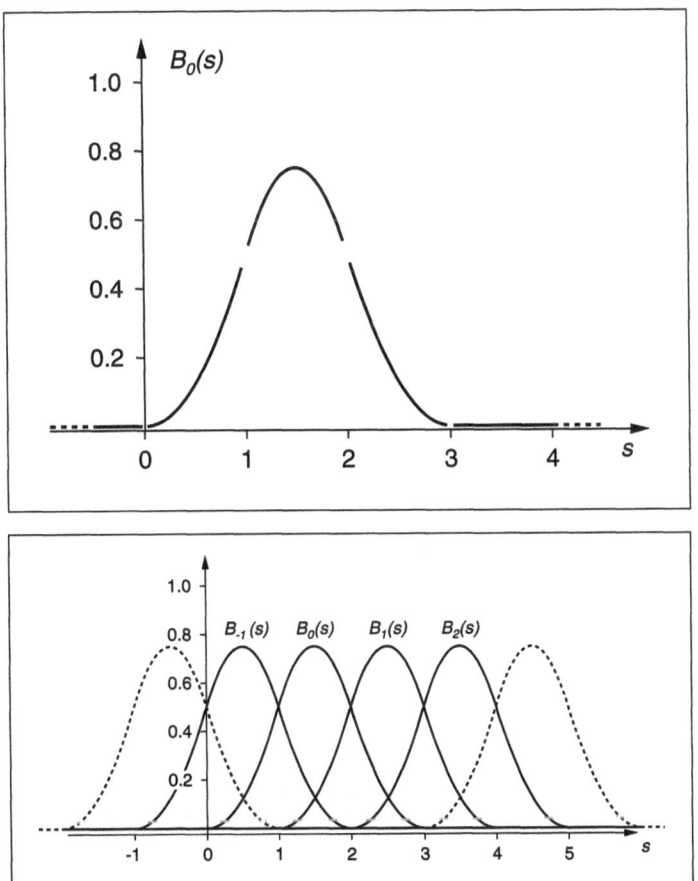

Figure 3.2: *(top) A single quadratic B-spline basis function $B_0(s)$. "Knots" at $s =, 0, 1, 2, 3, 4$ mark transitions between polynomial segments of the function. (bottom) In the regular case which has evenly spaced knots (at integral values of s), each B-spline basis function is a translated copy of the previous one.*

$$x(s) = \sum_{n=0}^{N_B-1} x_n B_n(s) \tag{3.1}$$

where x_n are the weights applied to the respective basis functions $B_n(s)$, as in figure 3.3. This can be expressed compactly in matrix notation as

$$x(s) = \mathbf{B}(s)^T \mathbf{Q}^x, \tag{3.2}$$

a matrix product between a vector of B-spline functions

$$\mathbf{B}(s) = (B_0(s), B_1(s), \dots, B_{N_B-1}(s))^T \tag{3.3}$$

and a vector of weights

$$\mathbf{Q}^x = \begin{pmatrix} x_0 \\ \vdots \\ x_{N_B-1} \end{pmatrix}. \tag{3.4}$$

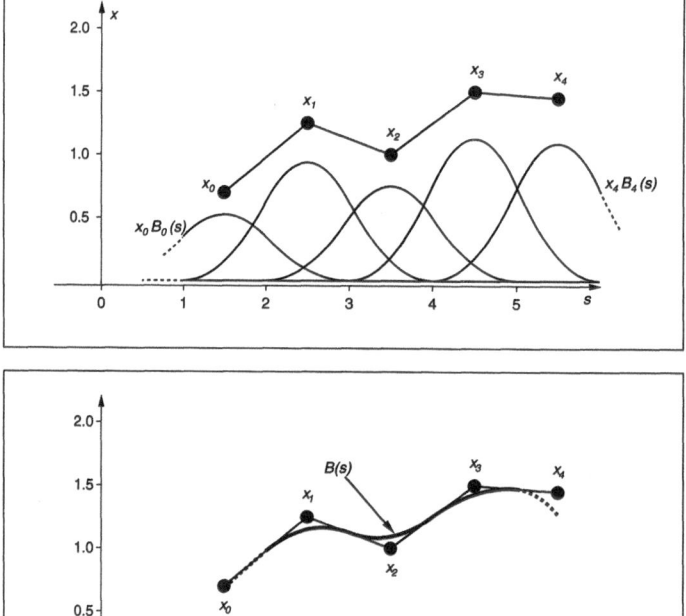

Figure 3.3: *B-spline basis functions $B_n(s)$ are weighted by coefficients x_n (top) and combined linearly to form a spline function $x(s)$ (bottom). Note that the spline function follows the "control polygon" closely.*

By convention, B-spline basis functions are constructed in such a way that they sum to 1 at all points:

$$\sum_{n=0}^{N_B-1} B_n(s) = 1 \text{ for all } s. \tag{3.5}$$

This summation or "convex hull" property is the underlying reason that the B-spline function in figure 3.3 follows the "control polygon," made up of the points $(\frac{3}{2}, x_0), (\frac{5}{2}, x_1), \ldots$ quite closely.

In the simple case of a quadratic B-spline with knots spaced regularly at unit intervals, the first B-spline basis function has the form

$$B_0(s) = \begin{cases} s^2/2 & \text{if } 0 \le s < 1 \\ \frac{3}{4} - (s - \frac{3}{2})^2 & \text{if } 1 \le s < 2 \\ (s-3)^2/2 & \text{if } 2 \le s < 3 \\ 0 & \text{otherwise} \end{cases} \tag{3.6}$$

and the others are simply translated copies:

$$B_n(s) = B_0(s - n).$$

However, this basis is bi-infinite: there are infinitely many B_n and the functions $x(s)$ they are used to construct are bi-infinite, extending to $s \to \pm\infty$. For practical applications finite bases are needed.

3.2 Finite bases

A finite spline basis can be either periodic (figure 3.4) or aperiodic (figures 3.5 and 3.6) over a closed interval $0 \le s \le L$. The periodic basis is simply the bi-infinite basis suitably wrapped around. For example, the basis functions for regular, quadratic splines are $B_0, \ldots, B_{L-1}$ ($N_B = L$), defined as above, but treated as periodic over the interval $0 \le s \le L$. The four periodic basis functions for the case that $L = 4$ are illustrated in figure 3.4. A non-periodic basis on a finite interval is more complex to construct, requiring so-called "multiple knots" (see later) at its endpoints. This allows full control over boundary conditions — the value of the function $x(s)$ and its

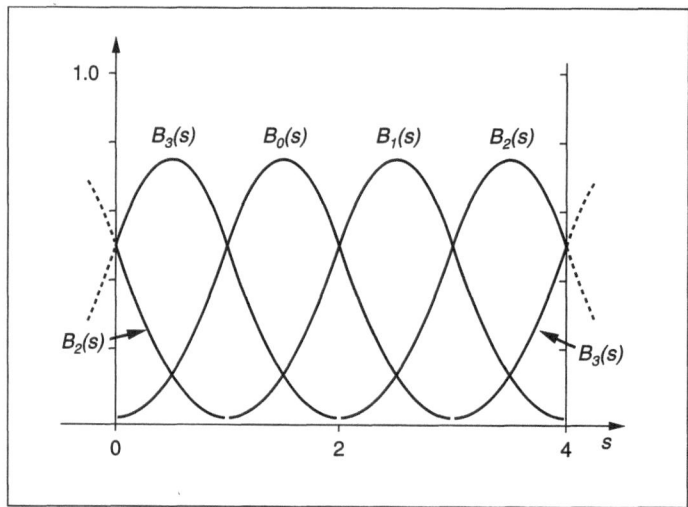

Figure 3.4: Periodic B-spline basis, *as figure 3.2 but for construction of functions that are periodic over the range $0 \leq s \leq 4$. Again each B-spline basis function is a translated copy of the previous one, but also wrapped around where the periodicity demands it.*

derivatives at the ends $s = 0, L$ of the interval. Details of the construction of the B_n for this case can be found in appendix A. It is no longer the case that the number of basis functions N_B is equal to the interval length L. Additional basis functions are needed to control boundary conditions (values of the spline function and its derivatives at $s = 0, L$). In the regular case, $d - 1$ extra functions are needed (d is the order of the polynomial) so that $N_B = L + d - 1$. In figure 3.5, for example, $L = 5$ and $d = 3$ (quadratic) so there must be $N_B = 7$ basis functions.

There is an efficient algorithm for generating spline functions from the weights x_n in which the basis functions are represented in terms of a matrix of polynomial coefficients. The standard method is given in appendix A.

3.3   Multiple knots

Sometimes it is desirable to allow a reduced degree of continuity at some point within the domain of a function $x(s)$. This can be achieved by forming a multiple knot, in which two knots in the B-spline basis approach one another and coincide. The spline then consists of a sequence of polynomial spans joined at *breakpoints*, some of

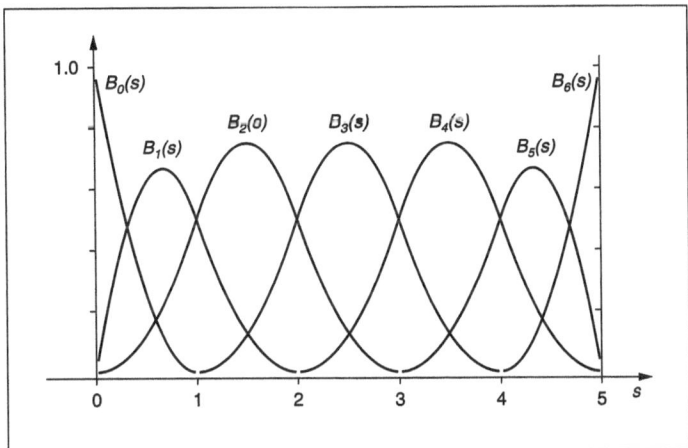

Figure 3.5: Spline basis over an interval. *Basis functions are not entirely composed of translated copies of one another, as they were in the bi-infinite case, but include special functions at the extremes of the interval for control of boundary conditions.*

which are single knots while others are multiple knots. At a regular breakpoint (single knot), the degree of smoothness is at its maximal value, that is C^{d-2} — continuity of all derivatives up to the $(d-2)$th. At a double knot, however, continuity is reduced to C^{d-3} and generally, continuity at a knot of multiplicity m is C^{d-m-1}. Forming a multiple knot is a limiting process in which m consecutive regular knots approach one another, as illustrated in figure 3.7 for the quadratic case. Once a double knot has been introduced into the basis, any constructed spline function generally loses one order of continuity at that breakpoint. In the quadratic case for instance, a spline function is C^0 at the knot: it remains continuous but its gradient becomes discontinuous — see figure 3.8 for an illustration. If a triple knot is introduced, the function becomes discontinuous, broken into two continuous pieces, one on each side of the knot. Hence triple knots are used to terminate a quadratic B-spline basis over a finite interval as in figure 3.5.

3.4   Norm and inner product for spline functions

It is very useful to be able to calculate the so-called "L_2-norm" $\|x\|$ of a function $x(s)$:

$$\|x\|^2 = \frac{1}{L} \int_0^L x(s)^2 \, ds. \tag{3.7}$$

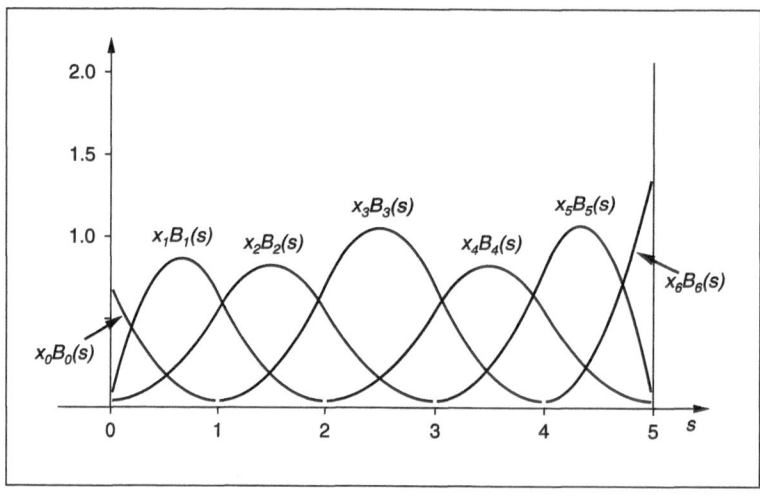

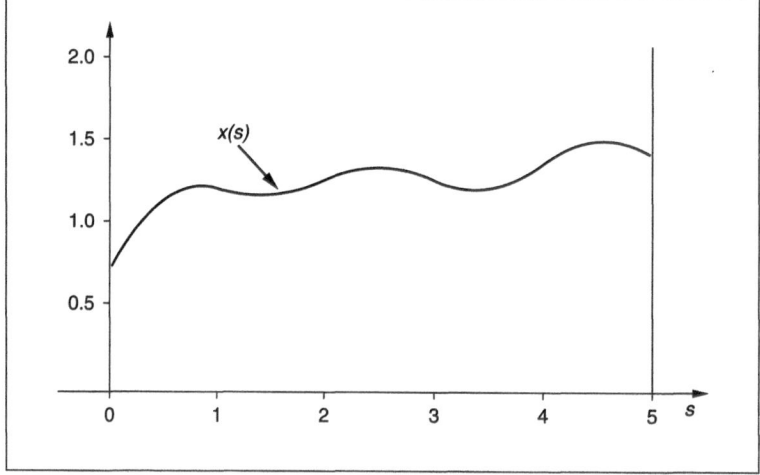

Figure 3.6: Splines over an interval. *Basis functions in figure 3.5 are blended to form a spline function $x(s)$. The choice of basis functions allows the boundary values $x(0), x(5)$ to be controlled directly by the weights x_0, x_6. Then weights x_1, x_5 control the values of derivatives at the extremes of the interval.*

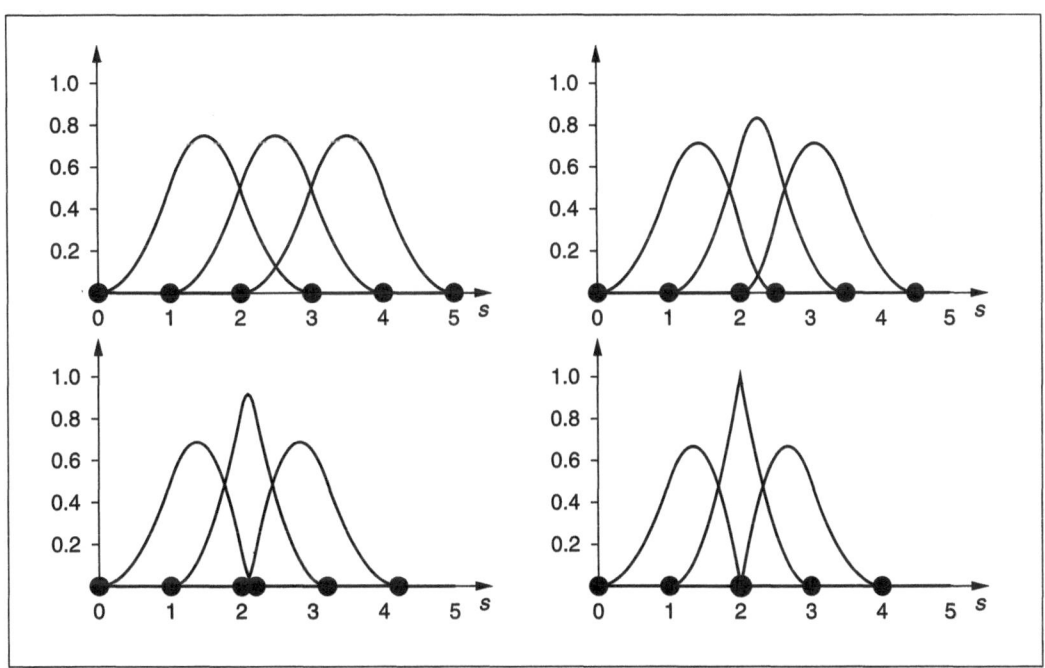

Figure 3.7: Forming multiple knots. *A double knot is introduced into a quadratic B-spline basis at $s = 2$. The resulting basis functions are the limit reached as the knot initially at $s = 3$ approaches $s = 2$.*

which is precisely the "root-mean-square" value[2] of $x(s)$ over the range $0 \leq s \leq L$. The functional norm is especially useful for measuring the difference between two functions $x_1(s), x_2(s)$ as $\|x_1 - x_2\|$, for instance when it is necessary to measure how closely a function x_1 is approximated by another function x_2.

The norm has a corresponding "inner product," denoted $< \cdot, \cdot >$, which is bilinear and is applied to a pair of functions x, y as $< x, y >$. The relationship between an inner product and a norm is that $< x, x >= \|x\|^2$, so in the L_2 case the inner product between two functions works out to be:

$$< x, y >= \frac{1}{L} \int_0^L x(s)y(s) \, ds. \tag{3.8}$$

The inner product will be used later to express function approximations concisely.

---

[2]Conventionally the $\frac{1}{L}$ scaling factor in the definition of the norm would be omitted; we include it so that $\|x\|$ is truly a root-mean-square measure.

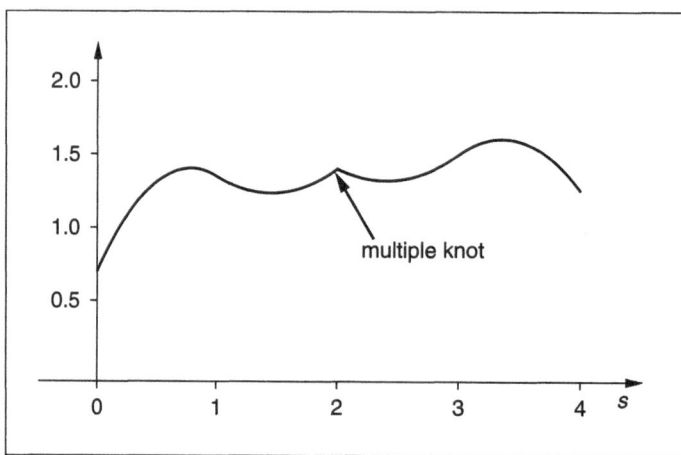

Figure 3.8: Reduced continuity. *Using a basis with a double knot as in figure 3.7, in which a double knot has been introduced at $s = 2$, constructed functions show a gradient discontinuity at $s = 2$ — compare with figure 3.6.*

Since we are representing functions compactly as vectors $\mathbf{Q}^x$ of spline weights, it is natural to express norms and inner products in terms of these vectors, that is, to define $\| \cdot \|$ for weight vectors such that

$$\|\mathbf{Q}^x\| = \|x\|.$$

Now from (3.2)

$$\|x\|^2 = (\mathbf{Q}^x)^T \frac{1}{L} \left(\int_0^L \mathbf{B}(s)\mathbf{B}(s)^T \, ds \right) \mathbf{Q}^x$$

so the norm must be defined as:

$$\|\mathbf{Q}^x\| = \sqrt{(\mathbf{Q}^x)^T \mathcal{B} \mathbf{Q}^x} \tag{3.9}$$

where

$$\mathcal{B} = \frac{1}{L} \int_0^L \mathbf{B}(s)\mathbf{B}(s)^T \, ds \tag{3.10}$$

defines the *metric matrix* for a given class of B-spline function, and the inner product is

$$< \mathbf{Q}_1^x, \mathbf{Q}_2^x > \equiv (\mathbf{Q}_1^x)^T \mathcal{B} \mathbf{Q}_2^x. \tag{3.11}$$

The $\mathcal{B}$-matrices are sparse. This reflects the fact that each weight Q_n^x affects the function $x(s)$ only over a short sub-interval — the "support" of the corresponding basis function B_n — and this lends efficiency to least-squares approximation algorithms. In the periodic, quadratic case the $\mathcal{B}$-matrices are sparse circulants of order 5, and in general they have order $2d - 1$ (the number of non-zero elements in each row). For a given polynomial order d, therefore, the sparsity is most significant when the number N_B of basis functions is large. For periodic quadratic splines with $N_B = 8$ the $\mathcal{B}$-matrix is:

$$\mathcal{B} = \frac{1}{8} \begin{pmatrix} 0.55 & 0.217 & 0.008 & 0.0 & 0.0 & 0.0 & 0.008 & 0.217 \\ 0.217 & 0.55 & 0.217 & 0.008 & 0.0 & 0.0 & 0.0 & 0.008 \\ 0.008 & 0.217 & 0.55 & 0.217 & 0.008 & 0.0 & 0.0 & 0.0 \\ 0.0 & 0.008 & 0.217 & 0.55 & 0.217 & 0.008 & 0.0 & 0.0 \\ 0.0 & 0.0 & 0.008 & 0.217 & 0.55 & 0.217 & 0.008 & 0.0 \\ 0.0 & 0.0 & 0.0 & 0.008 & 0.217 & 0.55 & 0.217 & 0.008 \\ 0.008 & 0.0 & 0.0 & 0.0 & 0.008 & 0.217 & 0.55 & 0.217 \\ 0.217 & 0.008 & 0.0 & 0.0 & 0.0 & 0.008 & 0.217 & 0.55 \end{pmatrix} \tag{3.12}$$

— note the circulant structure (repeating rows), characteristic of the periodic case. For a non-periodic quadratic function, $\mathcal{B}$ is still sparse, no longer a circulant, but now pentadiagonal, as shown here for the case $N_B = 6$ ($L = 4$):

$$\mathcal{B} = \frac{1}{4} \begin{pmatrix} 0.2 & 0.117 & 0.017 & 0.0 & 0.0 & 0.0 \\ 0.117 & 0.333 & 0.208 & 0.008 & 0.0 & 0.0 \\ 0.017 & 0.208 & 0.55 & 0.217 & 0.008 & 0.0 \\ 0.0 & 0.008 & 0.217 & 0.55 & 0.208 & 0.017 \\ 0.0 & 0.0 & 0.008 & 0.208 & 0.333 & 0.117 \\ 0.0 & 0.0 & 0.0 & 0.017 & 0.117 & 0.2 \end{pmatrix} \tag{3.13}$$

and would be heptadiagonal for cubic splines.

Armed with the inner product, some approximation problems become straightforward. The simplest tutorial example is the problem of approximating a spline function represented as $\mathbf{Q}^x$ in terms of two other spline functions $\mathbf{Q}_1^x, \mathbf{Q}_2^x$. The least-squares approximation can be expressed, using inner products, as

$$\hat{\mathbf{Q}}^x = \begin{pmatrix} \mathbf{Q}_1^x & \mathbf{Q}_2^x \end{pmatrix} \begin{pmatrix} \langle \mathbf{Q}_1^x, \mathbf{Q}_1^x \rangle & \langle \mathbf{Q}_1^x, \mathbf{Q}_2^x \rangle \\ \langle \mathbf{Q}_2^x, \mathbf{Q}_1^x \rangle & \langle \mathbf{Q}_2^x, \mathbf{Q}_2^x \rangle \end{pmatrix}^{-1} \begin{pmatrix} \langle \mathbf{Q}_1^x, \mathbf{Q}^x \rangle \\ \langle \mathbf{Q}_2^x, \mathbf{Q}^x \rangle \end{pmatrix} \tag{3.14}$$

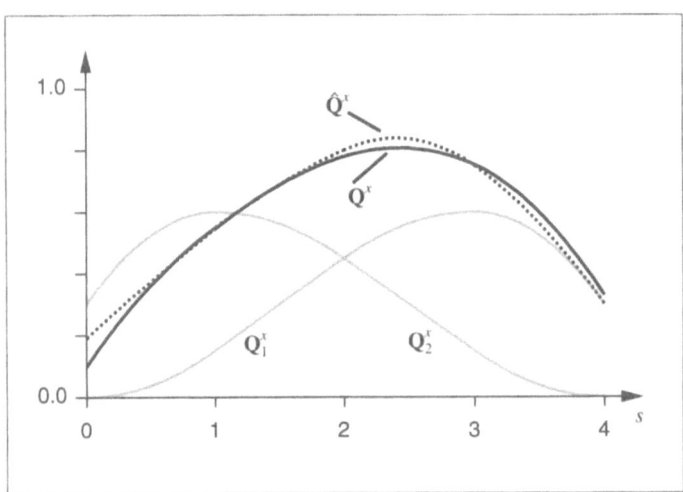

Figure 3.9: Spline approximation: $\hat{\mathbf{Q}}^x$ *is the least-squares approximation of a spline* $\mathbf{Q}^x$ *in terms of two other splines* $\mathbf{Q}_1^x, \mathbf{Q}_2^x$. *The solution can be concisely expressed in terms of inner products — see text.*

— see figure 3.9. A development of this method appears later when, in the interests of economy and of stability over time, it is required to express spline curves in terms of a relatively small number of parameters. The reduction of the parameter set is expressed as a projection operation rather like the approximation $\mathbf{Q}^x \to \hat{\mathbf{Q}}^x$ above, but for curves instead of functions.

Another important type of problem is to approximate some function $f(s)$, not necessarily a spline, as a spline function $x(s)$, represented, as usual, by weights $\mathbf{Q}^x$. Again, the solution can be derived neatly using inner products to give:

$$\mathbf{Q}^x = \mathcal{B}^{-1} \frac{1}{L} \int \mathbf{B}(s) f(s)\, ds \qquad (3.15)$$

and an example is shown in figure 3.10. Functional approximation of this kind can be developed to construct approximations of curves in image data, and this is close to what will be required for active contour algorithms. Of course it is not possible to evaluate integrals over data exactly, so in practice data must be sampled. The simplest approximation of a sampled function as a spline is:

$$\mathbf{Q}^x = \mathcal{B}^{-1} \frac{1}{N} \sum_{n=1}^{N} \mathbf{B}(s_n) f(s_n).$$

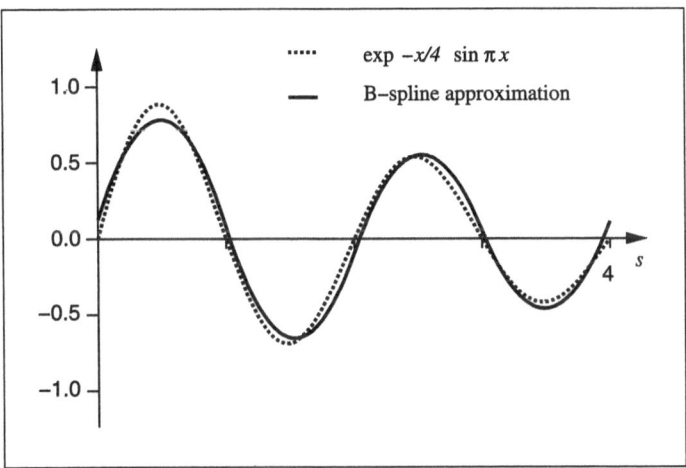

Figure 3.10: Spline approximation: *least-squares spline approximation to a damped sine wave (L = 4, $N_B = 6$).*

The application to image curves is discussed in chapter 6.

3.5 B-spline parametric curves

Spline functions were introduced to serve as a tool for constructing curves in the plane, which they do in the following manner. Parametric spline curves

$$\mathbf{r}(s) = (x(s), y(s))$$

have coordinates $x(s)$, $y(s)$ each of which is a spline function of the curve parameter s. First it is necessary to choose an appropriate interval $0 \le s \le L$ covering L spans and an appropriate basis $B_0, B_1, \ldots, B_{N_B-1}$ of N_B B-spline functions or basis functions. If the interval $[0, L]$ is taken to be periodic the resulting parametric curve will be closed. Alternatively, an open curve requires a B-spline basis over a finite interval as in figure 3.6. For each basis function B_n a *control point* $\mathbf{q}_n = (q_n^x, q_n^y)^T$ must now be defined and the curve is a weighted vector sum of control points

$$\mathbf{r}(s) = \sum_{n=0}^{N_B-1} B_n(s)\mathbf{q}_n \quad \text{for } 0 \le s \le L, \tag{3.16}$$

a smooth curve that follows approximately the "control polygon" defined by linking control points by lines (figure 3.11). The component functions of $\mathbf{r}(s)$ do, of course, turn out to be spline functions, for instance:

$$x(s) = \sum_{n=0}^{N_B-1} B_n(s)q_n^x \;\; \text{for} \;\; 0 \le s \le L, \tag{3.17}$$

— a weighted sum of basis functions with weights q_n^x. The example curve in figure 3.11 uses the basis functions B_n for regular, periodic, quadratic splines that were defined earlier in (3.6) on page 45 and, as before, $N_B = L$ so that the number of control points of the curve is equal to the number of its spans.

3.6   Curves with vertices

It is often necessary to introduce a vertex or hinge at a certain point along a parametric curve, to fit around sharp corners on an object outline. One straightforward way of doing this is to allow two or more consecutive control points to coincide to form a "multiple control point." When n consecutive control points coincide, the order of continuity of the curve is reduced by $n - 1$. A quadratic spline, for instance, has continuous first derivative but discontinuous second derivative when all control points are distinct. A hinge (discontinuous first derivative) is formed therefore when 2 consecutive control points coincide, as in figure 3.12. Unfortunately, introducing a hinge in this way generates spurious linearity constraints (see figure). What is more, the parameterisation of the curve behaves badly in the vicinity of the hinge in the sense that $\mathbf{r}'(s) = \mathbf{0}$ so that the parameter s is changing infinitely fast as the curve passes through the hinge. This would have the effect in the curve-fitting algorithms to be described in chapter 6 of giving undue weight to the region of the curve around the hinge. A good alternative is to use multiple knots — not quite as simple but having good geometric behaviour. The formation of multiple knots in a B-spline basis was explained earlier, in section 3.3. Parametric curves defined with the new basis inherit its reduced continuity. For example, a hinge can be formed in a quadratic, parametric curve by introducing a double knot into the underlying quadratic B-spline basis, as in figure 3.13. Alternatively a triple knot introduces a break in a quadratic B-spline curve.

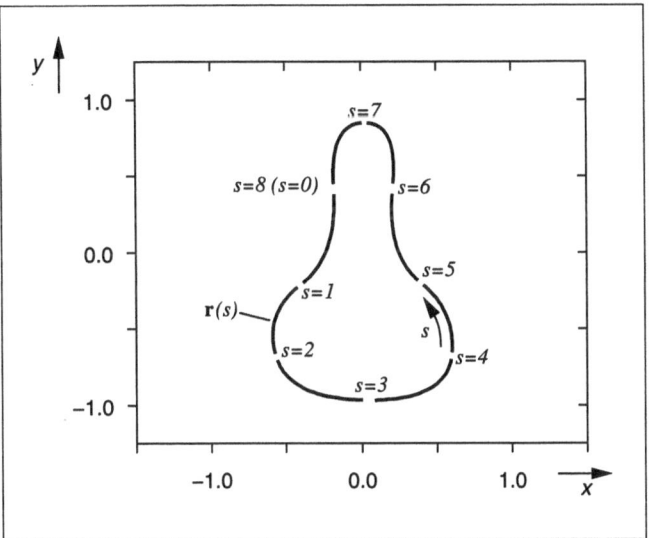

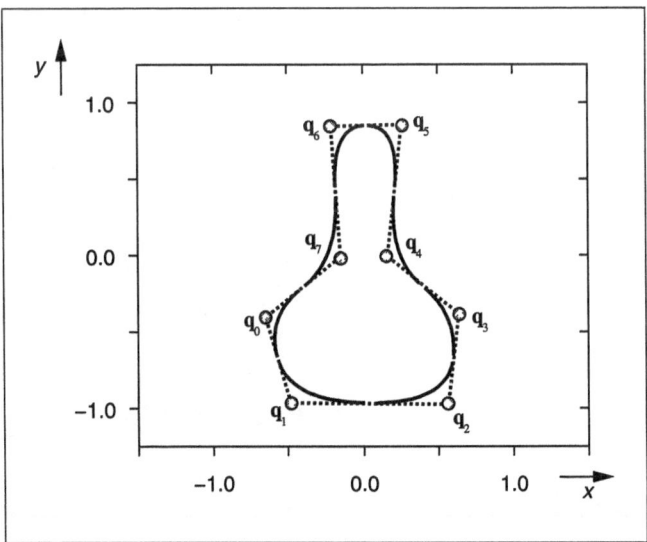

Figure 3.11: *A quadratic (d = 3), parametric spline curve* **r**(s) *is shown (top) that is regular, and closed. It has 8 knots, so L = 8, and* $0 \leq s \leq 8$, *and s is treated as periodic. The curve is a smooth approximation to its "control polygon" (bottom) formed from a sequence of control points* $\mathbf{q}_0, \mathbf{q}_1, \ldots, \mathbf{q}_7$. *Note that* $N_B = L$ *for regular closed spline curves — here* $N_B = L = 8$.

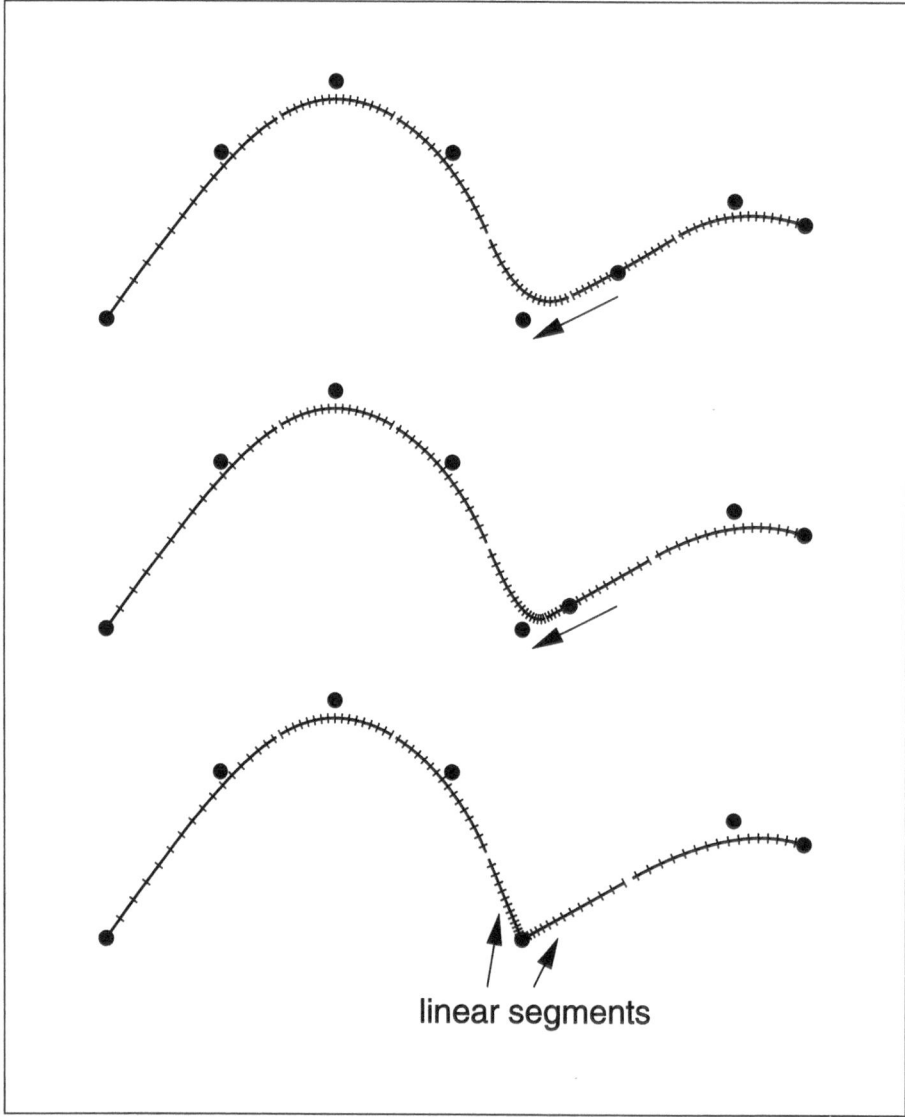

linear segments

Figure 3.12: Multiple control points. *On a quadratic, parametric spline curve, the curve tangent becomes discontinuous when two control points coincide, forming a "hinge." However, forming a hinge in this fashion turns out to be unsatisfactory, partly because of the segments on either side of the vertex which are constrained to be linear, and partly because of problems with uneven parameterisation — see text.*

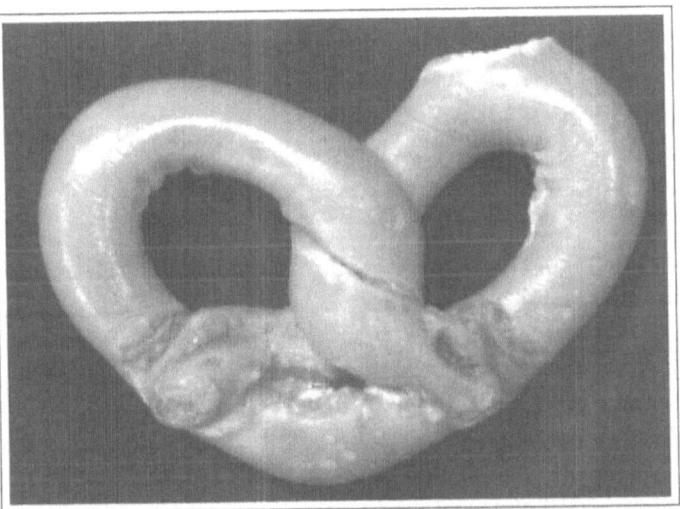

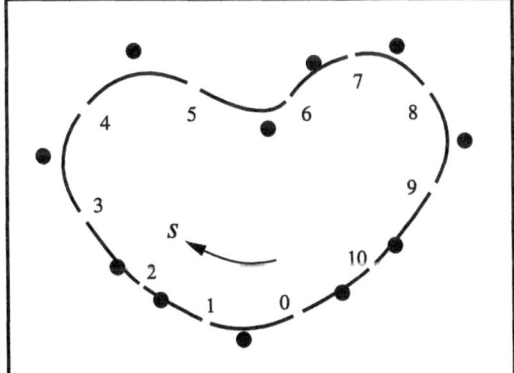 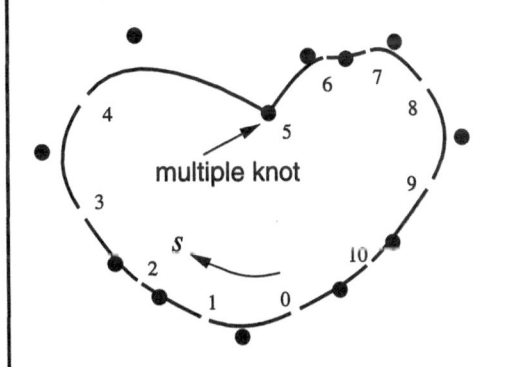

Figure 3.13: *The outline of the pretzel at the top is heart-shaped with a vertex at its apex. A regular, parametric spline curve (left) does not follow the outline as closely as one with a multiple knot (right). In the case of a quadratic spline curve, a tangent discontinuity forms at the control point corresponding to the double knot.*

3.7 Control vector

Dealing with control points explicitly is cumbersome so, as a first step towards a more compact notation, let us first define a space $\mathcal{S}_Q$ of *control vectors* $\mathbf{Q}$ consisting of control point coordinates, first all the x-coordinates, then all the y-coordinates:

$$\mathbf{Q} = \begin{pmatrix} \mathbf{Q}^x \\ \mathbf{Q}^y \end{pmatrix} \quad \text{where} \quad \mathbf{Q}^x = \begin{pmatrix} q_0^x \\ \cdots \\ \cdots \\ q_{N_B-1}^x \end{pmatrix} \tag{3.18}$$

and similarly for $\mathbf{Q}^y$. Then the coordinate functions can be written as

$$x(s) = \mathbf{B}(s)^T \mathbf{Q}^x,$$

where $\mathbf{B}(s)$ is a vector of B-spline basis functions as defined earlier, and similarly for $y(s)$, so that

$$\mathbf{r}(s) = U(s)\mathbf{Q} \quad \text{for} \ \ 0 \le s \le L \tag{3.19}$$

where

$$U(s) = I_2 \otimes \mathbf{B}(s)^T = \begin{pmatrix} \mathbf{B}(s)^T & \mathbf{0} \\ \mathbf{0} & \mathbf{B}(s)^T \end{pmatrix}, \tag{3.20}$$

a matrix of size $2 \times 2N_Q$. (Note that $\otimes$ denotes the "Kronecker product" of two matrices, a notation that will be used again. See appendix A for details. The matrix I_m denotes an $m \times m$ identity.)

3.8 Norm for curves

Now that we have set up a representation of curves as parametric splines, the next step is therefore to extend the norm and inner product to curves, for use in curve approximation. We can define a norm $\|\cdot\|$ for B-spline curves which is induced by the Euclidean distance measure in the image plane:

$$\|\mathbf{Q}\|^2 = \frac{1}{L} \int_{s=0}^{L} |\mathbf{r}(s)|^2 \, ds \tag{3.21}$$

or equivalently, from (3.19) and (3.10),

$$\|\mathbf{Q}\|^2 = \mathbf{Q}^T \mathcal{U} \mathbf{Q}, \tag{3.22}$$

where the metric matrix for curves $\mathcal{U}$ is defined in terms of the metric matrix $\mathcal{B}$ for B-spline functions:

$$\mathcal{U} \equiv \frac{1}{L} \int_0^L U(s)^T U(s) \, ds \qquad (3.23)$$

or, equivalently,

$$\mathcal{U} = I_2 \otimes \mathcal{B} = \begin{pmatrix} \mathcal{B} & 0 \\ 0 & \mathcal{B} \end{pmatrix}.$$

Of course, the norm also implies an inner product for curves:

$$\langle \mathbf{Q}_1, \mathbf{Q}_2 \rangle = \mathbf{Q}_1^T \mathcal{U} \mathbf{Q}_2.$$

The curve norm is particularly meaningful when used as a means of comparison between two curves, using the distance $\|\mathbf{Q}_1 - \mathbf{Q}_2\|$. This is the basis for approximation of visual data by curves, and is illustrated in figure 3.14.

There are potentially simpler norms than the one above, the obvious candidate being the Euclidean norm $|\mathbf{Q}|^2 \equiv \mathbf{Q}^T \mathbf{Q}$ of the control vector. Compared with the L_2 norm $\| \cdot \|$ defined above, the Euclidean norm is simpler to compute because the banded matrix $\mathcal{U}$ is replaced by the identity matrix. However, attractive as this short cut may be, the simpler norm does not work satisfactorily. This is made clear by the counter-example of figure 3.15 in which decreasing the displacement between two curves produces an increase in the Euclidean norm. This example makes it clear that the Euclidean norm is not suitable for ranking the closeness of curve approximations.

Invariance to re-parameterisation

It is important to note that the curve norm, as a measure of the difference between a pair of curves, does not allow for possible re-parameterisation of one of the curves. For example, a curve $\mathbf{r}(s)$, $0 \leq s \leq 1$ could be reparameterised to give a new curve $\mathbf{r}^*(s) = \mathbf{r}(1-s)$, $0 \leq s \leq 1$. Geometrically, the two curves are identical; the difference between them is simply a reversal of parameterisation. We would like ideally to measure shape differences in a way that is invariant to re-parameterisation, so that the vector difference function $\mathbf{r}^* - \mathbf{r}$ would ideally have a norm of zero. However, the L_2 norm will not behave in this way:

$$\|\mathbf{r}^*(s) - \mathbf{r}(s)\|^2 = \frac{1}{L} \int |\mathbf{r}(1-s) - \mathbf{r}(s)|^2 \, ds \neq 0,$$

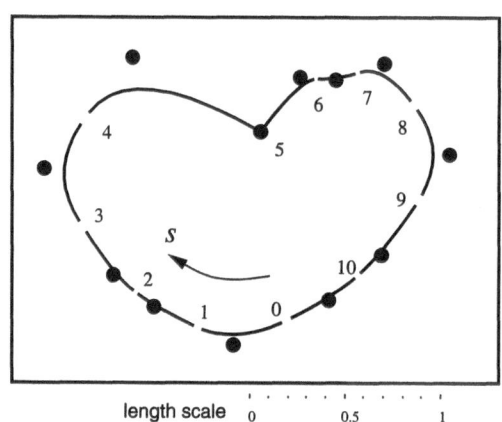

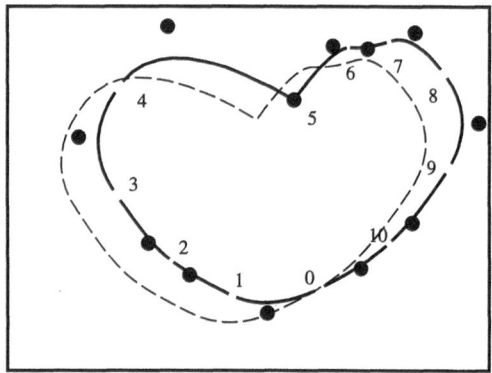
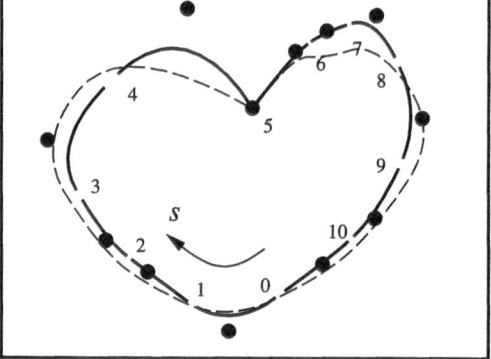

average displacement: norm= 0.22   average displacement: norm= 0.08

Figure 3.14: Measuring average displacement between curves. *The pretzel curve (top) is compared, using the norm as a measure, with two modified curves: a translation (left) and a modest distortion (right). The values given for the norm in each case represent average (root-mean-square) displacements for the curves, in length units as shown on the scale.*

in general. As a result, any curve-fitting algorithm that uses the norm will be disrupted if the parameterisation of the target curve fails to match that of the template, and this is illustrated in figure 3.16.

A general solution to the problem of parameterisation invariance would require a search over possible parameterisations. The proximity of a curve $\mathbf{r}(s)$ to a second

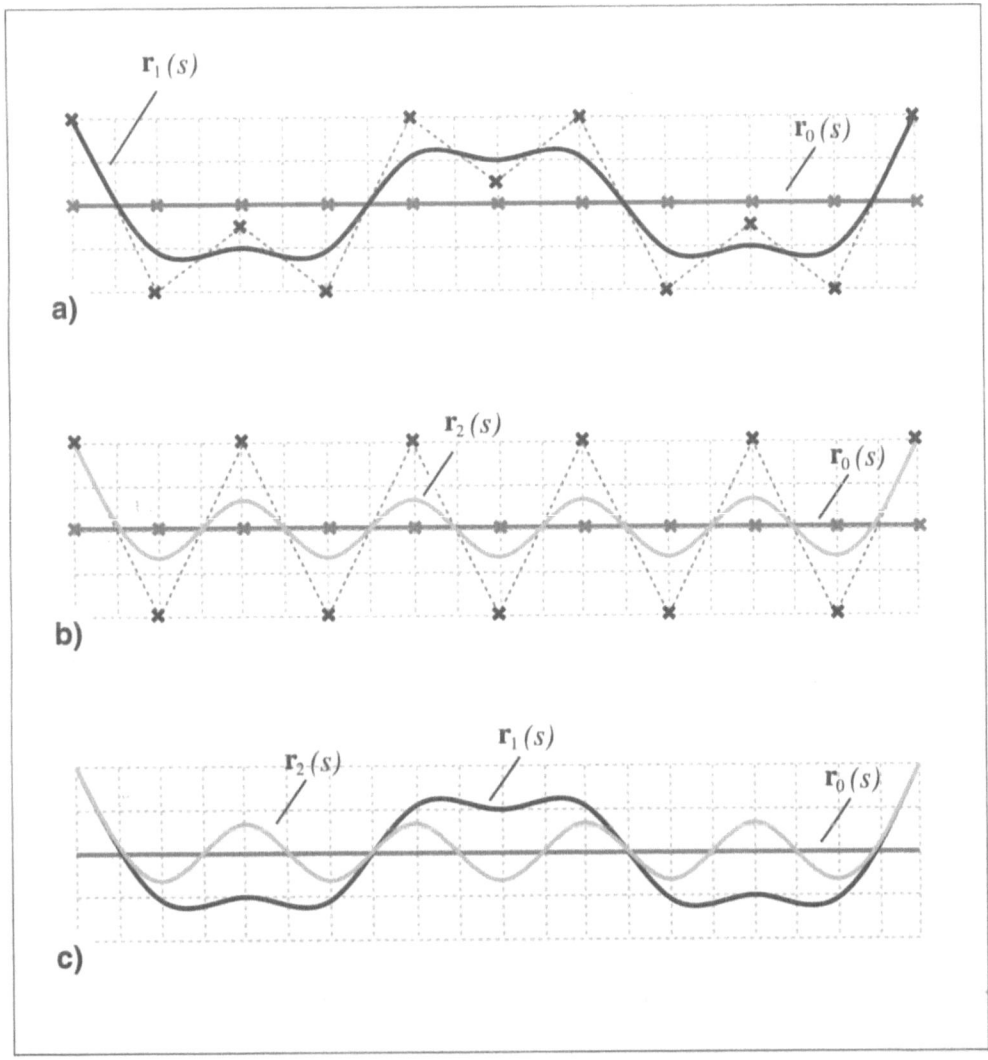

Figure 3.15: Euclidean distance between control vectors is a poor measure of curve displacement. *In a) and b) two cubic spline curves* $\mathbf{r}_1, \mathbf{r}_2$ *are shown together with their control points (crosses) and control polygons (dotted). Each is to be compared with a standard curve* $\mathbf{r}_0$. *It is clear that, pointwise,* $\mathbf{r}_2$ *is closer to* $\mathbf{r}_0$ *than* $\mathbf{r}_1$ *is. Any reasonable curve metric should reflect that and, sure enough, in the* L_2 *norm,* $\|\mathbf{r}_2 - \mathbf{r}_0\| < \|\mathbf{r}_1 - \mathbf{r}_0\|$. *However it is clear from the shape of the control polygons that, in Euclidean distance,* $\mathbf{r}_2$ *is actually further from* $\mathbf{r}_0$ *than* $\mathbf{r}_1$ *is. Euclidean distance between control vectors is therefore ruled out as a curve metric.*

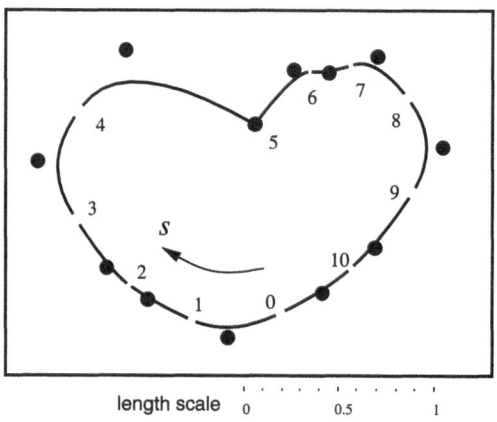

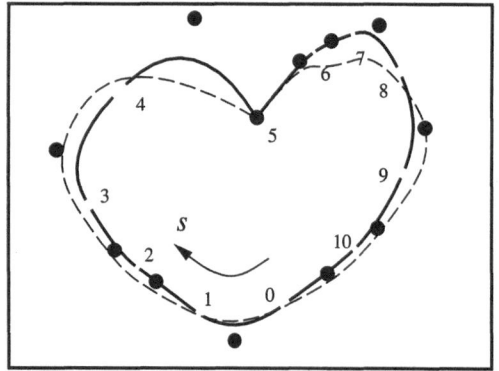
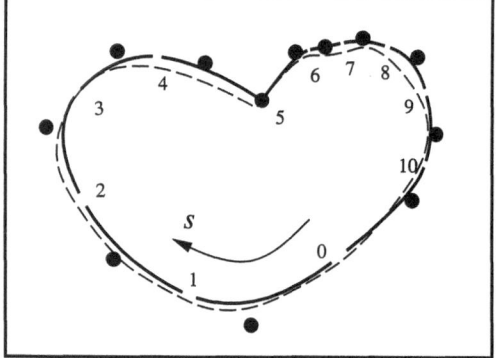

average displacement: norm= 0.08   average displacement: norm= 0.34

Figure 3.16: Norm-difference is sensitive to re-parameterisation. *A substantial change in parameterisation (right), though it gives rise to little change in shape, nonetheless registers a larger average displacement than the more distorted shape (left).*

curve $\mathbf{r}^*(s)$ could be evaluated as

$$\min_{g} \|\mathbf{r}(s) - \mathbf{r}^*(g(s))\|$$

where the minimum is explored over some space of re-parameterisation functions g, using an appropriate optimisation procedure. Suitably powerful optimisation procedures have been developed (see the bibliographic notes at the end of this chapter) but they are computationally costly. A more economical approach is to use a distance measure $d(\mathbf{r}, \mathbf{r}^*)$ that is invariant to *minor* re-parameterisation of the curve $\mathbf{r}$. This is developed later, in chapter 6. The distance measure is based on the *normal* displacement of $\mathbf{r}$ relative to $\mathbf{r}^*$, that is, omitting any tangential component of displacement — the sliding of one curve along the other. It will be shown that such a distance measure is approximately invariant and it remains to achieve a rough alignment of two curves sufficient for the invariant distance approximation to be effective. One way of doing this uses moments, as described below.

3.9   Areas and moments

Applications in computer vision often require the computation of gross properties of a curve. Curve moments — area, centroid, and higher moments are useful for computing approximate curve position and orientation, and to obtain gross shape information sufficient for some coarse discrimination between objects.

Generally, moments have two roles in active contours. The first is initialisation, in which a spline template is positioned sufficiently close to the tracked object to "lock" onto it. At this stage moments may also be used for coarse shape discrimination to confirm the identity of the object being tracked. The second role is in interpreting the position and orientation of a tracked object, for example the 3D visual mouse in figure 1.16 on page 20. For example, if hand motion is restricted to 3D translation and rotation in the image plane, those four parameters can be recovered from the zeroth moment (area) the first moment (centroid) and the second moment (inertia).

Centroid

A conventional definition for the centroid of a curve is

$$\bar{\mathbf{r}} = \frac{1}{L} \int_0^L \mathbf{r}(s)\, ds$$

which can be computed straightforwardly from the spline-vector $\mathbf{Q}$ using inner products:

$$\bar{\mathbf{r}} = \begin{pmatrix} \langle \mathbf{1}_x, \mathbf{Q} \rangle \\ \langle \mathbf{1}_y, \mathbf{Q} \rangle \end{pmatrix} \quad \text{where} \quad \mathbf{1}_x = \begin{pmatrix} 1 \\ 0 \end{pmatrix} \quad \text{and} \quad \mathbf{1}_y = \begin{pmatrix} 0 \\ 1 \end{pmatrix}. \tag{3.24}$$

This simple definition of centroid is computationally convenient but has the drawback that it is not invariant to re-parameterisation of the curve, as figure 3.17 shows. This

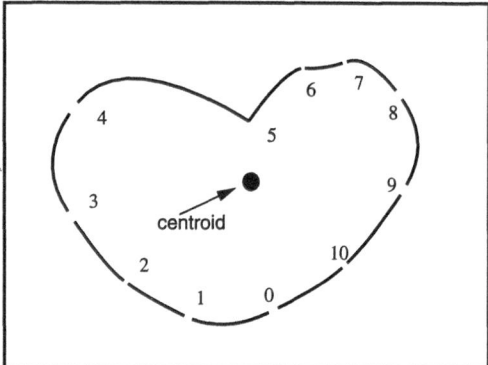 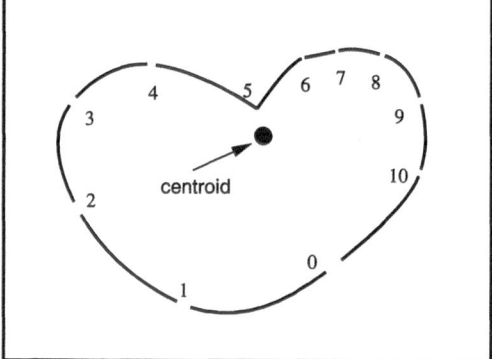

Figure 3.17: Simple centroid is not invariant. *The centroid of the pretzel curve (left), calculated as in (3.24), shifts substantially when the curve is reparameterised (right), even though there is almost no shape change.*

is because s is not generally true arclength; it is simply a convenient spline parameter. The length of an infinitesimal segment of curve is not ds but $|\mathbf{r}'(s)|ds$, so that an invariant centroid of the curve would be

$$\bar{\mathbf{r}} = \frac{\int_0^L \mathbf{r}(s)|\mathbf{r}'(s)|\,ds}{\int_0^L |\mathbf{r}'(s)|\,ds}.$$

The square root implicit in $|\mathbf{r}'(s)|$ means that this invariant centroid cannot be computed directly in terms of the spline-vector $\mathbf{Q}$. An alternative invariant centroid for closed curves is the centroid of area described below, which can be computed directly, as a cubic function of $\mathbf{Q}$. For many purposes, non-invariant moments are adequate. For example, in a hand-tracking application the parameterisation of the tracked curve is, typically, strongly stabilised by a template. The parameterisation of the tracked curve does not, in practice, deviate much from the standard parameterisation inherited from the template. In that case the 2D translational motion of the tracked object can be recovered satisfactorily from the non-invariant centroid.

Invariant moments

Suppose an active contour is to be initialised from an area of pixels detected by image processing based on brightness, colour or motion. The vector $\mathbf{Q}$ for the initial configuration of the tracked curve is set by manipulating it to bring the moments of the area enclosed by the curve into close agreement with the moments of the active area.

The simplest available parameterisation-invariant measure for a closed curve is the area

$$\int |\mathbf{r}(s), \mathbf{r}'(s)| \, ds$$

where $|\mathbf{x}, \mathbf{y}|$ denotes the determinant of the matrix whose columns are $\mathbf{x}, \mathbf{y}$. This is neatly expressible as a quadratic form in $\mathbf{Q}$:

$$A(\mathbf{Q}) = \mathbf{Q}^T \mathcal{A} \mathbf{Q}, \tag{3.25}$$

reminiscent of the norm in (3.22) but in place of the symmetric matrix $\mathcal{U}$ we have:

$$\mathcal{A} \equiv \begin{pmatrix} \mathcal{B}' & 0 \\ 0 & -\mathcal{B}' \end{pmatrix} \quad \text{where} \quad \mathcal{B}' = \int_0^L \mathbf{B}(s)\mathbf{B}'^T(s) \, ds. \tag{3.26}$$

(Details of efficient computation of $\mathcal{A}$ are given in appendix A.2.) As with the matrix $\mathcal{U}$, $\mathcal{A}$ is $2N_Q \times 2N_Q$ where

$$N_Q = 2N_B, \tag{3.27}$$

the dimension of the spline space. The matrix $\mathcal{A}$ is sparse which makes the computation of the area quadratic form relatively efficient. One direct application for curve area computation is in visual navigation, and a picturesque example is given in figure 3.18.

The centroid $\bar{\mathbf{r}}$ of the area enclosed by a closed B-spline curve, which is invariant to curve re-parameterisation, is given by

$$\bar{\mathbf{r}} = \frac{1}{A(\mathbf{Q})} \int_0^L |\mathbf{r}(s), \mathbf{r}'(s)| \, \mathbf{r}(s) \, ds. \tag{3.28}$$

In principle this is a useful measure for positioning, but is moderately costly — $O(N_Q^3)$ — to compute exactly. This improves to $O(N_X^3)$ if curves are restricted to a "shape-space" of reduced dimension N_X, and this is discussed in the next chapter. Similarly,

Figure 3.18: *While a travelling video camera approaches a car, a B-spline curve is locked onto the outline of the windscreen in successive video frames. The computed windscreen area $a(t)$, increasing over time, can be used to estimate time-to-collision as $a(t)/\dot{a}(t)$.* (Figure reproduced from (Cipolla and Blake, 1992a).)

the second moment

$$\mathcal{I} = \frac{1}{A(\mathbf{Q})} \int_0^L |\mathbf{r}(s), \mathbf{r}'(s)| \, \mathbf{r}(s)\mathbf{r}^T(s) \, ds \qquad (3.29)$$

is invariant and useful in principle for orienting a shape, but the computational cost is $O(N_Q^4)$, again reduced if a shape-space is used.

Bibliographic notes

This chapter has outlined a framework for representing curves in the image plane. It has been common both in robotics and in computer vision to represent curves algebraically as $f(x, y) = 0$ where f is a polynomial (Faverjon and Ponce, 1991; Petitjean et al., 1992; Forsyth et al., 1990). Although such representations are often attractive mathematically, for the purpose of constructing proofs, they are cumbersome from the computational point of view. Practical systems using curve approximation are better founded on B-splines. Tutorials on splines can be found in graphics books such as (Foley et al., 1990) or in books on computer-aided design such as (Faux and Pratt, 1979). Some essential details and algorithms are given also in the appendix of this book. A more complete book on splines, oriented toward computer graphics is (Bartels et al., 1987) and a mathematical source on spline functions (but not curves) is (de Boor, 1978).

Splines, common in computer graphics, have also been used in computer vision for some years, for shape-warping (Bookstein, 1988), representing corners and edges in static scenes (Medioni and Yasumoto, 1986; Arbogast and Mohr, 1990) and for shape approximation (Menet et al., 1990) and tracking (Cipolla and Blake, 1990).

Shape approximation using spline curves is an application of the "normal equations" for approximation problems (Press et al., 1988). Equivalently it uses a "pseudo-inverse" (Barnett, 1990) of which the B-spline metric matrix $\mathcal{B}$ is a component. Function norms are a standard mathematical tool for functional approximation (Kreysig, 1988) and in signal processing (Papoulis, 1991) and image processing (Gonzales and Wintz, 1987) for least-squares restoration problems.

Measures of curve difference that are more economical to compute than the L_2 norm can be made by replacing the metric matrix with the identity to give the Euclidean distance between control vectors, as done for polygons in (Cootes and Taylor, 1992). However, such measures do have some undesirable properties, as explained earlier. Curve matching using norms is not invariant to re-parameterisation; matching algorithms do exist that deal with re-parameterisation, for example ones developed for stereoscopic image matching (Ohta and Kanade, 1985; Witkin et al., 1986) but they are computationally expensive, too much so for use in real-time tracking systems. This is discussed again in chapter 6.

Chapter 4

Shape-space models

In practice, it is very desirable to distinguish between the *spline*-vector $\mathbf{Q} \in \mathcal{S}_Q$ that describes the basic shape of an object and the *shape*-vector which we denote $\mathbf{X} \in \mathcal{S}$, where $\mathcal{S}$ is a *shape-space*. Whereas $\mathcal{S}_Q$ is a vector space of B-splines and has dimension $N_Q = 2N_B$, the shape-space $\mathcal{S}_X$ is constructed from an underlying vector space of dimension N_X which is typically considerably smaller than N_Q. The shape-space is a linear parameterisation of the set of allowed deformations of a base curve. The necessity for the distinction is made clear in figure 4.1. To obtain a spline that does justice to the geometric complexity of the face shape, thirteen control points have been used. However, if all of the resulting 26 degrees of freedom of the spline-vector $\mathbf{Q}$ are manipulated arbitrarily, many uninteresting shapes are generated that are not at all reminiscent of faces. Restricting the displacements of control points to a lower-dimensional shape-space is more meaningful if it preserves the face-like quality of the shape. Conversely, using the unconstrained control-vector $\mathbf{Q}$ leads to unstable active contours and this was illustrated in figure 2.4 on page 31.

The requirement that a shape-space be a *linear* parameterisation is made for the sake of computational simplicity. The curve-fitting and tracking procedures described in the book are substantially simplified by linearity and in many cases exact algorithms are available only for linear parameterisations. Linearly parameterised, image-based models work well for rigid objects however, and for simpler non-rigid ones. Linearity can certainly be a limitation when the allowed motions of an object become more complex, for example a three-dimensional object with articulated parts. Articulation can in fact be dealt with in linearly parameterised, image-based models but only at the cost of relaxing certain geometric constraints. This is explored further in the

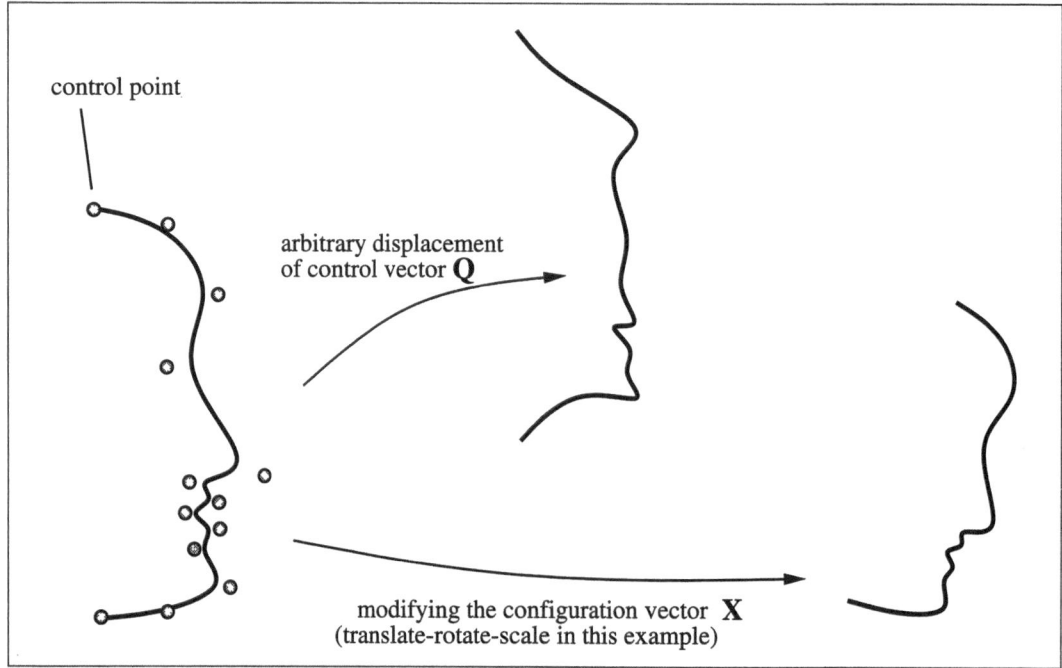

Figure 4.1: Configuration vector. *Arbitrary manipulation of the spline-vector* **Q** *of a spline curve is likely to be too general to be practically interesting. In this example a face curve ceases to look face-like. What is far more interesting is a restricted class S of transformations, parameterised by a relatively low-dimensional configuration vector* **X**. *In this case* **X** *is a Euclidean similarity transformation which does retain the face-like character.*

discussion of shape-spaces below. A more detailed discussion of the trade-off between image-based models and three-dimensional models is given in appendix C.

4.1   Representing transformations in shape-space

Rigid motion

A simple example of a shape-space is the space of Euclidean similarities of a template curve $\mathbf{r}_0(s)$. This is a space of dimension 4 corresponding exactly to the variation of an image curve as a camera with a zoom lens looks directly down on a planar object that is free to move on a table top. The effect on the curve is that it moves rigidly in the image plane and may also magnify or diminish in size, but its shape is preserved, as in figure 4.2. Alternatively it could be that the camera is able to translate in three

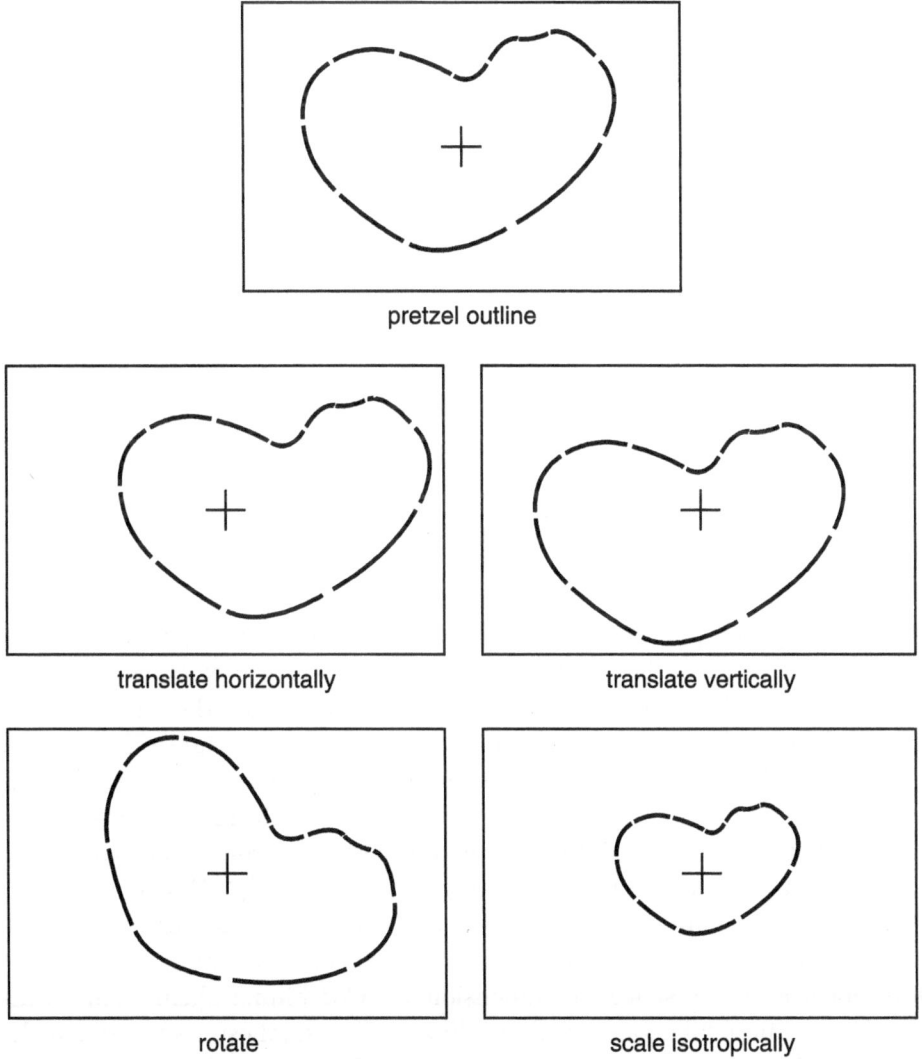

Figure 4.2: Euclidean similarities. *The shape-space of Euclidean similarities has 4 degrees of freedom. They are depicted here as applied to the outline of a pretzel (the pretzel from figure 3.13 on page 57 of the previous chapter).*

dimensions and to rotate about an axis perpendicular to the table top (something that occurs, for example, when a camera is mounted on a "SCARA" arm, popular in robot automation, which translates freely and rotates in the plane of the table). This combination also sweeps out a shape-space of Euclidean similarities.

Another important shape-space is the one that arises when a planar object has complete freedom to move in three dimensions. Its motion has 6 degrees of freedom, three for translation and three for rotation. Provided perspective effects are not too great, the image of a planar object contour is well described as a shape-space of planar affine transformations, a space with dimension 6. It can be thought of as the space of linear transformations of a template. Alternatively, it is the space of transformations which preserves parallelism between lines. The planar affine group of transformations is depicted in figure 4.3. Figure 4.4 illustrates how the planar affine shape-space can enhance active contours when used appropriately. The figure shows tracking of an outstretched hand which, being almost planar, is well modelled by a planar affine space. The increased degree of constraint enhances immunity to distraction from clutter in the background.

The planar affine and Euclidean similarity shape-spaces work efficiently in the sense that the dimension of the shape-space is exactly equal to the number (six/four respectively) of the degrees of freedom of camera movement. Unfortunately this happy state of affairs does not persist in general because transformation groups do not necessarily form vector spaces; it is not always possible to find a vector space which matches exactly the degrees of freedom of camera/object motion. Consider the case of a camera with fixed magnification, viewing a planar object moving rigidly on a table. The image curve translates and rotates rigidly without any change of size. This is now simply the planar Euclidean group which does not however form a vector space. To see this, consider the template $\mathbf{r}_0(s)$ and a copy of it rotated through 180° to give $-\mathbf{r}_0(s)$; when these two are added vectorially they give $\mathbf{r}_0(s) + (-\mathbf{r}_0(s)) = \mathbf{0}$ which is not a rotated version of the template at all. The rotation operation is therefore not "closed" under addition and therefore cannot form a vector space. Rotation and scaling taken *jointly* do form a vector space, of dimension 2. Combining them with translation gives the Euclidean similarities, of dimension 4. The smallest vector space that encompasses Euclidean transformations is therefore the space of Euclidean similarities. The price of insisting on a linear representation of the Euclidean transformations is that 4 dimensions are needed to represent 3 degrees of freedom; the resulting space is underconstrained by one degree of freedom.

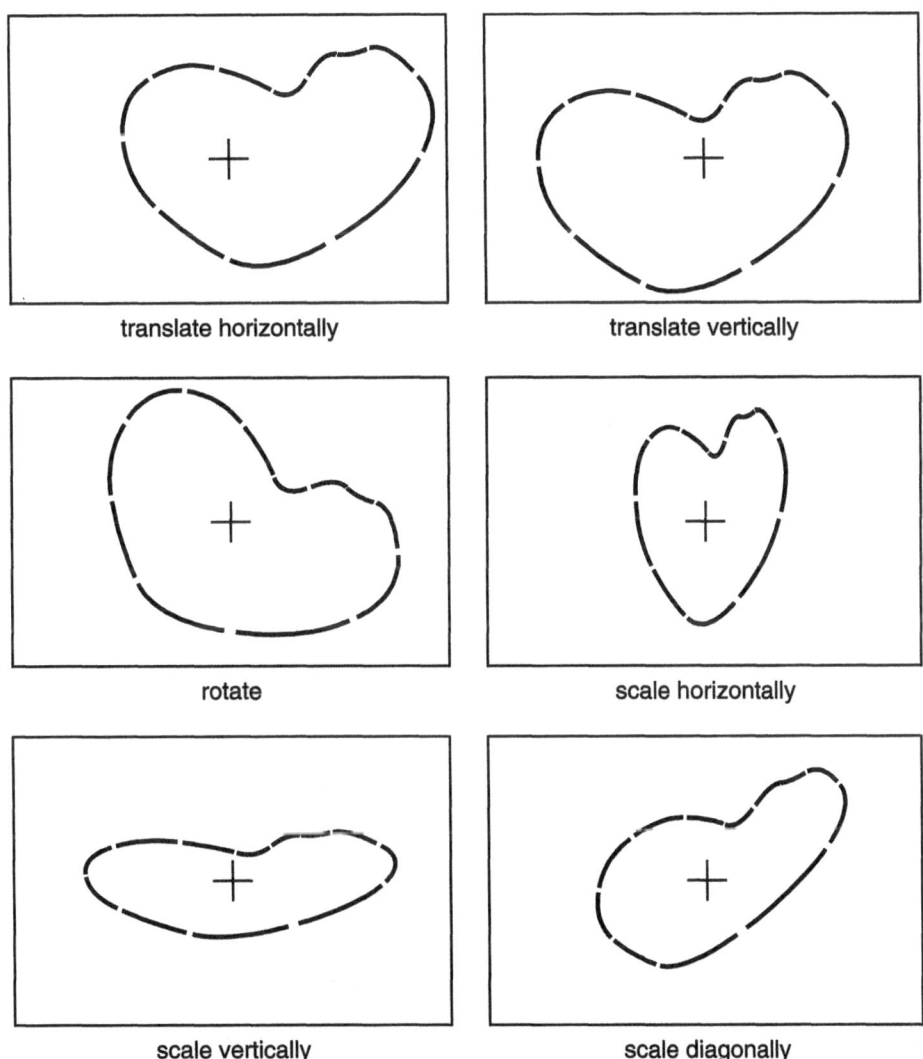

translate horizontally

translate vertically

rotate

scale horizontally

scale vertically

scale diagonally

Figure 4.3: Planar affine basis. *The planar affine transformation group has 6 degrees of freedom. A basis for them is depicted here, as applied to the pretzel outline from figure 4.2 on page 71. The first three elements of the basis correspond to the first three for the Euclidean similarities. The last three elements span a subspace that includes the fourth element — scaling — for the Euclidean similarities and two further degrees of freedom for directional scaling. Directional scaling occurs when a planar object, initially co-planar with the image, is allowed to rotate about an axis that lies parallel to the image plane.*

Spline space   Affine shape-space

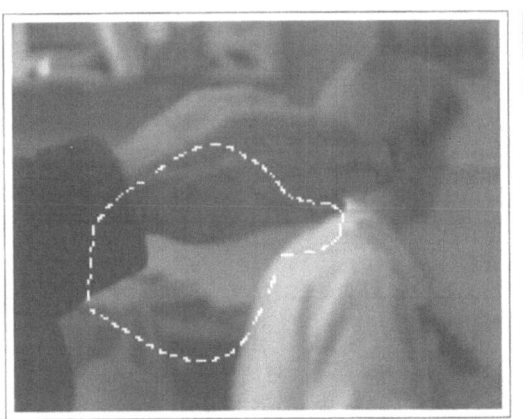 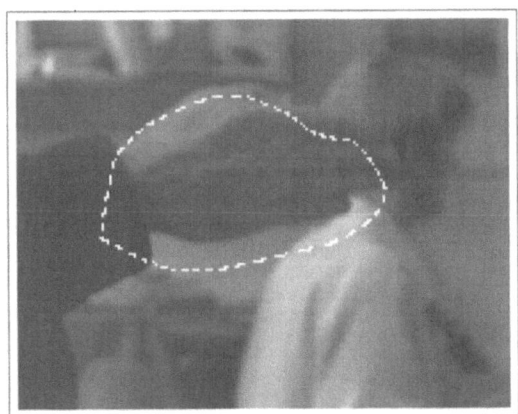

Figure 4.4: Shape-space can impose constraints that allow background clutter to be ignored. *A test sequence of hand motion against background clutter consists of vertical oscillation at around 0.5 Hz. The tracking algorithm developed later in the book (chapter 10) is applied over spline space and over an affine shape-space. The figure depicts snapshots after 9 seconds of tracking. It appears that the use of a planar affine shape-space confers enhanced immunity to background clutter.*

Definition of shape-space

At this stage, a more precise definition of shape-space is called for. A shape-space $\mathcal{S} = \mathcal{L}(W, \mathbf{Q}_0)$ is a linear mapping of a "shape-space vector" $\mathbf{X} \in \mathbb{R}^{N_X}$ to a spline-vector $\mathbf{Q} \in \mathbb{R}^{N_Q}$:

$$\mathbf{Q} = W\mathbf{X} + \mathbf{Q}_0, \tag{4.1}$$

where W is a $N_Q \times N_X$ "shape-matrix." The constant offset $\mathbf{Q}_0$ is a template curve against which shape variations are measured; for instance, a class of shapes consisting of $\mathbf{Q}_0$ and curves close to $\mathbf{Q}_0$ could be expressed by restricting the shape-space $\mathcal{S}$ to "small" $\mathbf{X}$. The image of $\mathbb{R}^{N_X}$ need not necessarily be a vector space itself but is a "coset" — an underlying vector space $\{W\mathbf{X}, \ \mathbf{X} \in \mathbb{R}^{N_X}\}$ plus an offset $\mathbf{Q}_0$. We talk of the "basis" $\mathcal{V}$ of a shape-space meaning a basis for the underlying vector space. The matrix W is comprised of columns which are the vectors of the basis $\mathcal{V}$. In fact the two spaces discussed in this chapter — Euclidean similarity and Affine, are vector spaces, because there exists an $\mathbf{X}$ for which $\mathbf{Q} = W\mathbf{X}$. In chapter 8 we encounter shape-spaces whose images are not vector spaces because the offset $\mathbf{Q}_0$ is linearly independent of the basis $\mathcal{V}$. In fact the simplest shape-space that is not a pure vector space is the space of translations of a template $\mathbf{Q}_0$.

4.2   The space of Euclidean similarities

The Euclidean similarities of a template curve $\mathbf{r}_0(s)$ represented by $\mathbf{Q}_0$ form a 4-dimensional shape-space $\mathcal{S}$ with shape-matrix

$$W = \begin{pmatrix} \mathbf{1} & \mathbf{0} & \mathbf{Q}_0^x & -\mathbf{Q}_0^y \\ \mathbf{0} & \mathbf{1} & \mathbf{Q}_0^y & \mathbf{Q}_0^x \end{pmatrix} \tag{4.2}$$

where the N_B-vectors $\mathbf{0}$ and $\mathbf{1}$ are:

$$\mathbf{0} = (0, 0, \dots, 0)^T, \quad \mathbf{1} = (1, 1, \dots, 1)^T.$$

The first two columns of W govern horizontal and vertical translations respectively. The third and fourth columns, made up from components of the spline-vector $\mathbf{Q}_0$ for the template, cover rotation and scaling. By convention, we choose $\mathbf{Q}_0$ to have its centroid at the origin ($< \mathbf{Q}_0^x, \mathbf{1} >=< \mathbf{Q}_0^y, \mathbf{1} >= 0$) so that the third and fourth columns are associated with *pure* rotation and scaling, free of translation. In practice the template is obtained by fitting a spline interactively around a standard view of the shape, and translating it so that its centroid lies over the origin.

Some examples of shape representations in the space of Euclidean similarities follow.

1. $\mathbf{X} = (0, 0, 0, 0)^T$ represents the original template shape $\mathbf{Q}_0$

2. $\mathbf{X} = (1, 0, 0, 0)^T$ represents the template translated 1 unit to the right, so that, from (4.1),

$$\mathbf{Q} = \mathbf{Q}_0 + \begin{pmatrix} \mathbf{1} \\ \mathbf{0} \end{pmatrix}$$

3. $\mathbf{X} = (0, 0, 1, 0)^T$ represents the template doubled in size

$$\mathbf{Q} = 2\mathbf{Q}_0$$

4. $\mathbf{X} = (0, 0, \cos\theta - 1, \sin\theta)^T$ represents the template rotated through angle θ:

$$\mathbf{Q} = \begin{pmatrix} \cos\theta \ \mathbf{Q}_0^x - \sin\theta \ \mathbf{Q}_0^y \\[1em] \sin\theta \ \mathbf{Q}_0^x + \cos\theta \ \mathbf{Q}_0^y \end{pmatrix}.$$

As an example, the lotion bottle in figure 4.5 moves rigidly from the template configuration $\mathbf{X} = 0$ in shape-space to the configuration

$$\mathbf{X} = (0.465, 0.047, -0.282, -0.698)^T$$

representing a translation through $(0.465, 0.047)^T$, almost horizontal as the figure shows, a magnification by a factor

$$\sqrt{(1 - 0.282)^2 + 0.698^2} = 1.001$$

and a rotation through

$$\arctan(1 - 0.282, -0.698) = -44.2^o$$

— all consistent with the figure.

4.3   Planar affine shape-space

It was claimed that for a planar shape just six *affine* degrees of freedom are required to describe, to a good approximation, the possible shapes of its bounding curve. The planar affine group can be viewed as the class of all *linear* transformations that can be applied to a template curve $\mathbf{r}_0(s)$:

$$\mathbf{r}(s) = \mathbf{u} + M\mathbf{r}_0(s), \tag{4.3}$$

where $\mathbf{u} = (u_1, u_2)^T$ is a two-dimensional translation vector and M is a 2×2 matrix, so that $M, \mathbf{u}$ between them represent the 6 degrees of freedom of the space. This class can be represented as a shape-space with template $\mathbf{Q}_0$ and shape-matrix:

$$W = \begin{pmatrix} 1 & 0 & \mathbf{Q}_0^x & 0 & 0 & \mathbf{Q}_0^y \\ 0 & 1 & 0 & \mathbf{Q}_0^y & \mathbf{Q}_0^x & 0 \end{pmatrix}. \tag{4.4}$$

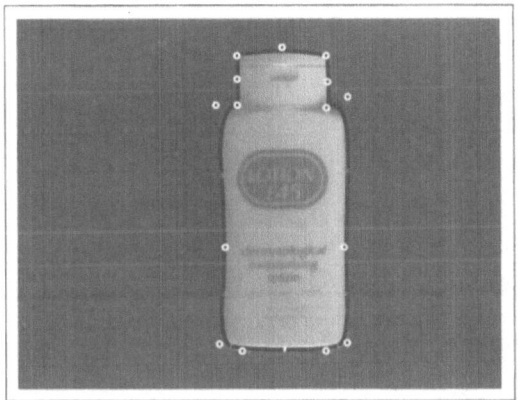
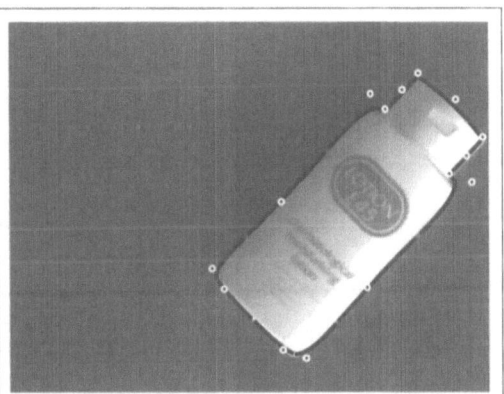
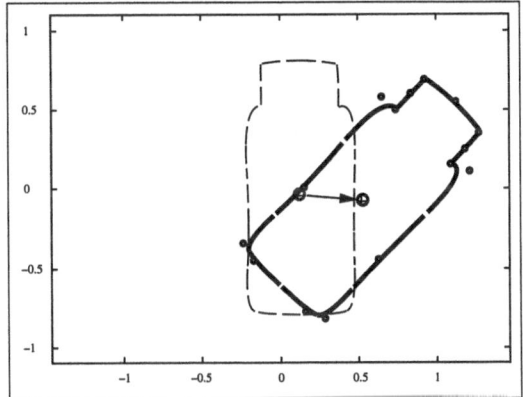

Figure 4.5: Euclidean similarities. *The outline of a bottle in a standard position (left) is taken as a template for a shape-space of Euclidean similarities in which any new outline (right) can be described. In this case the new outline is displaced and rotated relative to template (bottom) and this is apparent from the shape-space representation — see text.*

(A derivation is given below.) The first two columns of W represent horizontal and vertical translation. As before, by convention, the template $\mathbf{r}_0(s)$ represented by $\mathbf{Q}_0$ is chosen with its centroid at the origin. Then the remaining four affine motions (figure 4.3), which do not correspond one-for-one the last four columns of W, can however be expressed as simple linear combinations of those columns. Recall that the shape-space transformation is $\mathbf{Q} = W\mathbf{X} + \mathbf{Q}_0$ so that the elements of $\mathbf{X}$ act as weights on the columns of W. The interpretation of those weights in terms of planar transformations (4.3) of the template is:

$$\mathbf{X} = (u_1, u_2, M_{11} - 1, M_{22} - 1, M_{21}, M_{12})^T. \tag{4.5}$$

Some examples of transformations are:

1. $\mathbf{X} = (0,0,0,0,0,0)^T$ represents the original template shape $\mathbf{Q}_0$

2. $\mathbf{X} = (1,0,0,0,0,0)^T$ represents the template translated 1 unit to the right,

3. $\mathbf{X} = (0,0,1,1,0,0)^T$ represents the template doubled in size

4. $\mathbf{X} = (0,0,\cos\theta - 1, \cos\theta - 1, -\sin\theta, \sin\theta)^T$ represents the template rotated through angle θ

5. $\mathbf{X} = (0,0,1,0,0,0)^T$ represents the template doubled in width

In practice it is convenient to arrange for the elements of the affine basis to have similar magnitudes to improve numerical stability. If the control-vector $\mathbf{Q}_0$ is expressed in pixels, for computational simplicity, the magnitudes of the last four columns of the shape-matrix may be several hundred times larger than those of the first two, and it is then necessary to scale the translation columns to match.

Derivation of affine basis. Using (3.19), (4.3) can be rewritten:

$$\mathbf{r}(s) - \mathbf{r}_0(s) = \mathbf{u} + (M - I)U(s)\mathbf{Q}_0.$$

Now using the definition (3.20) of $U(s)$ and noting that $\mathbf{B}(s)^T\mathbf{1} = 1$ (3.5), this becomes:

$$\mathbf{r}(s) - \mathbf{r}_0(s) = \begin{pmatrix} u_1 \mathbf{B}^T(s)\mathbf{1} \\ u_2 \mathbf{B}^T(s)\mathbf{1} \end{pmatrix} + \begin{pmatrix} (M_{11}-1)\mathbf{B}^T(s)\mathbf{Q}_0^x + M_{12}\mathbf{B}^T(s)\mathbf{Q}_0^y \\ M_{21}\mathbf{B}^T(s)\mathbf{Q}_0^x + (M_{22}-1)\mathbf{B}^T(s)\mathbf{Q}_0^y \end{pmatrix}$$

$$= u_1 U(s)\begin{pmatrix} 1 \\ 0 \end{pmatrix} + u_2 U(s)\begin{pmatrix} 0 \\ 1 \end{pmatrix} + (M_{11}-1)U(s)\begin{pmatrix} \mathbf{Q}_0^x \\ 0 \end{pmatrix}$$

$$+ M_{12}U(s)\begin{pmatrix} \mathbf{Q}_0^y \\ 0 \end{pmatrix} + M_{21}U(s)\begin{pmatrix} 0 \\ \mathbf{Q}_0^x \end{pmatrix} + (M_{22}-1)U(s)\begin{pmatrix} 0 \\ \mathbf{Q}_0^y \end{pmatrix}.$$

From (3.19),

$$\mathbf{r}(s) - \mathbf{r}_0(s) = U(s)(\mathbf{Q} - \mathbf{Q}_0)$$

and comparing this with the expression for $\mathbf{r}(s) - \mathbf{r}_0(s)$ above shows that $\mathbf{Q} - \mathbf{Q}_0$ belongs to a vector space of dimension 6, for which the rows of W in (4.4) form a basis, and furthermore, given that $\mathbf{Q} = W\mathbf{X} + \mathbf{Q}_0$, $\mathbf{X}$ is composed of elements of M and $\mathbf{u}$ as in (4.5).

4.4   Norms and moments in a shape-space

Given that it is generally preferred to work in a shape-space $\mathcal{S}$, a formula for the curve norm is needed that applies to the shape-space parameter $\mathbf{X}$. We require a consistent definition so that, for a given space, $\|\mathbf{Q}_1 - \mathbf{Q}_2\| = \|\mathbf{X}_1 - \mathbf{X}_2\|$. The L_2 norm in shape-space $\mathcal{S}$ is said to be "induced" from the norm over $\mathcal{S}_Q$, which was in turn induced from the L_2 norm over the space of curves $\mathbf{r}(s)$. From (4.1), this is achieved by defining:

$$\|\mathbf{X}\| = \sqrt{\mathbf{X}^T \mathcal{H} \mathbf{X}}, \tag{4.6}$$

where

$$\mathcal{H} = W^T \mathcal{U} W. \tag{4.7}$$

The norm over $\mathcal{S}$ has a geometric interpretation:

$$\|\mathbf{X}\| = \|\mathbf{Q} - \mathbf{Q}_0\|$$

is the average displacement of the curve parameterised by $\mathbf{X}$ from the template curve. We can also now define a natural mapping from $\mathcal{S}_Q$ onto the shape-space $\mathcal{S}$. Of course there is in general no inverse of the mapping W in (4.1) from $\mathcal{S}_Q$ to $\mathcal{X}$ but, providing W has full rank (its columns are linearly independent), a pseudo-inverse W^+ can be defined:

$$\mathbf{X} = W^+(\mathbf{Q} - \mathbf{Q}_0) \quad \text{where} \quad W^+ = \mathcal{H}^{-1} W^T \mathcal{U}. \tag{4.8}$$

It turns out (see chapter 6) that W^+ can be naturally interpreted as an error-minimising *projection* onto shape-space.

In the case of spline space, it was argued in the previous chapter, the Euclidean norm $|\cdot|$ defined by $|\mathbf{Q}|^2 \equiv \mathbf{Q}^T \mathbf{Q}$ is not as natural as the L_2-norm $\|\mathbf{Q}\|$, although the two can have approximately similar values in practice, especially when curvature is small. Their approximate similarity derives from the fact that the metric matrix $\mathcal{U}$ is banded. However, the matrix $\mathcal{H}$ above is dense and so the Euclidean norm $|\mathbf{X}|$ in shape-space does not approximate the induced L_2-norm, and in fact $|\mathbf{X}|$ has no obvious geometric interpretation. Therefore, while one might get away with using the Euclidean norm in spline space, it is of no use at all in shape-space — only the L_2-norm will do.

Computing Area

As with the norm, the area form $A(\mathbf{X})$ can be expressed in a shape-space as a function

$$A(\mathbf{X}) = (W\mathbf{X} + \mathbf{Q}_0)^T \mathcal{A}(W\mathbf{X} + \mathbf{Q}_0)$$

that is quadratic in $\mathbf{X}$, and whose quadratic and linear terms involve just $N_X(N_X + 3)/2$ coefficients so that, in the case of Euclidean similarity, there are just 14 independent coefficients.

Centroid and inertia

The centroid $\bar{\mathbf{r}}$ of the area enclosed by a closed B-spline curve ((3.28) on page 65) is a symmetric cubic function of the configuration $\mathbf{X}$. Such a function has $O((N_X)^3)$ terms which is obviously large for larger shape-spaces, but works out to be just 20 terms in the case of Euclidean similarities — quite practical to compute. (The exact formula for the number of terms is $N_X(N_X + 1)(N_X + 2)/6$.) The invariant second moment or inertia matrix ((3.29) on page 67) could be expressed in terms of a symmetric quartic form which, surprisingly, has only 23 terms in the case of Euclidean similarity $(N_X = 4)$ but of course this number is $O((N_X)^4)$ in general. In cases where the size of the configuration space is too large for efficient computation of invariant moments, the alternative is to compute them by numerical integration.

Finally, note that there is an important special case in which invariant moments are easily computed. The special properties of the affine space mean that moments can be computed efficiently under affine transformations. For instance, although the invariant second moment for a 6-dimensional shape-space turns out generally to be a quartic polynomial with 101 terms, in the affine space it can be computed simply as a product of three 2×2 matrices:

$$\mathcal{I} = M\mathcal{I}_0 M^T$$

where $\mathcal{I}_0$ is the inertia matrix of the template, computed numerically from (3.28) on page 65. Similarly, for area:

$$A = (\det M)A_0.$$

Note that $\mathcal{I}$ represents only 3 constraints on M and one of those is duplicated by the area A. So $\mathcal{I}$, $\bar{\mathbf{r}}$ and A between them give only 5 constraints on the 6 affine degrees of freedom. It would be necessary to compute higher moments to fix all 6 constraints. On the other hand, those moments up to second-order are sufficient to fix a vector in the space of Euclidean similarities.

Using moments for initialisation

For the Euclidean similarities, a shape-vector $\mathbf{X}$ can be recovered from the moments up to second order, as follows.

1. The displacement of the centroid $\bar{\mathbf{r}}$ gives the translational component of $\mathbf{X}$.

2. The scaling is given by $\sqrt{A/A_0}$.

3. The rotation is the angle θ through which the largest eigenvector of $\mathcal{I}$ rotates.

This procedure is illustrated in figure 4.6 below. In the illustration given here, moments were computed over a foreground patch segmented from the background on the basis of colour. An effective method for certain problems such as vehicle tracking, in which the foreground moves over a stationary background, is to use image motion. So-called "optical flow" is computed over the image, and a region of moving pixels is delineated. Either a snake may be wrapped around this region directly or, in a shape-space, moments can be computed for initialisation as above. Background subtraction methods are also useful here — see chapter 5 for an explanation.

4.5   Perspective and weak perspective

The next task is to set up notation for perspective projection in order to show that, under modest approximations, the set of possible shapes of image contours do indeed form affine spaces. Standard camera geometry is shown in figure 4.7 and leads to the following relationship between a three-dimensional object contour $\mathbf{R}(s)$ and its

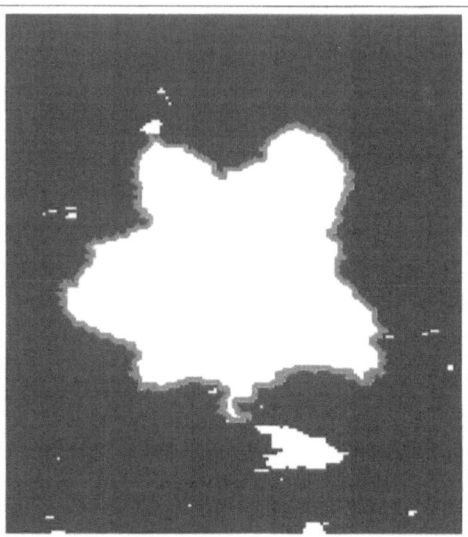

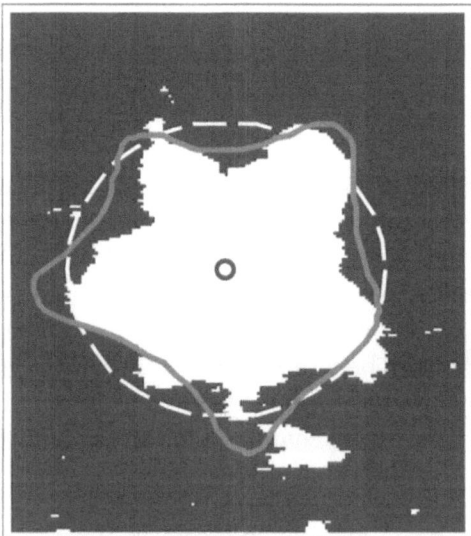

Figure 4.6: A contour is initialised using moments. *An image (left) (colour version in figure 1.6 on page 10) is processed using straightforward colour segmentation to obtain an interior region (right). Moments of the outline are calculated; the second moment is shown as an ellipse (bottom) and used to initialise a curve in a shape-space of Euclidean similarities (grey curve).*

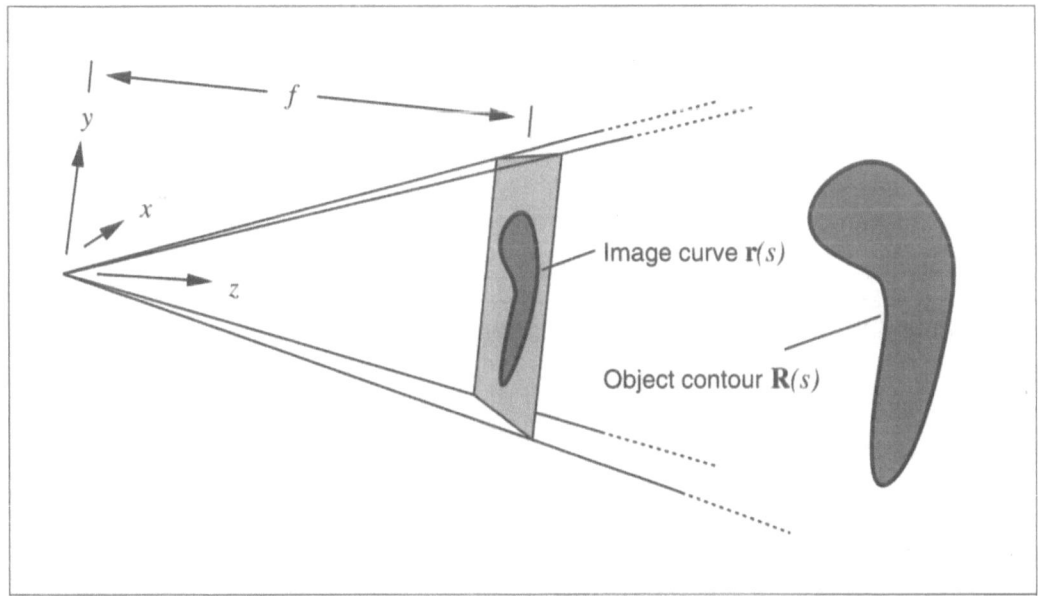

Figure 4.7: *It is a universally accepted fiction that, contrary to the layout of real cameras, the centre of projection in a mathematical camera is taken to lie* behind *the image plane. The focal length is f, measured in the same units as x, y and z, say mm for convenience.*

two-dimensional image $\mathbf{r}(s)$:

$$\mathbf{r}(s) = \frac{f}{Z(s)} \begin{pmatrix} X(s) \\ Y(s) \end{pmatrix} \quad \text{where} \quad \mathbf{R}(s) = \begin{pmatrix} X(s) \\ Y(s) \\ Z(s) \end{pmatrix}. \tag{4.9}$$

The $1/Z$ term is intuitively reasonable as it represents the tendency of objects to appear smaller as they recede from the camera. However, it makes the projection function non-linear which is problematic given that shape-spaces, being vector spaces, imply linearity. Fortunately there are well-established methods for making good linear approximations to perspective. The most general of these is the *weak perspective* projection.

Note that a good approximate value for f can be obtained simply by using the nominal value usually printed on the side of a lens housing, which we denote f_∞. To a first approximation, $f = f_\infty$, but a better one, taking into account the working distance Z_c, is

$$f = f_\infty \left(1 - \frac{f_\infty}{Z_c}\right)^{-1}. \tag{4.10}$$

Since image positions x, y available to a computer are measured in units of pixels relative to one corner of the camera array, a scale factor is needed to convert pixel units into length units (mm). This can be done quite effectively by taking a picture, as in figure 4.8, of a ruler lying in a plane parallel to the image plane. Moving a

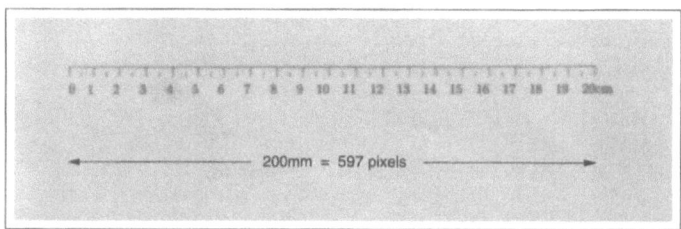

Figure 4.8: A simple camera calibration procedure. *The ruler is set at a distance 1060 mm from the camera iris.*

cursor over the image shows that 597 pixels corresponds to 200 mm at a distance of $Z_c = 1060$ mm from the camera iris, and the nominal focal length is $f_\infty = 25$ mm. From (4.10) we have $f = 25.6$ mm, and the scaling factor for distance x on the image plane is

$$\frac{200}{597} \frac{f}{Z_c} \text{ mm/pixel} = \frac{200}{597} \frac{25.6}{1060} \text{ mm/pixel} = 8.09 \times 10^{-3} \text{ mm/pixel}.$$

(This puts the width of the entire physical camera array of 768 pixels at $768 \times 8.09 \times 10^{-3}$ mm or 6.21 mm which is a very reasonable figure). Of course, for cameras whose pixels are not square, this procedure must be repeated in the vertical direction to calculate the vertical scaling factor.

Finally, note that more precise calculations, including allowances for minor mechanical defects such as asymmetry of lens placement with respect to the image array, can be made precisely using automatic but somewhat involved "camera calibration" procedures. However, the simple procedure above has proved sufficient for active contour interpretation, in most cases.

Weak Perspective

The weak perspective approximation is valid provided that the three-dimensional object is bounded so that its diameter is small compared with the distance from camera to object. Taking $\mathbf{R}_c = (X_c, Y_c, Z_c)^T$ to be a point close to the object — think of it as the object's centre, replace $\mathbf{R}(s)$ in the projection formula by $\mathbf{R}_c + \mathbf{R}(s)$ and then the assumption about object diameter can be written as

$$|\mathbf{R}(s)| \ll Z_c \; \forall s. \tag{4.11}$$

A useful alternative form of the assumption is that the subtended angle of the *image* contour is much less than 1 radian when viewed from any aspect.

That assumption can now be used in the perspective equation (4.9) to give the weak perspective projection:

$$\mathbf{r}(s) = \frac{f}{Z_c} \left[\begin{pmatrix} X_c \\ Y_c \end{pmatrix} + \begin{pmatrix} X(s) \\ Y(s) \end{pmatrix} - \frac{Z(s)}{Z_c} \begin{pmatrix} X_c \\ Y_c \end{pmatrix} \right] \tag{4.12}$$

which is linear in $\mathbf{R}(s)$ and approximates perspective projection to first order in $|\mathbf{R}(s)|/Z_c$. The tendency of image size to diminish as an object recedes is present in the f/Z_c term, now approximated to an "average" value for a given object. As individual points of $\mathbf{R}(s)$ recede they tend to move towards the centre of the image and the third term expresses this. In typical views, the approximation works well, as figure 4.9 shows. If, in addition to the camera having a large field of view, the object also fills that field of view, then errors in the weak perspective approximation become significant. That is not a situation that commonly arises in object tracking however. If the camera is mounted on a pan-tilt head, the camera's field of view is likely to be narrow in order to obtain the improved resolution that the movable head allows. Alternatively, when the camera is fixed, the image diameter is likely to be several times smaller than the field of view to allow for object movement. Since the field of view of a wide-angle camera lens is of the order of 1 radian, it follows that object diameter is likely to be considerably less than 1 radian, precisely the condition for the weak perspective approximation to hold good.

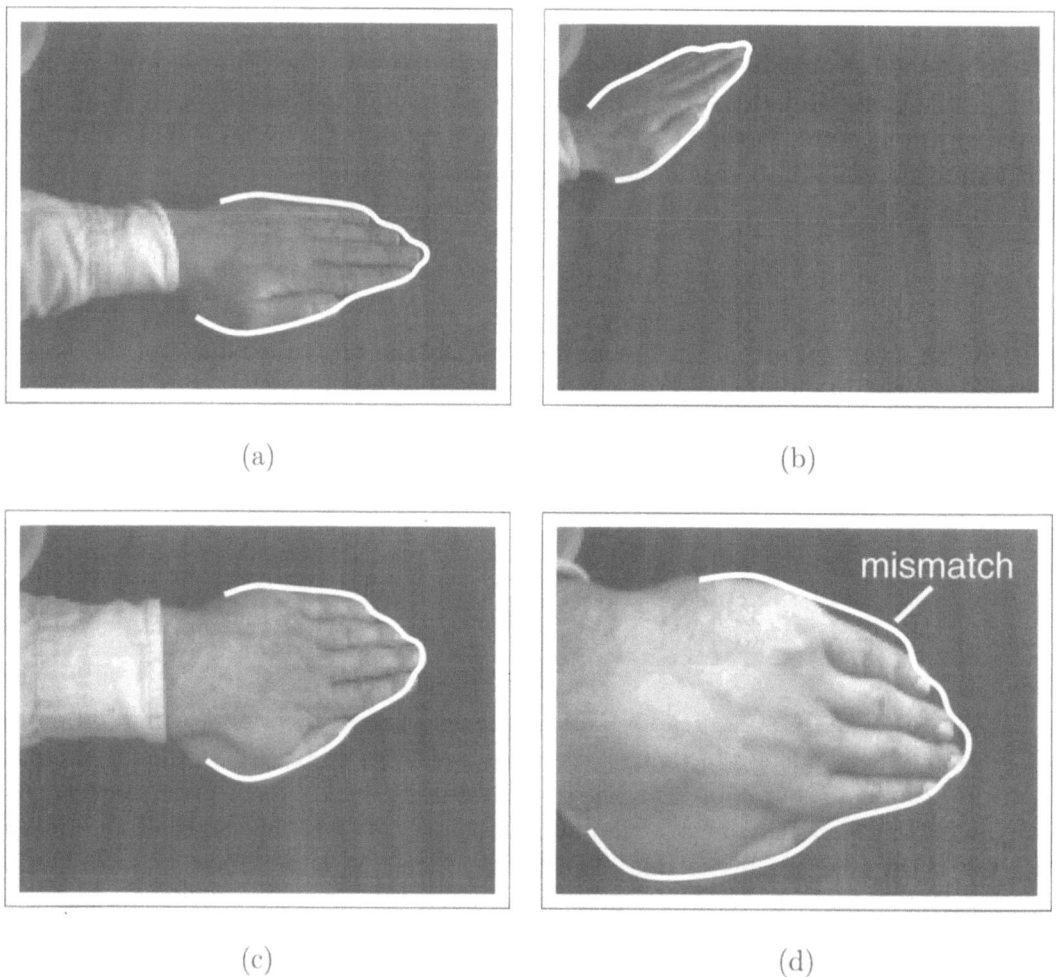

Figure 4.9: The weak perspective approximation is normally accurate. *(a–c) This image of a hand being tracked in a camera with a wide field of view shows that the weak perspective image of the hand outline closely approximates the true perspective image. (d) Under extreme viewing conditions, when perspective effects are strong, approximation error may be appreciable, visible here as the mismatch of the curve around the fingers.*

Orthographic projection

For a camera with a narrow field of view (substantially less than one radian) it can further be assumed, in addition to the assumption (4.11) about object diameter, that

$$|X_c| \ll Z_c \quad \text{and} \quad |Y_c| \ll Z_c \qquad (4.13)$$

— simply the condition that the contour centre is close enough to the centre of the image for the object actually to be visible. In that case, the third term in (4.12) is negligible and image perspective is well approximated by the *orthographic* projection

$$\mathbf{r}(s) = \frac{f}{Z_c} \left(\begin{array}{c} X_c + X(s) \\ Y_c + Y(s) \end{array} \right). \qquad (4.14)$$

Suppose the object contour $\mathbf{R}_c + \mathbf{R}(s)$ derives from a contour $\mathbf{R}_0(s)$ in a base coordinate frame which has then been rotated to give $\mathbf{R}(s) = R\mathbf{R}_0(s)$ and translated through $\mathbf{R}_c$, so that $R, \mathbf{R}_c$ are parameters for three-dimensional motion. Suppose also that the object is planar so (without loss of generality) $Z_0(s) = 0$. Then the orthographic projection equation becomes

$$\mathbf{r}(s) = \mathbf{u} + \frac{f}{Z_c} R_{2 \times 2} \left(\begin{array}{c} X_0(s) \\ Y_0(s) \end{array} \right)$$

where $\mathbf{u}$ is the orthographic projection of the three-dimensional displacement vector $\mathbf{R}_c$ and $R_{2 \times 2}$ is the upper-left 2×2 block of the rotation matrix R. Finally, take $M = (f/Z_c)R_{2 \times 2}$ and adopt the convention that $Z_c = f$ in the standard view so that

$$\mathbf{r}_0(s) = (X_0(s), Y_0(s))^T$$

is the image template. This gives a general planar affine transformation as in (4.3), so the image of a planar object moving rigidly in three dimensions does indeed sweep out a planar affine shape-space.

If the orthographic constraint (4.13) is relaxed again, to allow general weak perspective, it turns out that, when $\mathbf{R}(s)$ is planar, $\mathbf{r}(s)$ still inhabits the planar affine shape-space. Later we return to weak perspective for a general analysis of planar affine configurations, in particular to work out the three-dimensional pose of an object from its affine coordinates. This is used, for example, to calculate the three-dimensional position and attitude of the hand in the mouse application of figure 1.16 on page 20. That general method of pose calculation will work even when the camera is positioned obliquely relative to table-top coordinates and when the hand moves over the whole of a wide field of view.

4.6   Three-dimensional affine shape-space

Shape-space for a non-planar object is derived as a modest extension of the planar case. The object concerned should be visualised as a piece of bent wire, rather than a smooth three-dimensional surface. Smooth surfaces are, of course, of great interest but shape-space treatment is more difficult because of the complex geometrical behaviour of silhouettes. The bent wire model also implies freedom from hidden lines; the approach described here deals with parallax effects arising from three-dimensional shape but not with the problem of "occlusion" for which additional machinery is needed.

Clearly the 6-dimensional planar affine shape-space cannot be expected to suffice for non-planar surfaces and this is illustrated in figure 4.10. The new shape-space is "three-dimensional affine" with 8 degrees of freedom, made up of the six-parameter planar affine space and a two-parameter extension. Consider the object to be a three-dimensional curve

$$\mathbf{R}_0(s) = (X_0(s), Y_0(s), Z_0(s))^T$$

which is projected orthographically as in (4.14) to give an image curve

$$\mathbf{r}(s) = \mathbf{u} + \frac{f}{Z_c} R_{2\times3} \begin{pmatrix} X_0(s) \\ Y_0(s) \\ Z_0(s) \end{pmatrix}$$

and this can be expressed as the standard planar affine transformation $(\mathbf{u}, M)$ of (4.3) with an additional depth-dependent term:

$$\mathbf{r}(s) = \mathbf{u} + M\mathbf{r}_0(s) + \mathbf{v}Z_0(s) \tag{4.15}$$

where

$$R_{2\times3} = \frac{Z_c}{f} (M \mid \mathbf{v}). \tag{4.16}$$

Figure 4.10: The views of a general three-dimensional contour cannot be encompassed by a planar affine shape-space. *A planar affine space of contours has been generated from the outline of the first view of the cube. The outlines of subsequent views do not however lie in the space, as evidenced by the visible mismatch in the fitted contours. (Figure reprinted from (Curwen, 1993).) A similar effect is observed with the leaves.*

The three-dimensional shape-space therefore consists of the two-dimensional one for the planar affine space generated by template $\mathbf{Q}_0$, with two added components to account for the depth variation that is not visible in the template view. The two additional basis elements are:

$$\mathcal{V}' = \left\{ \begin{pmatrix} \mathbf{Q}_0^z \\ \mathbf{0} \end{pmatrix}, \begin{pmatrix} \mathbf{0} \\ \mathbf{Q}_0^z \end{pmatrix} \right\}.$$

The extra two elements are tacked onto the planar affine W-matrix (4.4) to form the W-matrix for the three-dimensional case:

$$W = \begin{pmatrix} 1 & 0 & \mathbf{Q}_0^x & 0 & 0 & \mathbf{Q}_0^y & \mathbf{Q}_0^z & 0 \\ 0 & 1 & 0 & \mathbf{Q}_0^y & \mathbf{Q}_0^x & 0 & 0 & \mathbf{Q}_0^z \end{pmatrix}. \tag{4.17}$$

Just as equation (4.5) provided a conversion from the planar affine shape-space to the real-world transformation, the three-dimensional affine shape-space components have the following interpretation:

$$\mathbf{X} = (u_1, u_2, M_{11} - 1, M_{22} - 1, M_{21}, M_{12}, v_1, v_2). \tag{4.18}$$

The expanded space now encompasses the outlines of all views of the three-dimensional outline as figure 4.11 shows. Automatic methods for determining $\mathbf{Q}_0^z$ from example views are discussed in chapter 7.

4.7   Key-frames

Affine spaces are appropriate shape-spaces for modelling the appearance of three-dimensional rigid body motion. In many applications, for instance facial animation, speech-reading and cardiac ultrasound, as described in chapter 1, motion is decidedly non-rigid. In the absence of any prior analytical description of the motion, the most effective strategy is to *learn* a shape-space from a training set of sample motion. A general approach to this, based on statistical modelling, is described in chapter 8. In the meantime, a simpler methodology is presented here, based on "key-frames" or representative image outlines of the moving shape. Often, an effective shape-space can be built by linear combination of such key-frames.

As an example, in figure 4.12, a sequence of three frames is shown which can be used to build a simple shape-space in which the first frame $\mathbf{Q}_0$ acts as the template and the shape-matrix W is constructed from the two key-frames $\mathbf{Q}_1', \mathbf{Q}_2'$:

$$W = \begin{pmatrix} \mathbf{Q}_1^x & \mathbf{Q}_2^x \\ \mathbf{Q}_1^y & \mathbf{Q}_2^y \end{pmatrix}. \tag{4.19}$$

where $\mathbf{Q}_i = \mathbf{Q}_i' - \mathbf{Q}_0$. This two-dimensional shape-space is sufficient to span all linear combinations of the three frames. What is more, the shape-space coordinates have clear interpretations, for example:

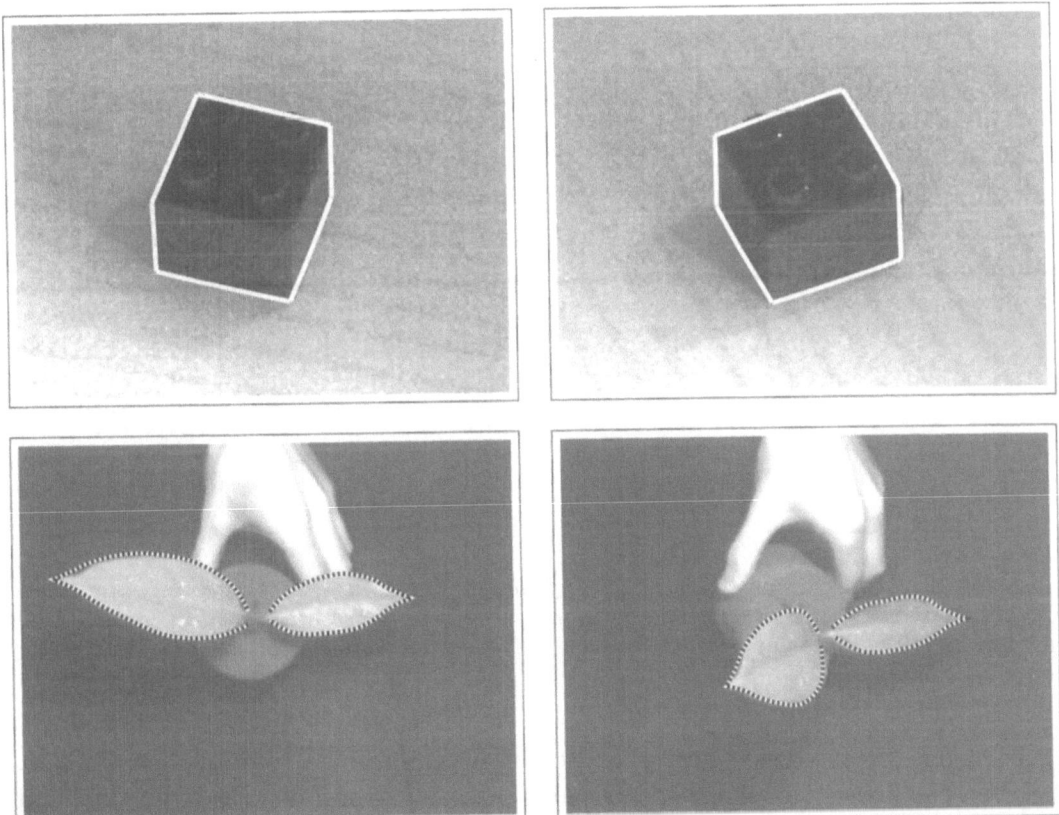

Figure 4.11: Three-dimensional affine shape-space. *The outlines of views of a cube which could not be contained in a planar affine shape-space now fall within a suitably constructed 3D affine space. (Figure reprinted from (Curwen, 1993).) Similarly, the non-planar arrangement of leaves is happily encompassed by a 3D affine space.*

- $\mathbf{X} = (0,0)^T$ represents the closed mouth;

- $\mathbf{X} = (1/2,0)^T$ represents the half-open mouth;

- $\mathbf{X} = (1/4,1/2)^T$ represents the mouth, half-protruding and slightly open.

A little more ambitiously, the same three frames can be used to build a more versatile shape-space that allows for translation, zooming and rotation of any of the expressions from the simple two-dimensional shape-space. Minimally, this should require

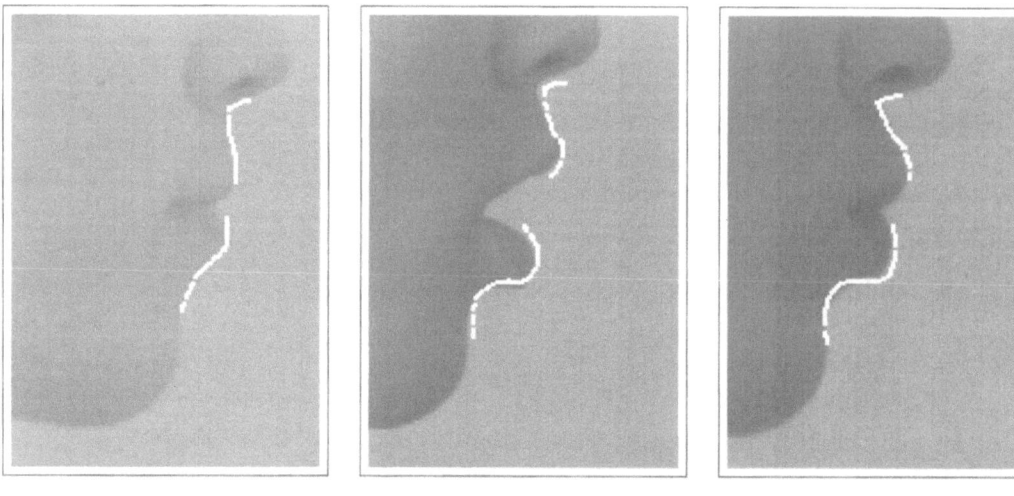

Template $\mathbf{Q}_0$ *Key-frame: opening* $\mathbf{Q}_1$ *Key-frame: protrusion* $\mathbf{Q}_2$

Figure 4.12: Key-frames. *Lips template followed by two key-frames, representing inter-actively tracked lips in characteristic positions. The key-frames are combined linearly with appropriate rigid degrees of freedom, to give a shape-space suitable for use in a tracker for non-rigid motion.*

2 parameters for expression plus 4 for Euclidean similarity, a total of 6 parameters. However, the linearity of shape-space leads to a wastage of 2 degrees of freedom and the shape-space is 8-dimensional with template $\mathbf{Q}_0$ as before and shape-matrix

$$
W = \begin{pmatrix} 1 & 0 & \mathbf{Q}_0^x & -\mathbf{Q}_0^y & \mathbf{Q}_1^x & -\mathbf{Q}_1^y & \mathbf{Q}_2^x & -\mathbf{Q}_2^y \\ 0 & 1 & \mathbf{Q}_0^y & \mathbf{Q}_0^x & \mathbf{Q}_1^y & \mathbf{Q}_1^x & \mathbf{Q}_2^y & \mathbf{Q}_2^x \end{pmatrix}. \tag{4.20}
$$

This is based on the shape-matrix for Euclidean similarities (4.2) on page 75, extended to all three frames. Again, expressions are naturally represented by the shape-vector, for example:

- $\mathbf{X} = (u, 0, 0, 0, 1, 0, 0, 0)^T$ represents the fully open mouth, shifted to the right by u;

- $\mathbf{X} = (0, 0, \cos\theta - 1, \sin\theta, 0, 0, \frac{1}{2}\cos\theta, \frac{1}{2}\sin\theta)^T$ represents the closed mouth, half-protruding and rotated through an angle θ.

Of course this technique, illustrated here for 2 key-frames under Euclidean similarity, does apply to an arbitrary number of key-frames, and a general space of rigid transformations spanned by a set $\{T^j, \ j = 1, \ldots, N_r\}$. In that case any contour corresponding to the appearance of ith key-frame is composed of a linear combination of contours

$$\mathbf{Q}_i^1, \mathbf{Q}_i^2, \ldots, \mathbf{Q}_i^{N_r},$$

for an N_r-dimensional space of rigid transformations. Then the W-matrix is composed of columns which are vectors $\mathbf{Q}_i^j, \ i = 0, 1, \ldots, \ j = 1, 2, \ldots, N_r$. To avoid introducing linear dependencies into the W-matrix, it is best to omit translation from the space of rigid transformations and treat it separately, as in the two key-frame example above. Then the W-matrix for the composite shape-space of rigid and non-rigid transformations is

$$W = \begin{pmatrix} 1 & 0 & & & & & \\ & & \mathbf{Q}_0^1 \ \mathbf{Q}_0^2 & \ldots & \mathbf{Q}_1^1 \ \mathbf{Q}_1^2 & \ldots \\ 0 & 1 & & & & & \end{pmatrix}. \qquad (4.21)$$

One final caveat is in order. With N_r degrees of transformational freedom (excluding translation) and N_k key-frames, there are a total of $N_r + N_k$ degrees of freedom in the system. However the linear representation as a shape-space with a W-matrix as above has dimension $N_r \times (N_k + 1)$, a "wastage" of $N_k(N_r - 1)$ degrees of freedom. The two key-frame example above has $N_r = 2, N_k = 2$ so the wastage is just 2 degrees of freedom, in a shape-space of total dimension 8 (including translation). With more key-frames and larger transformational spaces such as three-dimensional affine ($N_r = 6$), the wastage is more severe — 5 degrees of freedom per key-frame. In such cases, the constructed shape-space is likely to be too big for efficient or robust contour fitting. However, it is often possible to construct a smaller space by other means such as "PCA" (chapter 8) and use the large shape-space constructed as above for interpretation. In particular, shape displacements can be factored into components due to rigid and non-rigid transformations respectively, and this is explained at the end of chapter 7.

4.8   Articulated motion

When an object (e.g. a hand) is allowed, in addition to its freedom to move rigidly, to support articulated bodies (fingers), more general shape-spaces are needed. Clearly, one route is to take a kinematic model in the style used in robotics for multi-jointed arms and hands and use it as the basis of a configuration space. The advantage is that the resulting configuration space represents legal motions efficiently because the configuration space has minimal dimension. The drawback is that the resulting measurement models (see next chapter) are non-linear. This is due to trigonometric non-linearities as in the previous section on rigid motion but exacerbated by the hinges added onto the base body. The result is that classical linear Kalman filtering is no longer usable, though non-linear variants exist which are not however probabilistically rigorous. Furthermore, linear state-spaces admit motion models which apply globally throughout the space. In a non-linear space, motion models could perhaps be represented as a set of local linear models in tangent spaces placed strategically over a manifold. This is hard enough to represent and the task of learning such models seems a little daunting.

As with rigid motion, there is a way to avoid the non-linearities by generating appropriate shape-spaces. Again, there is some inefficiency in doing this and the resulting space underconstrains the modelled motion. The degree of wastage depends on the precise nature of the hinging of appendages, and this is summarised in the below. Proofs are not given here, but there is a more detailed discussion in appendix C.

Two-dimensional hinge

For a body in two dimensions, or equivalently a three-dimensional body constrained to lie on a plane, each additional hinged appendage increments the dimension of shape-space by 2, despite adding only one degree of kinematic freedom. Hence the wastage amounts to 1 degree of freedom per appendage.

Two-dimensional telescopic appendage

Still in two dimensions, each telescopic element added to the rigid body increments the shape-space dimension by 2, causing a wastage of one degree of freedom, as for the hinge.

Hinges on a planar body in three dimensions

The rigid planar body above, with its co-planar hinged appendages, is now allowed to move out of the ground plane, so that it can adopt any three-dimensional configuration. Each hinged appendage now adds 4 to the dimension of shape-space, resulting in the wastage of 3 degrees of freedom.

Universal joints on a rigid three-dimensional body

Given a three-dimensional rigid body, whose shape-space is 3D affine, each appendage attached with full rotational freedom (via a ball joint, for instance) increments the dimension of shape-space by 6. Such an appendage introduce 3 kinematic degrees of freedom, so the wastage is 3.

Hinges on a rigid three-dimensional body

For appendages attached to the three-dimensional body by planar hinges, with just 1 kinematic degree of freedom, the dimension of shape-space increases by 4, so again the wastage is 3 degrees of freedom per appendage.

Note that the above results hold regardless of how the appendages are attached — whether directly to the main body (parallel), or in a chain (serial) or a combination of the two.

Bibliographic notes

This chapter has explained how shape-spaces can be constructed for various classes of motion. The value of shape-spaces of modest dimensionality was illustrated in (Blake et al., 1993) as a cure to the instability that can arise in tracking with high-dimensional representations of curves such as the original finite-element snake (Kass et al., 1987) or unconstrained B-splines (Menet et al., 1990; Curwen and Blake, 1992). Shape-spaces are linear, parametric models in image-space, but non-linear models or *deformable templates* are also powerful tools (Fischler and Elschlager, 1973; Yuille, 1990; Yuille and Hallinan, 1992). Linear shape-spaces have been used effectively in recognition (Murase and Nayar, 1995). Shape-spaces discussed so far have been image-based but a related topic is the use of three-dimensional parametric models for tracking, either rigid (Harris, 1992b) or non-rigid (Terzopoulos and Waters, 1990; Terzopoulos and Metaxas, 1991; Lowe, 1991; Rehg and Kanade, 1994).

Initialisation from moments is discussed in (Blake and Marinos, 1990; Wildenberg, 1997) and the use of 3rd moments to recover a full planar affine transformation is described in (Cipolla and Blake, 1992a). In some circumstances, region-based optical flow computation (Buxton and Buxton, 1983; Horn and Schunk, 1981; Nagel, 1983; Horn, 1986; Nagel and Enkelmann, 1986; Enkelmann, 1986; Heeger, 1987; Bulthoff et al., 1989) can be used to define the region for snake initialisation. This has been shown to be particularly effective with traffic surveillance (Koller et al., 1994).

Shape-spaces are based on perspective projection and its linear approximations in terms of vector spaces (Strang, 1986). Mathematically, projective geometry is a somewhat old-fashioned topic and so the standard textbook (Semple and Kneebone, 1952) is rather old-fashioned too. More accessible, is a graphics book such as (Foley et al., 1990) for the basics of camera geometry and perspective transformations. Computer vision has been concerned with *camera calibration* (Tsai, 1987) in which test images of grids are analysed to deduce the projective parameters for a particular camera, including both *extrinsic* parameters (the camera-to-world transformation) and *intrinsic* parameters such as focal length.

The most general linear approximation to perspective is known variously as *paraperspective* (Aloimonos, 1990) or *weak perspective* (Mundy and Zisserman, 1992) and can be particularly effective if separate approximations are constructed for different neighbourhoods of an image (Lawn and Cipolla, 1994). The gamut of possible appearances of three-dimensional contours under a particular weak perspective transformation forms an affine space (Ullman and Basri, 1991; Koenderink and van Doorn, 1991). This idea led to a series of studies on using affine models to analyse motion, including (Harris, 1990; Demey et al., 1992; Bergen et al., 1992a; Reid and Murray, 1993; Bascle and Deriche, 1995; Black and Yacoob, 1995; Ivins and Porrill, 1995; Shapiro et al., 1995).

The first three chapters of (Faugeras, 1993) are an excellent introduction to projective and affine geometry and to camera calibration.

Articulated structures are most naturally described in terms of non-linear kinematics (Craig, 1986) in which the non-linearities arise from the trigonometry of rotary joints. Such a model has been incorporated into a hand tracker, for instance (Rehg and Kanade, 1994), in which the articulation the fingers is full treated. Articulated structures can be embedded in linear shape-spaces but this can be very "inefficient," in the sense of section 4.1, that kinematic constraints have to be relaxed — see appendix C.

Finally, smooth silhouette curves and their shape-spaces are beyond the scope of this book. However, it can be shown that a shape-space of dimension 11 is appropriate. This shape-space representation of the curve is an approximation, valid for sufficiently small changes of viewpoint. Its validity follows from results in the computer vision literature about the projection of silhouettes into images (Giblin and Weiss, 1987; Blake and Cipolla, 1990; Vaillant, 1990; Koenderink, 1990; Cipolla and Blake, 1992b).

Chapter 5

Image processing techniques for feature location

The use of image-filtering operations to highlight image features was illustrated in chapter 2. Figure 2.1 on page 27 illustrated operators for emphasising edges, valleys and ridges, and it was shown how the emphasised image could be used as a landscape for a snake. However, for efficiency, the deformable templates described in the next two chapters are driven towards a distinguished feature curve $\mathbf{r}_f(s)$ rather than over the entire image landscape F that is used in the snake model. This is rather like making a quadratic approximation to the external snake energy:

$$E_{\text{ext}} \propto -F(\mathbf{r}) \propto \int \left(\mathbf{r}(s) - \mathbf{r}_f(s) \right)^2 \, ds, \tag{5.1}$$

where $\mathbf{r}_f(s)$ lies along a ridge of the feature-map function F. The increase in efficiency comes from being able to move directly to the curve $\mathbf{r}_f$, rather than having to iterate towards it as in the original snake algorithm described in section 2.1.

It is therefore necessary to extract $\mathbf{r}_f(s)$ from an image. One way of doing this is to mark high strength values on the feature maps and group them to form point sets to which spline curves could be fitted. An example of feature curves grouped in this way was given, for edge-features, in figure 3.1 on page 42. However, the wholesale application of filters across entire images is excessively computationally costly. At any given instant, an estimate is available of the position of a tracked image-contour and this can be used to define a "search-region," in which the corresponding image feature is likely to lie. Image processing can then effectively be restricted to this search region,

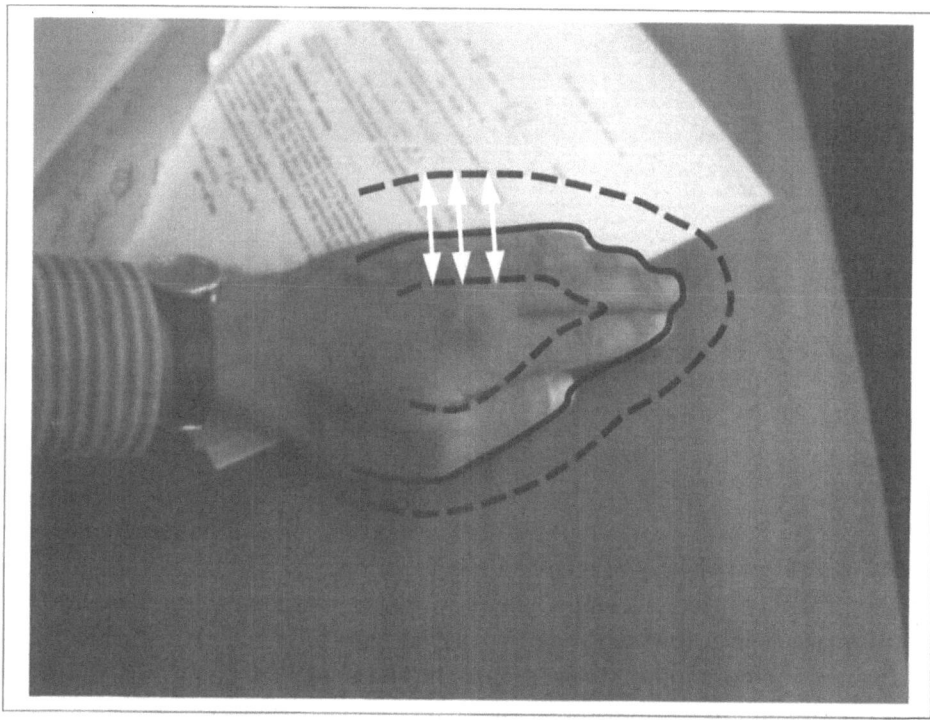

Figure 5.1: Search region. *It is computationally efficient to restrict image processing operations to lie within a "region of interest" (dashed lines), one either side of the currently estimated contour position (solid line). Image processing operations are then performed along certain lines passing through the estimated contour. In this example, the lines are normals to the estimated curve, three of which are shown as arrowed white lines.*

as in figure 5.1. The search region displayed in the figure is formed there by sweeping normal vectors of a chosen length along the entire contour. Features can then be detected by performing image filtering along each of the sampled normals, and this is very efficient. If normals are constructed at points $s = s_i$, $i = 1, \ldots, N$, along the curve $\mathbf{r}(s)$, this will give a sequence of sampled points $\mathbf{r}_f(s_i), i = 1, \ldots, N$ along the feature curve $\mathbf{r}_f(s)$. It is of course possible that more than one feature may be found on each normal, but for now it is assumed that just one — the favourite feature — is retained.

5.1   Linear scanning

In order to perform one-dimensional image processing, image intensity is sampled at regularly spaced intervals along each image normal. An arbitrarily placed normal line generally intersects image pixels in an irregular fashion, as in figure 5.2. This is well

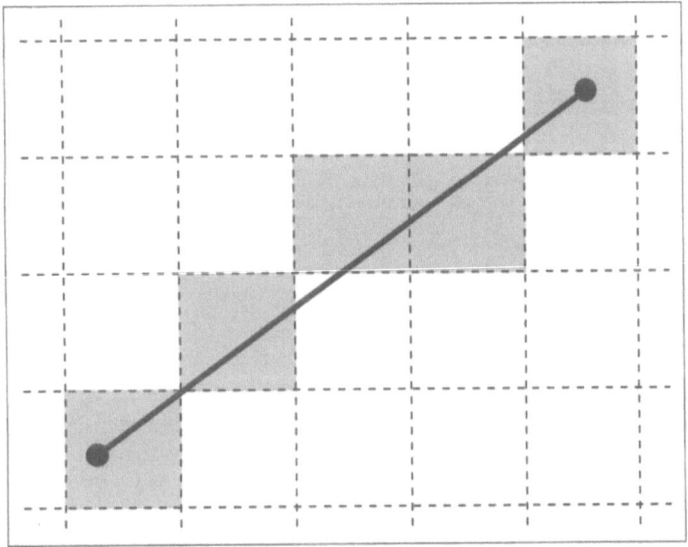

Figure 5.2: Irregular image sampling. *Listing the intensities of pixels crossed by a normal line would result in a non-uniform sampling of intensity that would suffer abrupt variations as the line moved over the image.*

known to produce undesirable artifacts in Computer graphics — "jaggies" in static images and twinkling effects in moving ones, for which the usual cure is "anti-aliasing." Unlike graphics, in which the task is to map from a mathematical line onto pixels, the problem here is to generate the opposite mapping, from image to line. This calls for a sampling scheme of its own.

An effective sampling scheme, spatially regular and temporally smooth (when the line moves) involves interpolation as follows. A sequence of regularly spaced sample points are chosen along the line. The intensity I at a particular sample point (x, y) is computed as a weighted sum of the intensities at 4 neighbouring pixels, as in figure 5.3. A pixel with centre sited at integer coordinates (i, j) has intensity $I_{i,j}$. The intensity

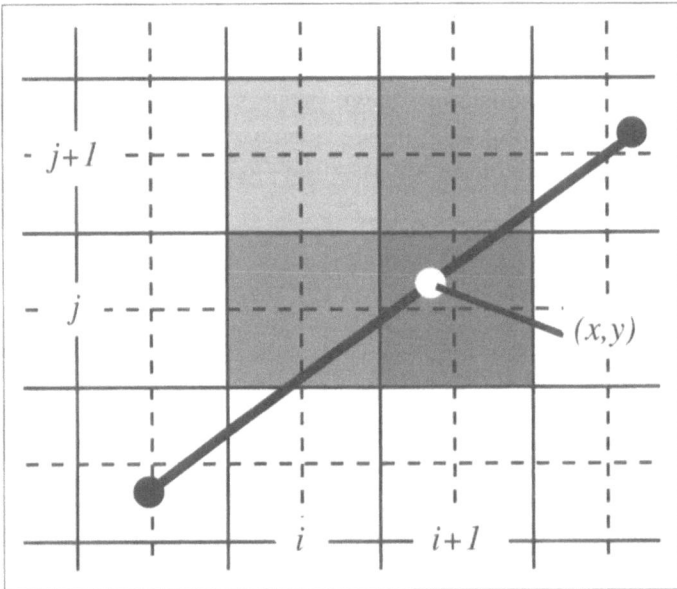

Figure 5.3: Interpolated image sampling. *The intensity at a chosen sampling point* (x, y) *is computed as a weighted sum of the intensities at the four immediately adjacent pixels.*

I at (x, y) is then computed by bilinear interpolation:

$$I = \sum_{i,j} w_{i,j} I_{i,j} \tag{5.2}$$

with weights

$$w_{i,j} = \begin{cases} (1 - |x - i|)(1 - |y - j|) & \text{if } |x - i| < 1 \ \text{ and } \ |y - j| < 1 \\ 0 & \text{otherwise} \end{cases} \tag{5.3}$$

so that at most four pixels, the ones whose centres are closest to (x, y), have non-zero weights, as the figure depicts.

5.2 Image filtering

Analysis of image intensities now concentrates on the one-dimensional signals along normals. The intensity $I(x)$ along a particular normal is sampled regularly at $x = x_i$

and intensities are stored in an array $I_i = I(x_i)$, $i = 1, \ldots, N$. A variety of feature detection operators can be applied to the line, popular ones being edges, valleys and ridges. Features are located by applying an appropriate operator or mask C_n, $-N_C \leq n \leq N_C$, by discrete convolution, to the sampled intensity signal I_n, $1 \leq n \leq N_I$, to give a feature-strength signal

$$E_n = \sum_{m=-N_C}^{N_C} C_m I_{n+m}.$$

Maxima of that signal are then located, and marked wherever the value at that maximum exceeds a preset threshold (chosen to exclude spurious, noise-generated maxima). This is illustrated for edges in figure 5.4 and for valleys in figure 5.5.

Corners

Effective operators for corners also exist and have been used for visual tracking. However, corners do not quite fit into the search paradigm described here. Being discrete points, a corner is likely to be missed by search along normals unless it happens to lie exactly on some normal; more generally it will be located in the gap between two adjacent normals. This problem does not arise with edges because they are extended and should generally intersect one or more of the normals. If corners are to be used they must be located by an exhaustive search over the region of interest, which is rather more expensive computationally than a search that is restricted to normals.

One operator, the "Harris" corner detector, works by computing a discrete approximation to the moment matrix

$$S(x,y) = \int G(x',y')[\nabla I(x+x', y+y')][\nabla I(x+x', y+y')]^T \, dx' \, dy'$$

at each image point (x, y), where $\nabla I = (\partial I/\partial x, \partial I/\partial y)^T$, the image-gradient vector at a point, and G is a two-dimensional Gaussian mask for smoothing, typically 2–4 pixels in diameter. The trace $\mathrm{tr}(S)$ and the determinant $\det(S)$ are examined at each point (x, y). Wherever $\mathrm{tr}(S)$ exceeds some threshold, signaling a significantly large image gradient, and also the ratio $\mathrm{tr}(S)/2\sqrt{\det(S)}$ is sufficiently close to its lower bound of 1, a corner feature is marked. The Harris detector responds reliably to objects that are largely polyhedral, marking corners that are likely to be stable to changing viewpoint. Natural shapes (figure 5.6) may however fire the corner detector in locations that are

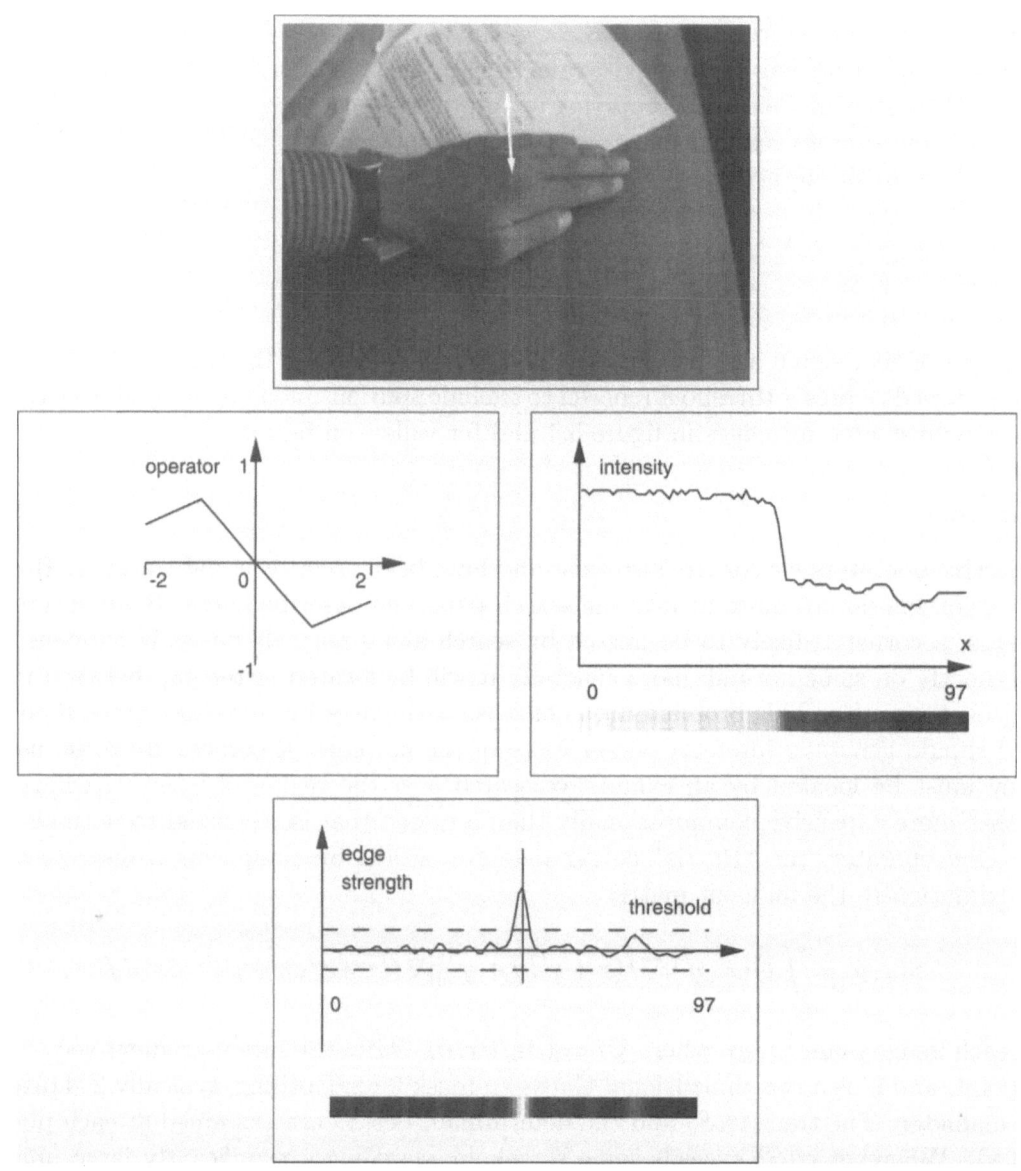

Figure 5.4: Operator for edge detection *The problem is to search along a line in an image (top) to find edges — locations where contrast is high. An operator (left, shown on an expanded length scale) is convolved with the image intensity function along the line (right). One edge is found, corresponding to a maximum of the feature-strength function (bottom).*

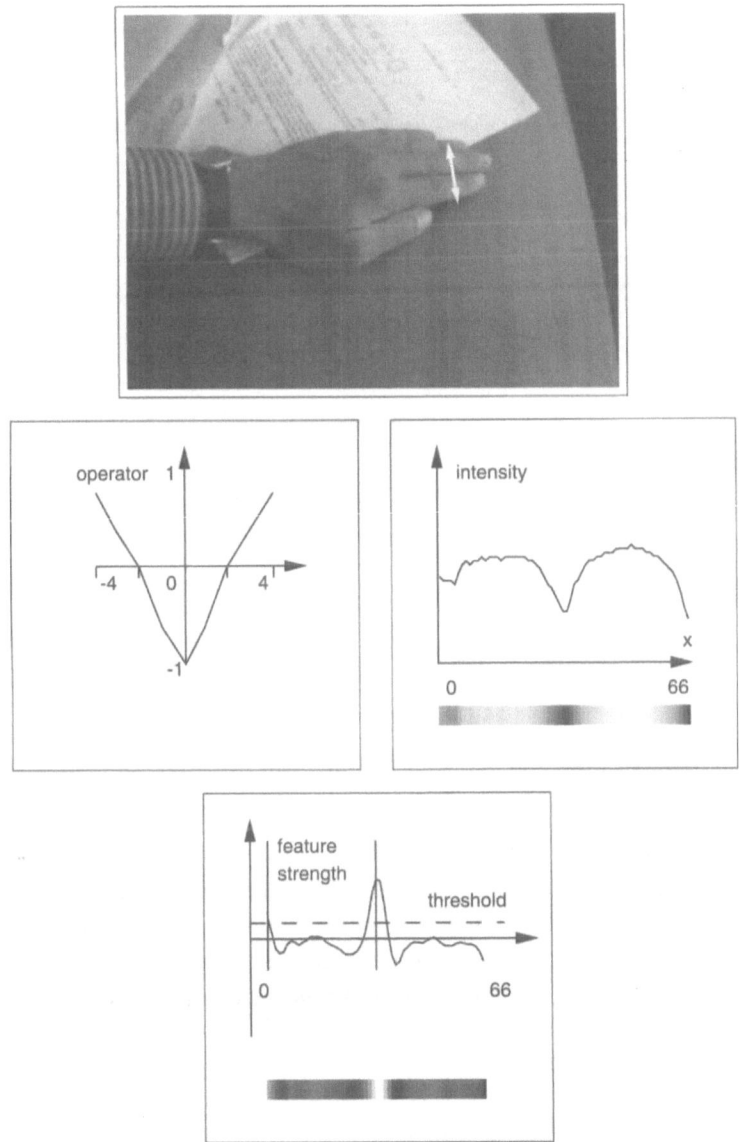

Figure 5.5: Operator for valley detection *The problem is to search along a line in an image (top) to find valleys — locations of minimum intensity. The valley operator (left) convolved with the intensity signal (right) produces a ridge (bottom) corresponding to the dark line between adjacent fingers.*

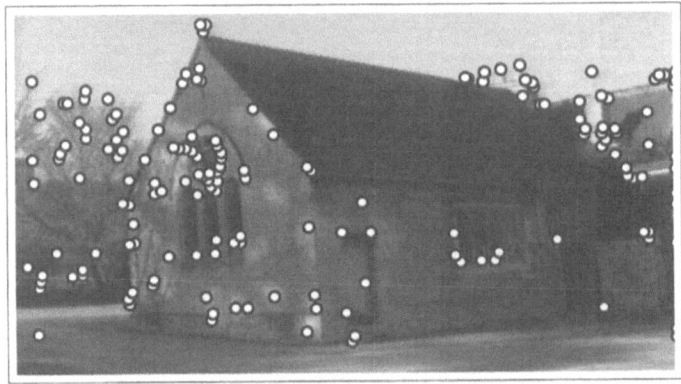

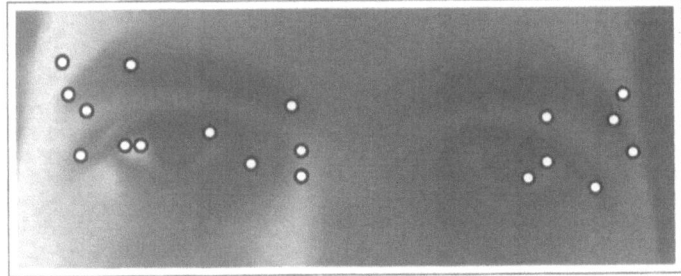

Figure 5.6: Corner detection. *The corner detector fires reliably at sharp, polyhedral corners and junctions. It also fires on certain other features such as tighter curves, but with less spatial accuracy and reliability. (Figures courtesy of Andrew Fitzgibbon).*

hard to predict and may be unstable under changing viewpoint. Not surprisingly, it works best where there are well-defined geometrical features, such as on buildings. On a natural object such as a face, the response is a mixture of reasonably reliable features at eye corners and other less reliable responses on curves.

5.3   Using colour

Colour in images is a valuable source of information especially where contrast is weak, as in discriminating lips from facial skin. Colour information is commonly presented as a vector $\mathbf{I} = (r, g, b)$ of red-green-blue values, at each pixel. The most economical way to treat the colour vector is to reduce it to a single value by computing a suitable scalar function of $\mathbf{I}$. The scalar I can then be treated as a single intensity value and subjected to the same image-feature operators as were used above, for processing monochrome images. Two scalar functions are described here: one that is general, the *hue* function and one that is customised — learned from training data, the *Fisher linear discriminant*.

The hue function is used commonly to represent colour in graphics, vision and colorimetry and corresponds roughly to the polar angle for polar coordinates in r, g, b-space. It separates out what is roughly a correlate of spectral colour — i.e. colour wavelength — from two other colour components: intensity (overall brightness) and saturation ("colouredness" as opposed to whiteness). This explains why it is particularly effective when contrast is low so that changes in intensity $(r + g + b)$ are hard to perceive. Hue is defined as follows

$$\text{hue}(r, g, b) = \arctan\left((2r - g - b), \sqrt{3}(g - b)\right). \tag{5.4}$$

Note that there is a linear "hexcone" approximation to the hue function for applications where it is important to compute it fast, for example to convert entire video images.

The Fisher linear discriminant is an alternative scalar function that attempts to represent as much of the relevant colour information as possible in a single scalar value. Its efficiency is optimal in certain sense, but this comes at the cost of re-deriving the function, in a learning step, for each new application. Learning is straightforward and works as follows. First, a foreground area F and a background area B are delineated in a training image. For instance, in figure 5.7, F would be the lip region and B would be part of the immediate surround. The Fisher discriminant function is defined as a simple scalar product:

$$\text{fisher}(\mathbf{I}) = \mathbf{f} \cdot \mathbf{I} \quad \text{where} \quad \mathbf{I} = (r, g, b)^T. \tag{5.5}$$

The function is learned from the foreground and background areas F and B by the algorithm of figure 5.8, which determines the coefficient vector $\mathbf{f}$. The effect of the algorithm is to choose the vector $\mathbf{f}$ in colour (r, g, b) space which best separates the background and foreground populations of colours. When the Fisher function is used in place of intensity, feature detection works effectively, as figure 5.7 shows.

5.4 Correlation matching

The oldest idea in visual matching and tracking, and one that is widely used in practical tracking systems, is correlation. Given a template T in the form of a small

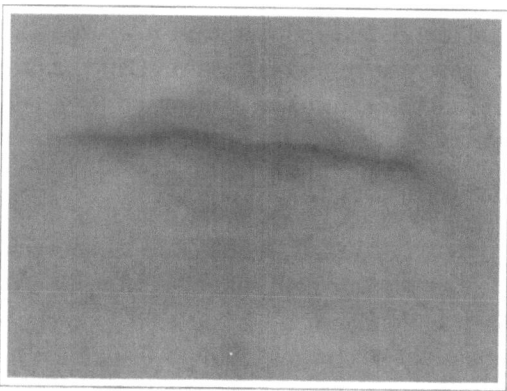

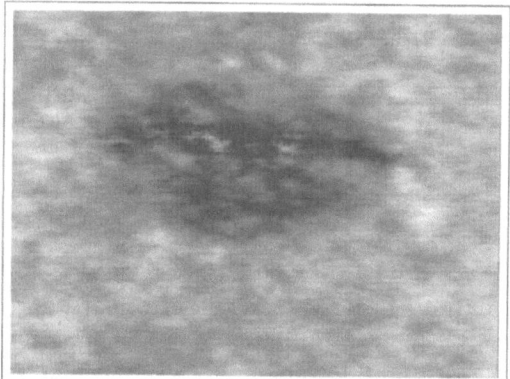
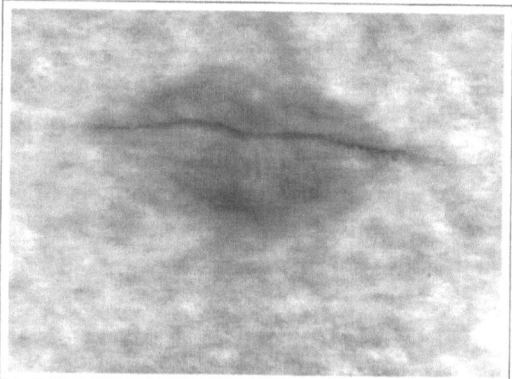

Figure 5.7: Detecting colour boundaries. *Contrast between lip and skin can be lost if only intensity (top) is extracted from a colour image; this is especially so for the lower lip in this example. The hue function (left) shows improved contrast, but is rather noisy, and this tends to disturb the operation of an active contour by generating spurious clutter. A better solution is the Fisher discriminant function (right). In this example, colour is to be used to locate lip boundaries. A colour training image is used to learn the Fisher discriminant function. (Figures courtesy of Robert Kaucic.)*

array of image intensities, the aim is to find the likely locations of that template in some larger test image I. In one dimension for example, with image $I(x)$ and template $T(x)$, $0 \leq x \leq \Delta$, the problem is to find the offset x' for which $T(x)$ best matches $I(x + x')$ over the range $0 \leq x \leq \Delta$ of the template. This is most naturally done by

1. **Calculate** the mean pixel values in each class

$$\bar{I}_F = \frac{1}{N_F} \sum_{(x,y)\in F} \mathbf{I}(x,y)$$

$$\bar{I}_B = \frac{1}{N_B} \sum_{(x,y)\in B} \mathbf{I}(x,y)$$

2. **Determine** the within class scatter matrices

$$S_F = \sum_{(x,y)\in F} (\mathbf{I}(x,y) - \bar{\mathbf{I}}_F)(\mathbf{I}(x,y) - \bar{\mathbf{I}}_F)^T$$

$$S_B = \sum_{(x,y)\in B} (\mathbf{I}(x,y) - \bar{\mathbf{I}}_B)(\mathbf{I}(x,y) - \bar{\mathbf{I}}_B)^T$$

3. **Find** the Fisher discriminant vector

$$\mathbf{f} = S^{-1}(\bar{\mathbf{I}}_F - \bar{\mathbf{I}}_B) \quad \text{where} \quad S = S_F + S_B.$$

Figure 5.8: Learning the Fisher discriminant function. *A discriminant vector* **f** *is computed from the foreground* F *and background* B *populations of pixels, aiming to separate foreground optimally from background.*

minimising a difference measure such as

$$M(x') = \int_{x=0}^{\Delta} (I(x + x') - T(x))^2 \, dx \tag{5.6}$$

with respect to x'. In practice this theoretical measure must be computed as a sum over image pixels.

An illustration is given in figure 5.9 in which the problem is to locate the position of the eyebrow along a particular line. A template T is available that represents an ideal distribution of intensity along a cross-section located in a standard position. The task is to find the position on the line which best corresponds to the given intensity template, and this is achieved by minimising $M(x)$ as above. The nomenclature

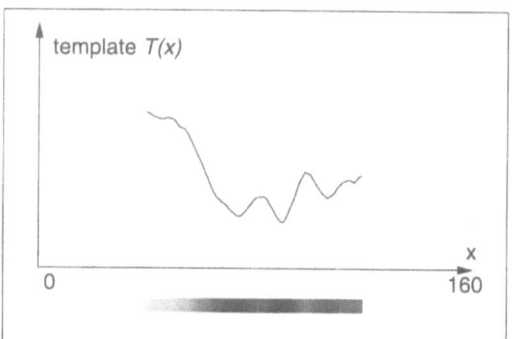

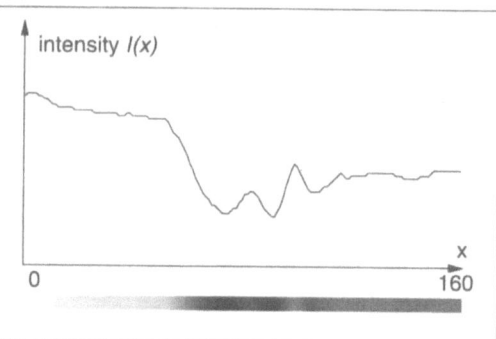

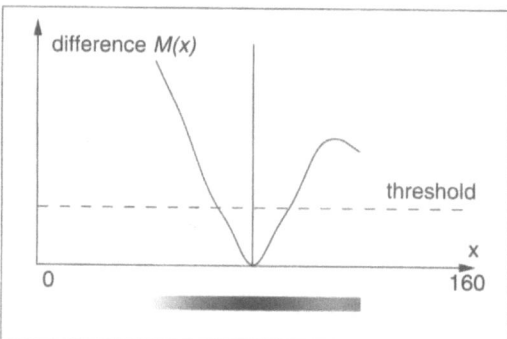

Figure 5.9: One-dimensional correlation matching. *The problem is to search along a line in an image (top) to find the position of an eye. A template (left) of the cross-section of an eye is to be matched with the image intensity function along the line (right). The correct position is marked by the minimum of the matching function (bottom). Figure courtesy of Robert Kaucic.*

"correlation" derives from the idea that in the case that $T(x)$ and $I(x)$ span the same x-interval, and are periodic, minimising $M(x)$ is equivalent to maximising the "mathematical correlation"

$$\int_{x=0}^{\Delta} I(x+x')T(x)\,dx. \tag{5.7}$$

Correlation matching can be used to considerable effect as a generalised substitute for edge and valley location. Position of the located feature along each normal can be reported in the same way as for edges and used for curve matching. This can be particularly effective in problems where image contrast is poor. In the lip-tracking application of figure 1.10 on page 14, tracking without lip make-up proved possible only when edge detection was replaced by correlation.

There are numerous variations on the basic theme (see bibliographic notes, at the end of the chapter). One variation is to pre-process $I(x)$, for example to emphasise its spatial derivative, which tends to generate a sharper valley in $M(x')$, which can then be located more accurately. More modern approaches dispense with intensities altogether, representing I and T simply as lists of the positions of prominent features. The problem then is to match those lists, using discrete algorithms. This has the advantage of efficiency because of the data-compression involved in reducing the intensity arrays to lists. It is also more robust for two reasons. First, intensity profiles vary as ambient illumination changes whereas the locations of features are approximately invariant to illumination. Secondly, there is the additional flexibility that different amounts of offset x' can be associated with different features, for example when performing stereo matching over a large image region, whereas in the correlation framework x' is fixed. For these reasons, feature-based matching is considered superior to correlation for many problems.

Correlation matching is often used in two dimensions with a template $T(x,y)$ matched to an offset image $I(x+x',y+y')$, as in figure 5.10. This is considerably more computation-intensive than in one dimension as it involves a double integral over x,y and also a two-dimensional search to minimise M with respect to x',y'. Various techniques such as Discrete Fourier Transforms and "pyramid" processing at multiple spatial scales can be used to improve efficiency. In higher dimensions than two, for example when rotation and scaling are to be allowed in addition to translation, exhaustive correlation becomes prohibitively expensive and alternative algorithms are needed. One approach is to generate the offsets in higher dimensions in a more sparing fashion, using gradient descent for instance. Numerous authors have shown that this

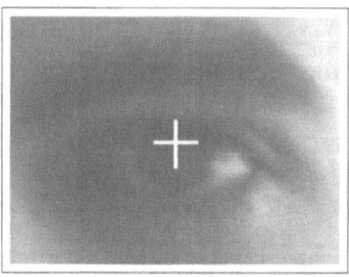

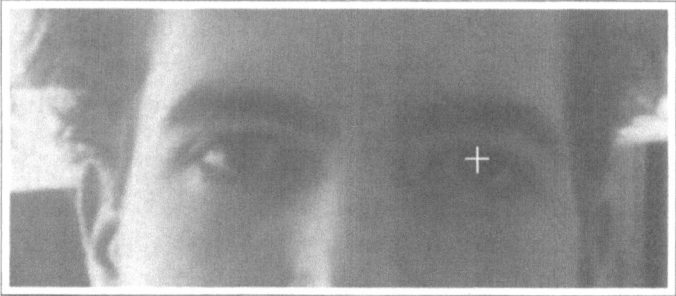

Figure 5.10: Matching eyes by image correlation. *The template (top) is a reversed copy of the right eye in the face image (bottom). When the template is correlated with the image, its centre collocates accurately with the centre of the left eye. (Figures courtesy of Steve Smith.)*

can be very successful.

5.5   Background subtraction

A widely used technique for separating moving objects from their backgrounds is based on subtraction. It is used as a pre-process in advance of feature detection to suppress background features to prevent them distracting fitting and tracking processes. It is particularly suited for applications such as surveillance where the background is often largely stationary.

An image of $I_B(x, y)$ of the background is stored before the introduction of a foreground object. Then, given an image $I(x, y)$ captured with the object present, feature detection is restricted to areas of $I(x, y)$ that are labelled as foreground because they satisfy

$$|I(x, y) - I_B(x, y)| > \sigma,$$

where σ is a suitable chosen noise threshold. As figure 5.11 shows, background features tend to be successfully inhibited by this procedure. Cancellation can disrupt the

Background $I_B(x, y)$ Image $I(x, y)$

Background suppressed

Figure 5.11: Background subtraction. *The difference between an image (right) and a stored background (left) is computed to suppress background features (bottom), though some background features do "print through" the foreground.*

foreground, as the figure shows, where the background intensity happens to match the foreground too closely. This results in some loss of genuine foreground features, a cost which is eminently justified by the effectiveness of background suppression.

Finally, it should be noted that the expense of computing the entire difference image ΔI can be largely saved by computing differences "lazily" just of those pixels actually required for interpolation along normals in (5.2). This is an important consideration for real-time tracking systems.

Bibliographic notes

The Bresenham algorithm (Foley et al., 1990) is routinely used in graphics to convert a mathematical line to a sequence of pixels and "anti-aliasing" is employed to achieve an interpolated pattern of pixel intensities which varies smoothly, without flicker, as the line is moved. The interpolated sampling scheme described in this chapter is similar in spirit, but uses a different sampling pattern which performs the inverse function of mapping from an array of pixels to an arbitrary point on a mathematical line.

A general reference on feature detection and the use of convolution masks is (Ballard and Brown, 1982). A much-consulted study of trade-offs in the design of operators for edge detection is (Canny, 1986) and the design of operators for ridges and valleys is described in (Haralick, 1980); the discussions relate to two-dimensional image processing whereas in this chapter the simpler one-dimensional problem is addressed. Effective detectors for corners exist (Kitchen and Rosenfeld, 1982; Zuniga and Haralick, 1983; Noble, 1988) and have been used to good effect in motion analysis, e.g. (Harris, 1992a) and tracking (Reid and Murray, 1996). Operators that respond to regions rather than curves are also important, for example texture masks (Jain and Farrokhnia, 1991) which can be used effectively in snakes that settle on texture boundaries (Ivins and Porrill, 1995).

Correlation matching is based on the idea of "mathematical correlation" which is central to the processing of one-dimensional signals (Bracewell, 1978). It is also used in two-dimensional processing of images to locate patterns (Ballard and Brown, 1982), track motion (Bergen et al., 1992b) and register images for stereo vision (Lucas and Kanade, 1981). Two-dimensional correlation can be computed efficiently using pyramid architectures (Burt, 1983). A notable variation on the correlation approach is to allow the correlation offset to vary spatially, adding considerable flexibility at some computational expense (Witkin et al., 1986). Successful applications of correlation in higher dimensions have used gradient descent (Sullivan, 1992) which may also be coupled with feature detection (Bascle and Deriche, 1995) for added robustness to illumination variations and specularity. Very efficient algorithms can be constructed in the case of affine shape-spaces, by pre-processing image deformation maps (Hager and Belhumeur, 1996). One-dimensional correlation along normals has proved a useful tool in matching contours (Cootes et al., 1993; Rowe and Blake, 1996a).

The hue model for colour is used in vision (Ballard and Brown, 1982) and in graphics (Foley et al., 1990) in the form of a linear "hexcone" approximation which can be computed efficiently. The Fisher discriminant function is a general technique in pattern recognition (Bishop, 1995) that has also been used to good effect in vision to discriminate faces from one another (Belhumeur et al., 1996). A Bayesian treatment of colour segmentation for tracking is given in (Crisman, 1992). With careful modelling of the physics of reflection, colour segmentation can be even made robust in an outdoor environment with its varying illumination (Plá et al., 1993).

Background subtraction/cancellation assists greatly in the generation of benign training sets, by suppressing clutter, and is a valuable technique in learning dynamics (chapter 11). A variety of techniques exists based on linear and non-linear (morphological) filtering, and on statistical hypothesis testing (Baumberg and Hogg, 1994; Murray and Basu, 1994; Koller et al., 1994; Rowe and Blake, 1996b).

Chapter 6

Fitting spline templates

Chapters 3 and 4 dealt with the geometry and representation of curves and classes of curves — the shape-spaces. Now it is time to look at some image data to see how shapes can be approximated by members of those classes. The norm and inner product machinery developed earlier proves useful together with the image-processing techniques of chapter 5. Curve approximation techniques are built up step by step in this chapter until the necessary tools are assembled for basic B-spline snakes and deformable templates.

6.1 Regularised matching

Generally, measurements made from images are "noisy" — prone to unpredictable variations from a number of sources. At the finest grain there is the effect of electrical noise on the video signal from the camera and optical noise such as the flickering of fluorescent lights. Coarser effects are the interactions of lighting with object surfaces, causing specularities or highlights and shadows which vary in a manner that is too hard to model. Since we have to live with such disturbances it is imperative that algorithms for analysing images are designed to be intrinsically robust to them. This is standard practice in image processing, where "regularisation" is used to clean poor quality images by imposing prior constraints on the likely appearance of valid images. This section describes the application of regularisation to curves reconstructed from image data. First basic curve-fitting machinery is developed.

Projection

Shape-space was introduced as a means of reducing shape-variability and, in the context of the problem of fitting an image-feature curve, acts as a way of encouraging smoothness. Suppose the image feature were expressed in the form of a spline curve $\mathbf{r}_f$ where $\mathbf{r}_f(s) = U(s)\mathbf{Q}_f$. If the fitted spline $\mathbf{Q}$ is restricted to shape-space, and using the energy landscape for a quadratically approximated feature map ((5.1) on page 97), the fitting problem is

$$\min_{\mathbf{X}} \; \|W\mathbf{X} + \mathbf{Q}_0 - \mathbf{Q}_f\|^2.$$

The solution $\mathbf{X} = \hat{\mathbf{X}}$ is given (proof below) by the pseudo-inverse $W^+ = \mathcal{H}^{-1}W^T\mathcal{U}$ that was defined in the previous chapter:

$$\hat{\mathbf{X}} = W^+(\mathbf{Q}_f - \mathbf{Q}_0). \tag{6.1}$$

The fact that $\hat{\mathbf{X}}$ minimises the error in approximating the curve $\mathbf{Q}$ within shape-space $\mathcal{S}$ motivates the interpretation of W^+ as an operator for *projection* onto shape-space. It also explains why, in the example of figure 6.1, this operator manages to smooth noisy data while staying close to the gross shape of that data.

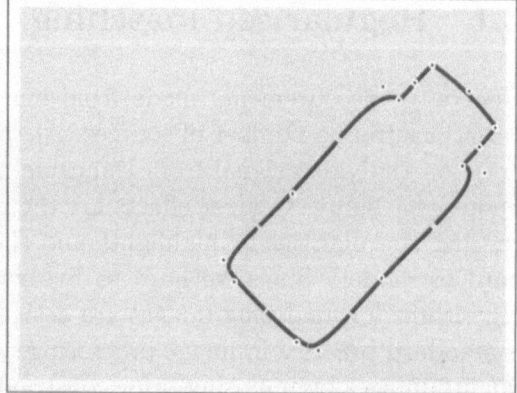

Figure 6.1: Curve approximation in shape-space. A distorted bottle shape (left, black curve) does not fit the bottle outline at all well. Projecting the distorted curve onto the lotion bottle shape-space of figure 4.5 on page 77, finds the closest "valid" curve shape (right). Clearly this removes the distortion.

Regularisation

Further tolerance to noise is procured by biasing the fitted curve towards a mean shape $\bar{\mathbf{r}}(s)$ to a degree determined by a regularisation constant α. In the simplest form of the problem, the fitted curve is the solution of

$$\min_{\mathbf{r}(s)} \alpha\|\mathbf{r} - \bar{\mathbf{r}}\|^2 + \|\mathbf{r} - \mathbf{r}_f\|^2 \tag{6.2}$$

where $\mathbf{r}_f$ is a member of a class $\mathcal{S}_Q$ of B-spline curves, and the possible fitted curves $\mathbf{r}$ are constrained to lie in some shape-space $\mathcal{S} \subset \mathcal{S}_Q$. The problem can be expressed conveniently as

$$\min_{\mathbf{X}} \alpha\|\mathbf{X} - \overline{\mathbf{X}}\|^2 + \|\mathbf{Q} - \mathbf{Q}_f\|^2 \quad \text{with} \quad \mathbf{Q} = W\mathbf{X} + \mathbf{Q}_0.$$

The idea of the regularising term is that it tends to pull any fitted curve towards the mean $\bar{\mathbf{r}}$ but this is rarely satisfactory as it stands. For example, it may be desirable in practice for $\bar{\mathbf{r}}$ to influence the *shape* of the fitted curve but not its position or orientation. A more general regulariser is needed therefore, using a weight matrix $\overline{S}$ (which must be positive semi-definite), so that the fitting problem becomes:

$$\min_{\mathbf{X}} (\mathbf{X} - \overline{\mathbf{X}})^T \overline{S}(\mathbf{X} - \overline{\mathbf{X}}) + \|\mathbf{Q} - \mathbf{Q}_f\|^2 \quad \text{with} \quad \mathbf{Q} = W\mathbf{X} + \mathbf{Q}_0. \tag{6.3}$$

For instance, the regulariser $\alpha\|\mathbf{X} - \overline{\mathbf{X}}\|^2$ would be obtained by setting $\overline{S} = \alpha\mathcal{H}$. To achieve the desired invariance of the regulariser over some subspace $\mathcal{S}_s \subset \mathcal{S}$ of transformations, for instance the Euclidean similarities, $\overline{S}$ must be restricted by means of a projection operation E^d to operate over deformations outside the invariant subspace $\mathcal{S}_s$:

$$\overline{S} = \alpha E^{d^T} \mathcal{H} E^d. \tag{6.4}$$

The projection operator can be expressed in terms of the shape-matrix W_s for the subspace and its pseudo-inverse shape-matrix W_s^+:

$$E^d = I - E^s \quad \text{where} \quad E^s = W^+ W_s W_s^+ W. \tag{6.5}$$

The solution $\mathbf{X} = \hat{\mathbf{X}}$ to the fitting problem (6.3) is obtained in two stages, a projection onto shape-space

$$\mathbf{X}_f = W^+(\mathbf{Q}_f - \mathbf{Q}_0), \tag{6.6}$$

followed by weighted summation

$$\hat{\mathbf{X}} = \left(\overline{S} + \mathcal{H}\right)^{-1} \left(\overline{S}\,\overline{\mathbf{X}} + \mathcal{H}\mathbf{X}_f\right).$$

(6.7)

This is illustrated in the example of figure 6.2, in which the shape-space $\mathcal{S}$ is taken to be the entire spline space $(\mathcal{S} = \mathcal{S}_Q)$ and the invariant subspace $\mathcal{S}_s$ is the space of Euclidean similarities which allows the *shape* of $\overline{\mathbf{X}}$ to influence the fit while its position and orientation are ignored. As $\alpha \to 0$, the influence of the template diminishes and the fitted curve moves closer to the data, as expected.

Proof that projection is realised by the pseudo-inverse. Writing $\mathbf{Q}' = \mathbf{Q}_f - \mathbf{Q}_0$, the fitting problem above is to minimise

$$
\begin{aligned}
\|W\mathbf{X} - \mathbf{Q}'\|^2 &= (W\mathbf{X} - \mathbf{Q}')^T \mathcal{U}(W\mathbf{X} - \mathbf{Q}') \\
&= \mathbf{X}^T W^T \mathcal{U} W \mathbf{X} - \mathbf{X}^T W^T \mathcal{U} \mathbf{Q}' - \mathbf{Q}'^T \mathcal{U} W \mathbf{X} + \text{const} \\
&= (\mathbf{X} - \hat{\mathbf{X}})^T W^T \mathcal{U} W (\mathbf{X} - \hat{\mathbf{X}}) + \text{const},
\end{aligned}
$$

(completing the square) in which

$$\hat{\mathbf{X}} = (W^T \mathcal{U} W)^{-1} W^T \mathcal{U} \mathbf{Q}' = \mathcal{H}^{-1} W^T \mathcal{U} \mathbf{Q}',$$

which gives the required result (6.1).

Derivation of the shape-space regularisation formula. First a "projection lemma" is needed, essentially Pythagoras' theorem applied to shape-space, that

$$\|\mathbf{Q} - \mathbf{Q}_f\|^2 = \|\mathbf{X} - \mathbf{X}_f\|^2 + \|W\mathbf{X}_f + \mathbf{Q}_0 - \mathbf{Q}_f\|^2,$$

(6.8)

and this is illustrated in figure 6.3. (Note that $\|W\mathbf{X} - W\mathbf{X}_f\| = \|\mathbf{X} - \mathbf{X}_f\|$ by definition of the norm over $\mathcal{S}$). Now the problem of (6.3) becomes the minimisation of

$$(\mathbf{X} - \overline{\mathbf{X}})^T \overline{S}(\mathbf{X} - \overline{\mathbf{X}}) + (\mathbf{X} - \mathbf{X}_f)^T \mathcal{H}(\mathbf{X} - \mathbf{X}_f)$$

and this is simplified by "completing the square" to give

$$(\mathbf{X} - \hat{\mathbf{X}})^T (\overline{S} + \mathcal{H})(\mathbf{X} - \hat{\mathbf{X}}) + c$$

where $\hat{\mathbf{X}}$ is as defined above and c is a constant independent of $\mathbf{X}$, so that the minimising shape is $\mathbf{X} = \hat{\mathbf{X}}$ as in 6.7.

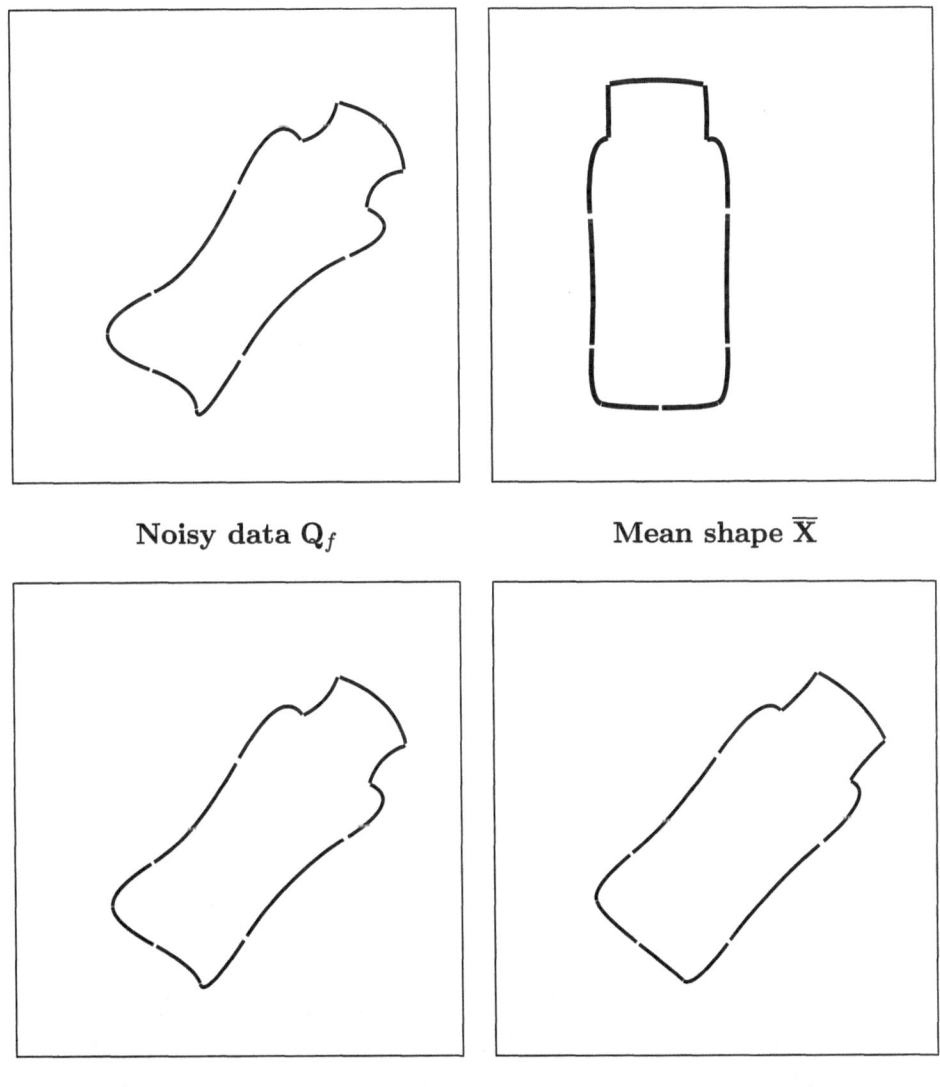

Noisy data $\mathbf{Q}_f$ Mean shape $\overline{\mathbf{X}}$

Fitted curve: $\alpha = 0.5$ Fitted curve: $\alpha = 3.0$

Figure 6.2: Curve-fitting with regularisation. *A regularisation parameter α controls the trade-off from high noise resistance but biased towards a mean shape (α large) to more accurate fitting but with greater sensitivity to noisy data (alpha small). The regulariser is constructed to be invariant to Euclidean similarity transformations, so the position and orientation of the mean shape have no influence on the fitted curve.*

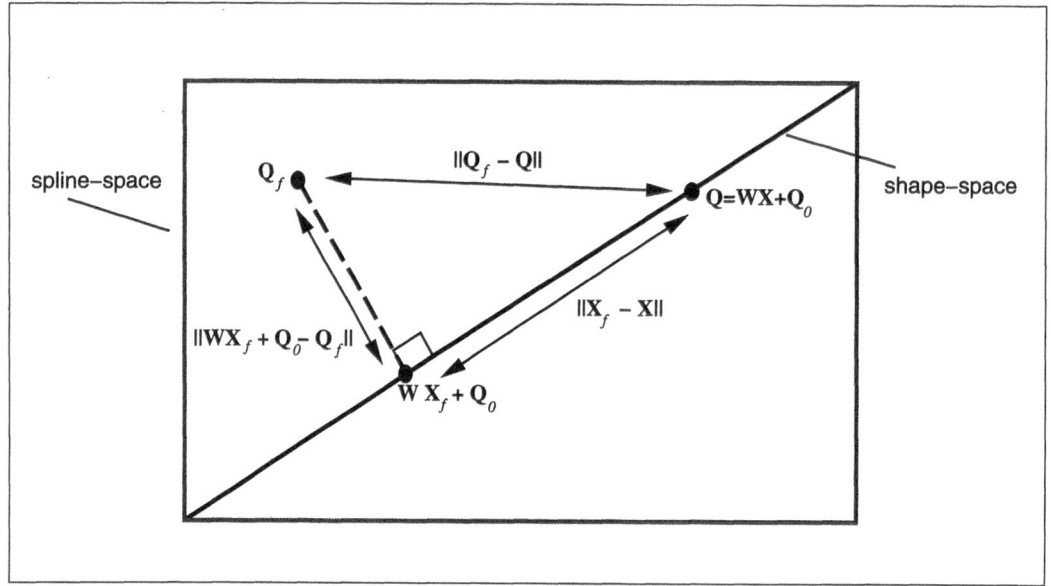

Figure 6.3: *Projection lemma.*

6.2 Normal displacement in curve fitting

In chapter 3 we saw that the measure of curve difference using the norm is sensitive to parameterisation. Two curves with similar shapes will nonetheless register a substantial difference according to the norm unless the parameterisations of those curves also match. One example, figure 3.16 on page 62, showed two similarly shaped curves with a norm-difference that was substantial simply because the parameterisation of one of the curves had been shifted.

Curve fitting based on the norm will therefore suffer from sensitivity to parameterisation. This is particularly a problem when the data-curve is derived from an image because, as the previous section showed, the parameterisation of an image curve is inherited from a spline curve $\mathbf{r}(s)$, either the template curve itself or some initial estimate of the image curve. This anchors the parameterisation of the image curve $\mathbf{r}_f(s)$ close to that of $\mathbf{r}(s)$. The result is that the fitted curve exhibits "reluctance" to move away from $\mathbf{r}(s)$, as figure 6.4 illustrates.

A solution to this problem, proposed in chapter 3, is to redefine the fitting problem to take re-parameterisation explicitly into account when measuring the difference

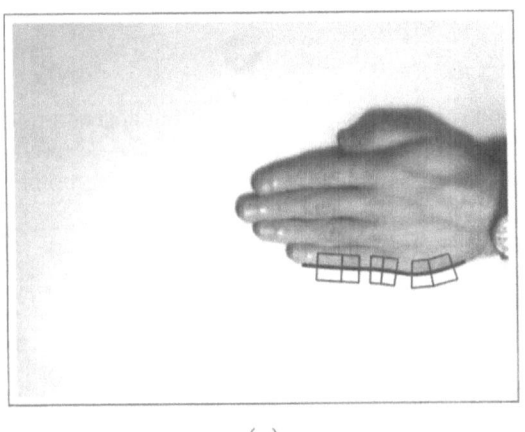 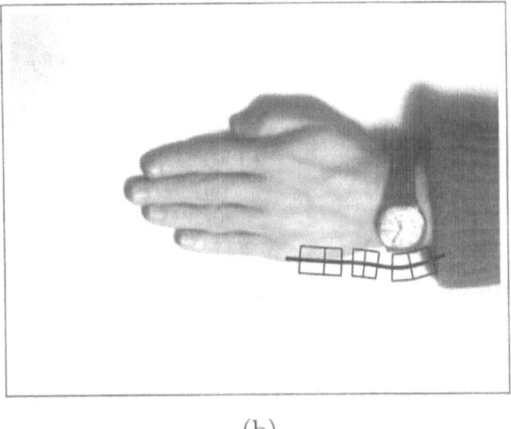

(a)   (b)

Figure 6.4: Inherited parameterisation leads to reluctance in curve fitting. *Curve fitting using norm-difference is applied repeatedly to a sequence of images starting with (a) and ending with (b). In each case, the fitted curve obtained from the previous image is used as the initial estimate for fitting in the current image. At each step, the inherited parameterisation of the feature curve follows closely the parameterisation of the estimated curve. The result is a reluctance to move, as illustrated.*

between the fitted curve $\mathbf{r}$ and the data $\mathbf{r}_f$. A re-parameterisation function $g(s)$ is defined such that $g(s) = s$ gives the original inherited parameterisation, mapping points on curve $\mathbf{r}$ to points on data-curve $\mathbf{r}_f$. Then the image-data curve $\mathbf{r}_f(s)$ is to be compared with the re-parameterised version of the curve $\mathbf{r}(g(s))$. In place of $\|\mathbf{r} - \mathbf{r}_f\|$ in (6.2), the alternative curve-displacement measure $d(\mathbf{r}, \mathbf{r}_f)$ is:

$$d^2 = \min_g \frac{1}{L} \int D^2 \, ds \ \text{ where } \ D^2(s) = (\mathbf{r}(g(s)) - \mathbf{r}_f(s))^2, \tag{6.9}$$

defined as a minimum over all possible reparameterisations.

Local invariance

Computation of the curve-displacement measure $d(\mathbf{r}, \mathbf{r}_f)$ is feasible if we are content with local, rather than global minimisation. According to the rules of the "calculus of variations", a local minimum of d is achieved when

$$\frac{\partial D^2}{\partial g} = 0 \ \text{ for all } s,$$

that is, when

$$[\mathbf{r}_f(s) - \mathbf{r}(g(s))] \cdot \mathbf{r}'(g(s)) = 0.$$

This says that, for the optimal parameterisation, the vector $\mathbf{r}_f(s) - \mathbf{r}(g(s))$ joining corresponding points on the two curves is perpendicular to the tangent vector $\mathbf{r}'(g(s))$; it is in the direction of the *normal* vector $\mathbf{n}(g(s))$ to the curve $\mathbf{r}$, so that the distance between corresponding points is

$$D(s) = |\mathbf{r}_f(s) - \mathbf{r}(g(s))| = [\mathbf{r}_f(s) - \mathbf{r}(g(s))] \cdot \mathbf{n}(g(s)).$$

Now, assuming that the curves are sufficiently similar that the extent of re-parameterisation is small $(g(s) \approx s)$, it follows first that

$$\mathbf{n}(g(s)) \approx \mathbf{n}(s)$$

since the implicit smoothness of the B-spline ensures the curvature is also small. In addition, using a first-order Taylor expansion,

$$[\mathbf{r}(s) - \mathbf{r}(g(s))] \cdot \mathbf{n}(s) \approx (g(s) - s)\mathbf{r}'(s) \cdot \mathbf{n}(s) = 0,$$

giving finally

$$D(s) \approx [\mathbf{r}_f(s) - \mathbf{r}(s)] \cdot \mathbf{n}(s),$$

 the "normal displacement" between corresponding points on the two curves (with the original parameterisation).

The use of normal displacement, which is a standard technique from Computer Vision, can be explained intuitively. The total displacement $\mathbf{r}_f(s) - \mathbf{r}(s)$ at a point can be expressed as the vector sum of components along the curve tangent and normal respectively (figure 6.5). The tangential component corresponds approximately to displacement *along* the curve $\mathbf{r}(s)$ without actually travelling any distance away from the curve. It reflects the variation of parameterisation between curves. If the tangential component is eliminated, what remains of the total displacement is the normal component, representing purely the distance between curves.

In terms of the original problem of finding a measure of distance between the fitting curve $\mathbf{r}(s)$ and a feature curve $\mathbf{r}_f(s)$, we have shown that the distance

$$d(\mathbf{r}, \mathbf{r}_f) \approx \frac{1}{L} \int [(\mathbf{r}(s) - \mathbf{r}_f(s)) \cdot \mathbf{n}(s)]^2 \, ds \qquad (6.10)$$

is a suitably invariant measure provided that the displacement between the two curves is small. This is now used to define a new norm $\| \cdot \|_{\bar{\mathbf{n}}}$ with respect to an estimated or

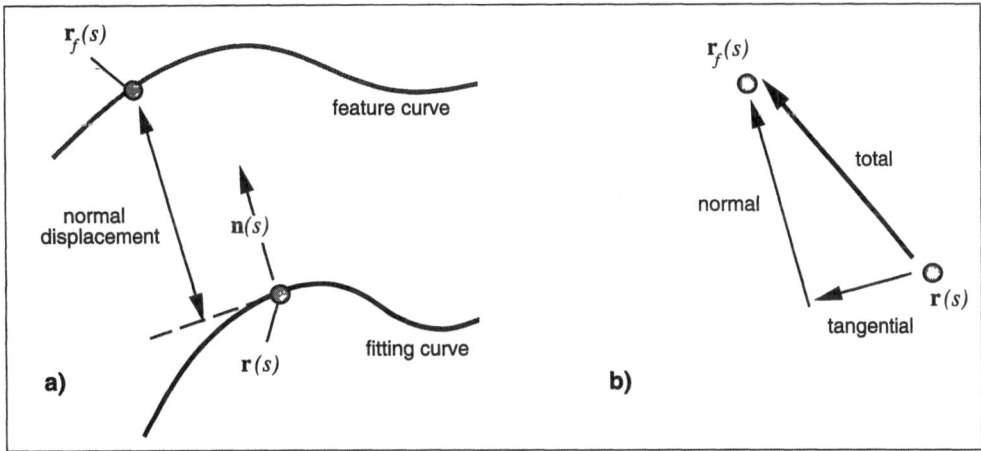

Figure 6.5: Normal Displacement. *a) Displacement along the normal from one curve to another, as shown, forms the basis for a measure of difference between curves that is approximately invariant to re-parameterisation. b) Total displacement can be factored vectorially into two components, tangential and normal.*

template curve $\bar{\mathbf{r}}(s)$, whose normals are $\bar{\mathbf{n}}(s)$:

$$\|\mathbf{r}\|_{\bar{\mathbf{n}}}^2 \equiv \frac{1}{L} \int [\mathbf{r}(s) \cdot \bar{\mathbf{n}}(s)]^2 \; ds. \tag{6.11}$$

It has the property that the norm-difference approximates the invariant distance measure, that is

$$\|\mathbf{r} - \mathbf{r}_f\|_{\bar{\mathbf{n}}} \approx d(\mathbf{r}, \mathbf{r}_f),$$

provided both curves $\mathbf{r}$ and $\mathbf{r}_f$ are sufficiently close to the estimated curve $\bar{\mathbf{r}}$. Norms $\|\mathbf{Q}\|_{\bar{\mathbf{n}}}$ in spline space and $\|\mathbf{X}\|_{\bar{\mathbf{n}}}$ in shape-space can be defined by inducing them from the curve norm $\|\mathbf{r}\|_{\bar{\mathbf{n}}}$, just as $\|\mathbf{Q}\|$ and $\|\mathbf{X}\|$ were induced from $\|\mathbf{r}\|$ originally.

The next step is to take account of the practicalities of image measurement by expressing the invariant difference discretely. Suppose normal vectors are sampled at regularly spaced points $s = s_i$, $i = 1, \ldots, N$, with inter-sample spacing h, along the entire curve $\mathbf{r}(s)$, so that, in the case of an open curve,

$$s_1 = 0, \; s_{i+1} = s_i + h, \text{ and } s_N = L.$$

Then the norm-difference (6.10) can be approximated as a sum:

$$\|\mathbf{r} - \mathbf{r}_f\|_{\overline{\mathbf{n}}}^2 \approx \frac{1}{N} \sum_{i=1}^{N} [(\mathbf{r}_f(s_i) - \mathbf{r}(s_i)) \cdot \overline{\mathbf{n}}(s_i)]^2 . \tag{6.12}$$

Recall that, in shape-space, $\mathbf{r}(s_i)$ can be expressed explicitly in terms of the shape-space vector $\mathbf{X}$:

$$\mathbf{r}(s_i) = U(s_i)(W\mathbf{X} + \mathbf{Q}_0)$$

so the norm-difference can be expressed explicitly in terms of the shape-space vector $\mathbf{X}$ as

$$\|\mathbf{r} - \mathbf{r}_f\|_{\overline{\mathbf{n}}}^2 \approx \frac{1}{N} \sum_{i=1}^{N} \left(\nu_i - \mathbf{h}(s_i)^T [\mathbf{X} - \overline{\mathbf{X}}] \right)^2 \tag{6.13}$$

where

$$\nu_i = (\mathbf{r}_f(s_i) - \overline{\mathbf{r}}(s_i)) \cdot \overline{\mathbf{n}}(s_i), \tag{6.14}$$

is the "innovation" — the displacement measured relative to the mean shape and resolved along the normal, and

$$\mathbf{h}(s)^T = \overline{\mathbf{n}}(s_i)^T U(s_i) W. \tag{6.15}$$

Details of an algorithm to solve the fitting problem, using the invariant norm, are developed next. In the meantime, note (figure 6.6) that the new algorithm solves the reluctance problem demonstrated earlier arising from the use of norm-difference with the inherited parameterisation (figure 6.4).

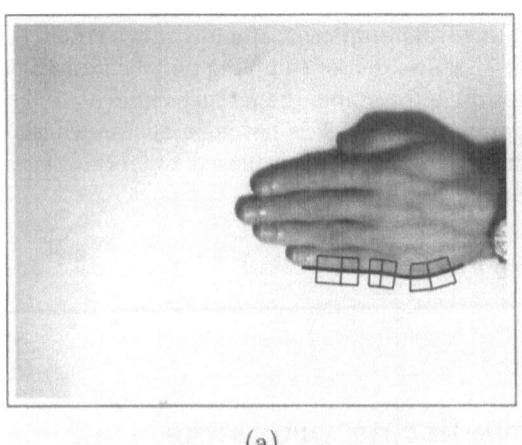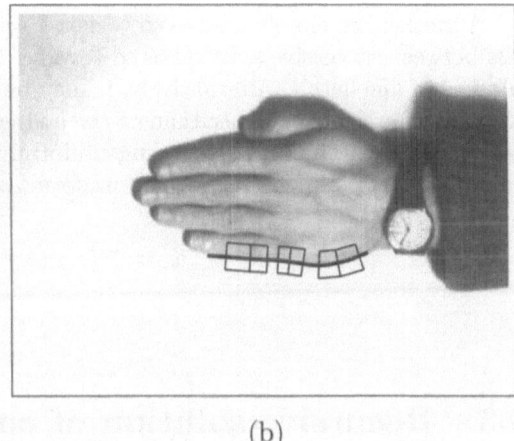

(a) (b)

Figure 6.6: Normal displacement solves the reluctance problem. *Curve fitting using normal displacement is applied repeatedly to a sequence of images starting with (a) and ending with (b). The fitted curve follows the moving image feature, without the reluctance associated with norm-difference and inherited parameterisation (figure 6.4).*

Weighted norm

A more general sampled form for the norm, in place of (6.13), incorporates weights w_i:

$$\|\mathbf{r} - \mathbf{r}_f\|_{\overline{\mathbf{n}}}^2 = \left(\sum_{i=1}^{N} w_i\right)^{-1} \sum_{i=1}^{N} w_i \delta_i^2 \qquad (6.16)$$

where

$$\delta_i = \nu_i - \mathbf{h}(s_i)^T [\mathbf{X} - \overline{\mathbf{X}}].$$

For instance, setting

$$w_1 = w_N = \frac{1}{2} \quad \text{and} \quad w_i = 1, \ 1 < i < N$$

implements the trapezium rule which correctly takes account of the ends of an open curve. For closed curves, in which $i = 0$ and $i = N$ represent the same physical point which must be counted exactly once, appropriate weights are

$$w_i = 1 \quad \text{for} \quad 1 \le i < N \quad \text{and} \quad w_N = 0.$$

Weights can also be used to implement modified norms which accentuate areas of the image-feature curve. One good reason for doing this is to allow increased influence around areas of fine detail, such as the ends of the fingers in figure 5.1.

A mechanism closely related to variable weighting is non-uniform sampling in which intervals between successive s_i are allowed to vary. For instance, denser sampling may be desirable in areas of fine detail. Alternatively, it may be desirable to sample the feature curve at equal intervals of arclength, rather than of the spline parameter s which is not generally arclength. Lastly, an effect similar to sampling uniformly in arclength can be achieved by appropriate weighting that compensates for non-uniform sampling:

$$w_i = |\bar{\mathbf{r}}'(s_i)| \quad \text{where} \quad \bar{\mathbf{r}}'(s) \equiv \frac{d\bar{\mathbf{r}}}{ds}.$$

6.3   Recursive solution of curve-fitting problems

We already saw one solution to the regularised curve-fitting problem (6.3), using projection followed by weighted summation. The new fitting problem based on the normal displacement measure is best solved in a rather different style. The new algorithm is recursive, working by traversing the data-curve once, updating the estimated shape $\hat{\mathbf{X}}$ as it does so. (In fact the original problem (6.3) could also be solved concisely in the recursive manner.) The complete fitting algorithm is given in figure 6.7. For the regularised fitting problem above, we set the "measurement error" constant $\sigma_i = \sigma = \sqrt{N}$, but in subsequent chapters, other values of σ_i will be used. The first two steps of the algorithm establish feature points $\mathbf{r}_f(s_i)$ which serve as the data for curve fitting. Step 3 initialises the "information matrix" S_i which is a measure of the "strength" of each intermediate estimate $\hat{\mathbf{X}}_i$, taking account of the first i data points. Step 3 also initialises the "information weighted sum" $\mathbf{Z}_i$ which accumulates the influence of the mean shape and the individual measurements, each with its proper weight. In step 4, the measurements $\mathbf{r}_f(s_i)$ are assimilated in turn. Note that, as expected, it is only the normal component ν_i of each measurement that is used. [Note the effect of the weighted norm (6.16) is to choose a variable "measurement error" $\sigma_i = \sqrt{w_i^{-1} \sum_j w_j}$.] In step 5, the "aggregated" observation vector $\mathbf{Z}$ is defined, together with S, its statistical information as an estimator of $\mathbf{X}$. In fact $\mathbf{Z}$ is an unbiased estimate not of $\mathbf{X}$ directly, but of $S\mathbf{X}$. (This is much safer than trying to deal with an estimate of $\mathbf{X}$ itself, given that the inverse of S need not exist.) In step 6, the aggregated measurement $\mathbf{Z}$ that incorporates the influence of all data points, is finally combined in what is, in fact, an information weighted sum (see derivation below) to give the estimated shape-space vector $\hat{\mathbf{X}}$.

Curve-fitting problem

Given an initial shape estimate $\bar{\mathbf{r}}(s)$ (or $\overline{\mathbf{X}}$ in shape-space) with normals $\bar{\mathbf{n}}(s)$, and a regularisation weight matrix $\overline{S}$, solve:

$$\min_{\mathbf{X}} T \quad \text{where} \quad T = (\mathbf{X} - \overline{\mathbf{X}})^T \overline{S} (\mathbf{X} - \overline{\mathbf{X}}) + \sum_{i=1}^{N} \frac{1}{\sigma_i^2} \left(\nu_i - \mathbf{h}(s_i)^T [\mathbf{X} - \overline{\mathbf{X}}] \right)^2 .$$

Algorithm

1. Choose samples $s_i, i = 1, \ldots, N$, s.t. $s_1 = 0$, $s_{i+1} = s_i + h$, $s_N = L$.

2. For each i, apply some image-processing filter along a suitable line (e.g. curve normal) passing through $\bar{\mathbf{r}}(s_i)$, to establish the position of $\mathbf{r}_f(s_i)$.

3. Initialise
$$\mathbf{Z}_0 = 0, \quad S_0 = 0.$$

4. Iterate, for $i = 1, \ldots, N$:
$$
\begin{aligned}
\nu_i &= (\mathbf{r}_f(s_i) - \bar{\mathbf{r}}(s_i)) \cdot \bar{\mathbf{n}}(s_i); \\
\mathbf{h}(s_i)^T &= \bar{\mathbf{n}}(s_i)^T U(s_i) W; \\
S_i &= S_{i-1} + \frac{1}{\sigma_i^2} \mathbf{h}(s_i) \mathbf{h}(s_i)^T; \\
\mathbf{Z}_i &= \mathbf{Z}_{i-1} + \frac{1}{\sigma_i^2} \mathbf{h}(s_i) \nu_i.
\end{aligned}
$$

5. The aggregated observation vector is
$$\mathbf{Z} = \mathbf{Z}_N \text{ with associated statistical information } S = S_N.$$

6. Finally, the best fitting curve is given in shape-space by:
$$\hat{\mathbf{X}} = \overline{\mathbf{X}} + (\overline{S} + S)^{-1} \mathbf{Z}.$$

Figure 6.7: Recursive algorithm for curve fitting.

Example 1

A very simple fitting problem for tutorial purposes is illustrated in figure 6.8. It

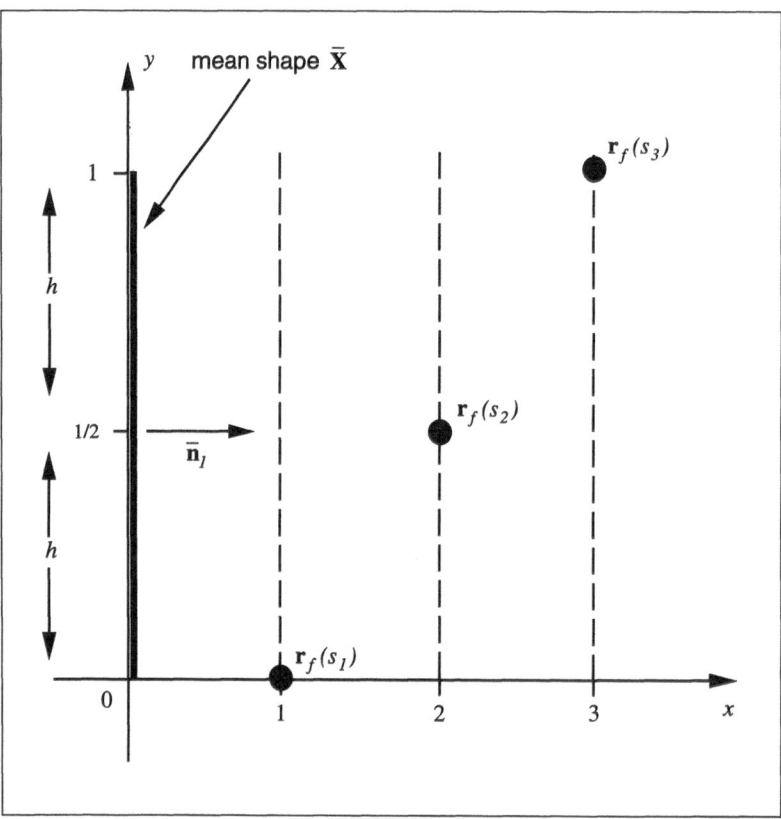

Figure 6.8: *Example fitting problem for the recursive algorithm — see text.*

involves a family of vertical line segments

$$\mathbf{r}(s) = (x, s)^T \quad \text{for} \ \ 0 \le s \le L$$

(a single-span linear spline!) with length $L = 1$. Take the template $\mathbf{Q}_0$ to correspond to a vertical line at the origin:

$$\mathbf{r}_0(s) = (0, s)^T \quad \text{for} \ \ 0 \le s \le L$$

so that shape-space is then parameterised by the single-component shape-vector $\mathbf{X} = x$, and

$$U(s)W = (1, 0)^T,$$

so that $\mathbf{X}$ is transformed to spline space as

$$U(s)W\mathbf{X} + \mathbf{r}_0 = (x, s)^T,$$

as required. The mean shape is also taken to be the line through the origin, so $\overline{\mathbf{X}} = 0$. Measurements will be made at

$$s_1 = 0, \ s_2 = \frac{1}{2}, \ s_3 = 1$$

so that $h = 1/2$, as illustrated in figure 6.8,

$$\sigma^2 = N = \frac{L}{h} + 1 = 3, \ \text{ and } \ \overline{\mathbf{n}}(s_i) = (1,0)^T \ \text{ for } \ i = 1, 2, 3.$$

Now we can also calculate

$$\mathbf{h}(s_i)^T = \overline{\mathbf{n}}(s_i)^T U(s_i) W = 1 \ \text{ for } \ i = 1, 2, 3.$$

The data points shown in the figure are

$$\mathbf{r}_f(s_1) = (1, 0)^T, \ \mathbf{r}_f(s_2) = (2, 1/2)^T, \ \mathbf{r}_f(s_3) = (3, 1)^T.$$

It is not difficult to show that the metric matrix for this shape-space is $\mathcal{H} = 1$.

First, take the case of fitting without regularisation, so $\alpha = 0$. Following the steps of the algorithm gives

i	ν_i	S_i	$\mathbf{Z}_i$
0		0	0
1	1	$\frac{1}{3}$	$\frac{1}{3}$
2	2	$\frac{2}{3}$	1
3	3	1	2

The aggregated measurement is then $\mathbf{Z} = 2$ with information $S = 1$, and since $\overline{S} = 0$ (because $\alpha = 0$), the estimate is $\hat{\mathbf{X}} = S^{-1}\mathbf{Z} = 2$, simply the average value of the x-coordinates of the three data points, as might be expected.

Example 2

If regularisation is added with $\alpha = 1$, the main part of the algorithm proceeds as above, but since now $\overline{S} = 1$, the final step is

$$\hat{\mathbf{X}} = \overline{\mathbf{X}} + \left(\overline{S} + S\right)^{-1} \mathbf{Z} = 0 + \frac{1}{2} \cdot 2 = 1,$$

which has been pulled down towards $\overline{\mathbf{X}} = 0$ by regularisation, as expected.

Proof of correctness: the algorithm is of a standard type for recursive solution of least-squares problems. Defining the partial sum

$$T_i = \frac{1}{\sigma^2} \sum_{j=1}^{i} \left(\nu_j - \mathbf{h}(s_j)^T [\mathbf{X} - \overline{\mathbf{X}}]\right)^2.$$

it is straightforward to prove by induction on i that

$$T_i = (\mathbf{X} - \overline{\mathbf{X}})^T S_i (\mathbf{X} - \overline{\mathbf{X}}) - (\mathbf{X} - \overline{\mathbf{X}})^T \mathbf{Z}_i - \mathbf{Z}_i^T (\mathbf{X} - \overline{\mathbf{X}}) + c_i \quad \text{for} \ \ i = 1, \dots, N,$$

where c_i is a constant, independent of $\mathbf{X}$. When $i = N$, this gives

$$T_N = (\mathbf{X} - \overline{\mathbf{X}})^T S (\mathbf{X} - \overline{\mathbf{X}}) - (\mathbf{X} - \overline{\mathbf{X}})^T \mathbf{Z} - \mathbf{Z}^T (\mathbf{X} - \overline{\mathbf{X}}) + c_N,$$

where $S = S_N$ and $\mathbf{Z} = \mathbf{Z}_N$, as in the algorithm. Now

$$T = (\mathbf{X} - \overline{\mathbf{X}})^T \overline{S} (\mathbf{X} - \overline{\mathbf{X}}) + T_N$$

and completing the square gives

$$T = (\mathbf{X} - \hat{\mathbf{X}})^T (\overline{S} + S)(\mathbf{X} - \hat{\mathbf{X}}) + c,$$

where c is independent of $\mathbf{X}$, and $\hat{\mathbf{X}}$ is as defined in the algorithm, so $\mathbf{X} = \hat{\mathbf{X}}$ optimises T, as required.

Validation gate

The innovation ν_i in (6.14) represents the difference between the actual measurement and the measurement that would be predicted at $s = s_i$ based on the mean shape. It is potentially useful for detecting and deleting rogue data or "outliers". In the context of tracking, outliers arise when the object being tracked is partially obscured, and the edge of some background object is then detected and masquerades as $\mathbf{r}_f(s_i)$, for some i. The resulting error in position may be considerable, well outside the normal range attributable to random factors such as electrical and optical noise. Such an "outlier" should be signaled by an unusually large magnitude $|\nu_i|$ of innovation. If this occurs, the ith measurement can be discarded, and the ith iteration in the algorithm altogether omitted. Otherwise, the ith data point is said to be validated and the ith iteration proceeds as normal.

It remains to choose a threshold — at what level is it considered that $|\nu_i|$ is sufficiently large to signal an outlier? A principled answer to this question emerges from statistical models described in chapter 8, leading to the "validation-gate" for outlier removal. The width of the validation gate effectively fixes the width of the search region for image processing along normals (figure 5.1 on page 98). In fact the threshold on $|\nu_i|$ determines the length of the search segment, the interval on the normal that lies within the search region.

Fitting corners

In some applications it may be desired to fit to corner features rather than edge-features, or to a mixture of corners and edges. The recursive fitting algorithm extends in a straightforward manner. Suppose the measured feature $\mathbf{r}_f(s_i)$ is a corner or other distinguished point, so that its full displacement $\nu_i^r = \mathbf{r}_f(s_i) - \mathbf{r}(s_i)$ must be taken into account, rather than just the normal component. Then step 4 of the recursive algorithm in figure 6.7 must be modified to take account of this, as in figure 6.9.

Alternative recursive solution to the fitting problem

For completeness, an alternative to the fitting algorithm described above in figure 6.7 is given here. It solves exactly the same minimisation problem, but using an alternative set of variables. Where the algorithm above was based on "information" S, the algorithm here is based on covariance P. This algorithm will be readily recognised as a special case of a "Kalman filter" by those who are already familiar with them. This alternative curve-fitting algorithm can also be explained intuitively. In place of the information matrix S_i, this algorithm is expressed in terms of its inverse $P_i = S_i^{-1}$ (to be interpreted in later chapters as a statistical covariance). The weighted sum variable $\mathbf{Z}_i$ is no longer needed; instead successive estimates $\hat{\mathbf{X}}_i$, based on the first i data points, are generated directly. The variable ν_i' is a "successive" form of innovation, the difference between the actual value of the ith measurement and its expected value based on extrapolation from the previous $i-1$ measurements, and it is only this difference

---

Modified iterative step:

4 Iterate, for $i = 1, \ldots, N$:

$$\boldsymbol{\nu}_i^r = \mathbf{r}_f(s_i) - \bar{\mathbf{r}}(s_i);$$

$$h^r(s_i)^T = U(s_i)W;$$

$$S_i = S_{i-1} + \frac{1}{\sigma_i^2} h^r(s_i) h^r(s_i)^T;$$

$$\mathbf{Z}_i = \mathbf{Z}_{i-1} + \frac{1}{\sigma_i^2} h^r(s_i) \boldsymbol{\nu}_i^r.$$

Figure 6.9: Recursive fitting algorithm: modification for corner features.

that generates any change in successive estimates $\hat{\mathbf{X}}_i$. The proof of equivalence of the two algorithms is omitted here, but could be found in a standard textbook on statistical filtering.

Computational cost

Both algorithms have computational complexity $O(NN_X^2)$, where N_X is the dimension of shape-space, as usual. This is based on the assumption that, for a given shape-space, $\mathcal{H}^{-1}$ and the products $U(s_i)W$ can be calculated off-line and need not be counted towards the total computational cost. It is further assumed that $N \geq N_X$ which is normally the case as this is the minimum value for which S can have full rank. The original algorithm has only one $O(NN_X^2)$ operation, as opposed to two in the alternative algorithm. However, the original algorithm involves additional matrix inversions which have complexity $O(N_X^3)$. This means that for small $N_X \ll N$ the original algorithm is more efficient, but as $N_X \to N$ the alternative algorithm becomes the more efficient.

Example 3

Example 2 is repeated here using the alternative algorithm; of course it should achieve the same result. The estimate is $\hat{\mathbf{X}} = \hat{\mathbf{X}}_3 = 1$, in agreement with the original algorithm.

Algorithm:

1. Obtain measurements $\mathbf{r}_f(s_i)$ as before.

2. Initialise

$$
\begin{aligned}
P_0 &= \overline{S}^{-1} \\
\hat{\mathbf{X}}_0 &= \overline{\mathbf{X}}
\end{aligned}
$$

3. Iterate, for $i = 1, \ldots, N$:

$$
\begin{aligned}
\nu_i &= (\mathbf{r}_f(s_i) - \overline{\mathbf{r}}(s_i)) \cdot \overline{\mathbf{n}}(s_i); \\
\mathbf{h}(s_i)^T &= \overline{\mathbf{n}}(s_i)^T U(s_i) W; \\
\nu_i' &= \nu_i - \mathbf{h}(s_i)^T (\hat{\mathbf{X}}_{i-1} - \overline{\mathbf{X}}); \\
K_i &= P_{i-1}\mathbf{h}(s_i)(\mathbf{h}(s_i)^T P_{i-1}\mathbf{h}(s_i) + \sigma_i^2)^{-1} \\
\hat{\mathbf{X}}_i &= \hat{\mathbf{X}}_{i-1} + K_i \nu_i'; \\
P_i &= (I - K_i \mathbf{h}(s_i)^T) P_{i-1};
\end{aligned}
$$

4. The best fitting curve is given in shape-space by:
$$\hat{\mathbf{X}} = \mathbf{X}_N.$$

Note that intermediate estimates $\hat{\mathbf{X}}_i$ are automatically generated by this algorithm.

Figure 6.10: Alternative recursive fitting algorithm.

i	ν_i	ν_i'	K_i	$\hat{\mathbf{X}}_i$	P_i
0				0	1
1	1	1	$\frac{1}{4}$	$\frac{1}{4}$	$\frac{3}{4}$
2	2	$\frac{7}{4}$	$\frac{1}{5}$	$\frac{3}{5}$	$\frac{3}{5}$
3	3	$\frac{12}{5}$	$\frac{1}{6}$	1	$\frac{1}{2}$

Note also that, as expected, the final value of covariance P_3, satisfies $P_3 = S_3^{-1}$ where S_3 is the final information from example 2.

6.4   Examples

B-spline snake

An example of fitting a B-spline snake is given in figure 6.11. Working in a spline

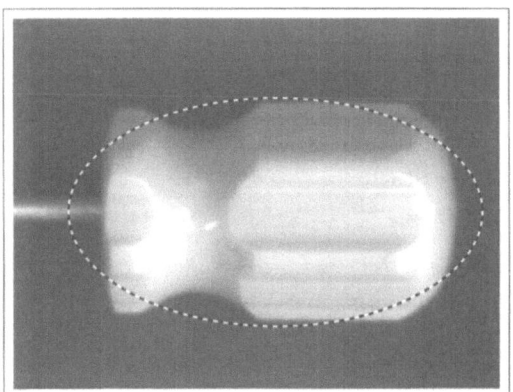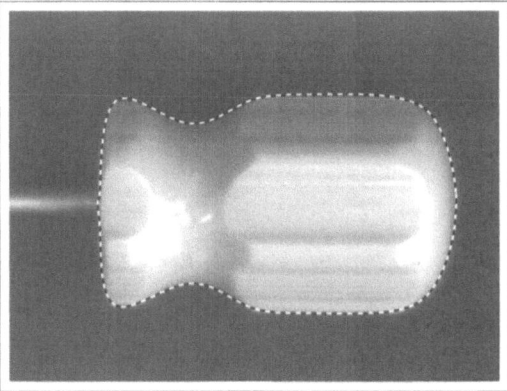

Figure 6.11: Using a snake to capture an outline. *For automated planning of robot grasps (see chapter 1) it is necessary to capture the outline of the part to be grasped. This can be done even without much prior knowledge of part shape, using a B-spline snake initialised as an ellipse (left), achieving a good fit (right). (Figure courtesy of Colin Davidson.)*

space is appropriate when no strong assumption can be made about the expected shape. The snake is initialised using moments (as explained in figure 4.6 on page 82) to obtain a coarse elliptical approximation to the outline. Weak regularisation is used, just sufficient to stabilise the computation but without unduly biasing the fit. The low level of regularisation suffices in this case because the background is clean and featureless.

More stable behaviour may be needed when the background is more cluttered or the foreground partly obscured or simply outside the search region. Stability can be achieved by increasing the strength of regularisation as in figure 6.12. In that example, missing measurements around the fingertips mean that the shape of the fitted curve is underconstrained. (In fact a minimal regulariser with $\alpha = 0.001$ is applied simply to ensure numerical stability). When a significant degree of regularisation ($\alpha = 0.1$) is applied, the missing data is satisfactorily interpolated by defaulting to the template (initial) shape. Regularisation is set to be invariant to Euclidean similarities, to allow the snake to rotate and translate freely, while maintaining shape constraints. Even

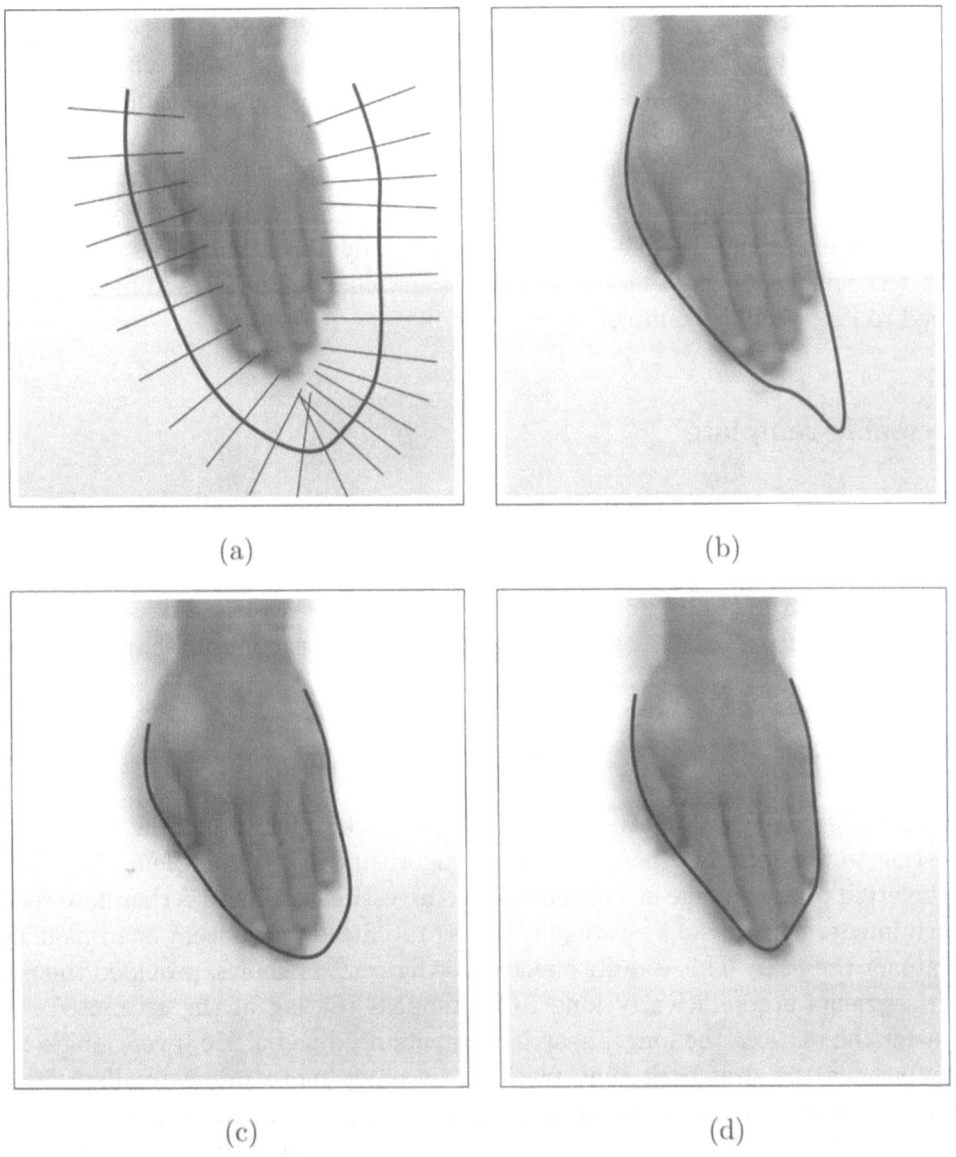

(a)   (b)

(c)   (d)

Figure 6.12: Regularisation stabilises snake fitting. *Given an initial snake curve (a) in which some features (fingertips) fall outside the search region, unregularised fitting gives poor results (b). Introducing regularisation (with invariance to Euclidean similarities) produces a good fit (c). This is refined further after iterating the fitting cycle (d).*

more accurate fitting is achieved by iterating the fitting process to convergence. At each successive iteration the fitted shape from the previous iteration is used as an estimate from which image measurements are made. However the stabilising template remains constant throughout.

An alternative method of stabilisation is to restrict the displacement of points $\mathbf{r}(s)$ to run along a family of straight lines such as normals, or parallel lines in a chosen direction. It is simple and effective, but usable only for applications where motion can be restricted to a fixed family of lines. Surveillance of railed vehicles or traffic confined to lanes is one example.

Deformable template

At the expense of needing more specific prior knowledge about shape, deformable templates running in a suitably constrained shape-space generally behave more stably than snakes. An example of fitting a deformable template in affine shape-space is shown in figure 6.13. The initial estimated contour was obtained using moments. Then recursive fitting in affine spaces considerably refines contour shape, as shown.

Rudimentary tracking

Much of the second part of the book concerns tracking shapes in motion and it is interesting to try applying the curve-fitting algorithm to that problem. Now a curve must be fitted to each image in a sequence, and an estimated curve is therefore required for each image. The obvious strategy is to use the fitted curve from one image as the estimate for the next. This is quite effective, as figure 6.14 shows, provided the normal search segments are sufficiently long to encompass the lag of the estimated curves. The faster the motion, the longer search segments need to be. However, longer search segments have the drawback that tracking becomes more prone to distraction by background clutter, as the figure shows. There is therefore a trade-off between agility of motion and density of clutter. The capacity to deal with this trade-off is greatly expanded by dynamical modelling. If each estimated curve, rather than being a mere copy of the previous fitted curve, is actually an extrapolation of the motion to date, tracking performance can be greatly improved, and this is the subject of chapters 9 and 10.

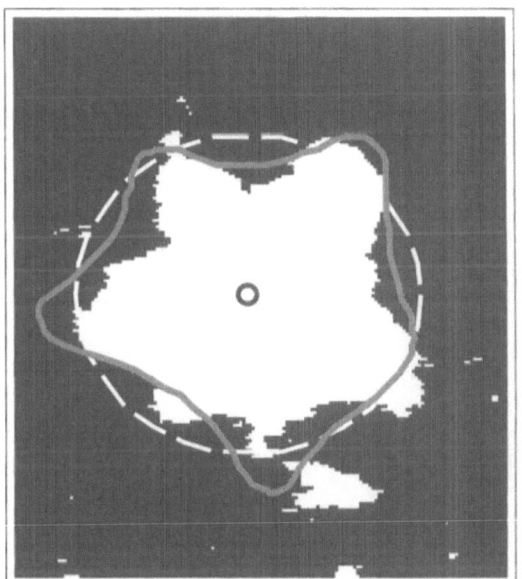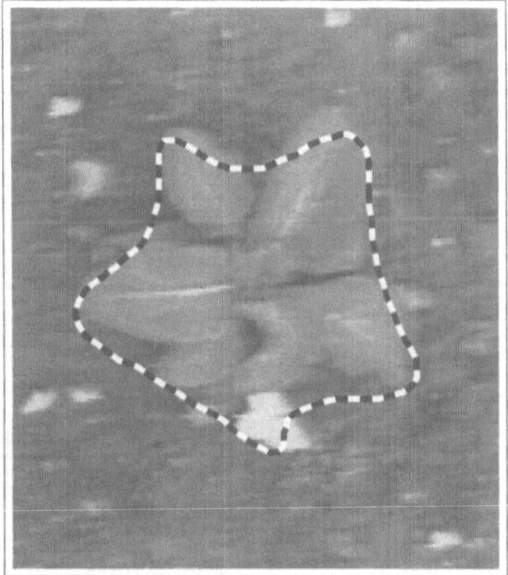

Figure 6.13: Deformable template in affine space. *From an initial estimate (grey curve, left) obtained using moments, a cabbage template is fitted to image data over an affine space to obtain the fit shown (right).*

Bibliographic notes

Regularisation, used in this chapter to bias fitted curves towards a default shape, is a standard algorithmic device for stabilising otherwise unstable or underconstrained systems of equations (Press et al., 1988). Regularisation has been used a good deal in Computer Vision (Horn and Schunk, 1981; Grimson, 1981; Ikeuchi and Horn, 1981; Poggio et al., 1985; Terzopoulos, 1986) and in image processing for "constrained restoration" of degraded images (Gonzales and Wintz, 1987).

An important aspect of the curve fitter described in the chapter is that it avoids the very considerable computational expense of applying filters to entire images. Feature detection is restricted to be one-dimensional, along normal curves (Harris and Stennett, 1990; Lowe, 1991), and within a region of interest (Inoue and Mizoguchi, 1985) or search region.

The need to use normal displacement, factoring out the spurious tangential displacement is known in vision as the "aperture problem," is the basis of the visual

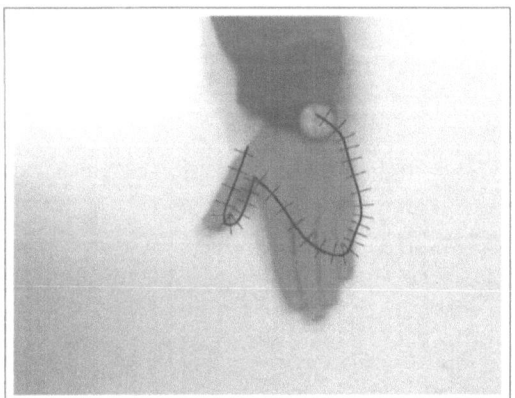
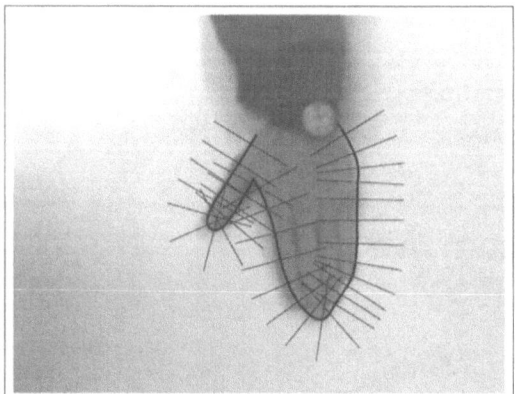
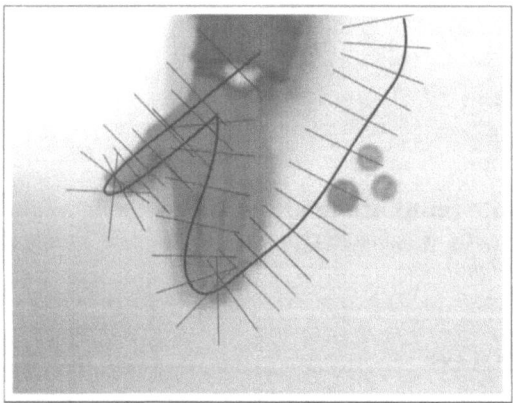

Figure 6.14: From fitting to tracking. *Recursive fitting is applied to successive images in a sequence in which a hand moves from right to left. The fitted curve from one image is used as the initial estimate in the next. With short search segments (left), even slow motion causes estimated shape to lag sufficiently that the hand falls outside the search region. Longer search segments (right) may cure this problem, but make tracking more prone to distraction by clutter (bottom).*

"barber's pole" illusion and has received much attention in the design of algorithms for analysing visual motion (Horn and Schunk, 1981; Hildreth, 1983; Horn, 1986; Murray and Buxton, 1990). However, normal displacement alone underconstrains the motion of a contour — the "aperture problem in the large" (Waxman and Wohn, 1985), though this defect is somewhat mitigated by regularisation.

Curve fitting is done efficiently here by a recursive algorithm, and this is based on recursive least-squares estimation (Bar-Shalom and Fortmann, 1988) which in turn is an application of the "dynamic programming" principle (Bellman and Dreyfus, 1962) for optimisation, applied to quadratic, multi-variate minimisation (Jacobs, 1993). Dynamic programming was first used with snakes by (Amini et al., 1988), in a discrete form as opposed to the continuous form described in this chapter. The original snake paper (Kass et al., 1987) used sparse matrix methods to solve least-squares curve fitting and that is closely related to the recursive algorithm in the case that shape-space is the spline space $\mathcal{S}_Q$. The validation of data by checking the absolute value of innovations is used in a simple form of robust estimator known as an m-estimator (Hampel et al., 1995). It is a special case from the family of "robust" estimators known as M-estimators which allow some flexibility in the way outliers are treated. General M-estimators can be expensive to compute, requiring repeatedly refined regressions to form the final robust estimate. The simple validation mechanism described in the chapter, the validation gate (Bar-Shalom and Fortmann, 1988), has the great virtue of low computational cost and this will be essential for real-time processing of image sequences in later chapters.

Chapter 7

Pose recovery

In certain three-dimensional applications (chapter 1), such as the 3D mouse in figure 1.16 on page 20, a shape-vector $\mathbf{X}$ is used to compute pose, in that case the position and attitude of the hand. Similarly, in facial animation, it is desirable to compute the attitude of the head, independently of expression if possible. The problem is to convert a shape-vector $\mathbf{X}$, from a planar or three-dimensional shape-space respectively, into three-dimensional translation $\mathbf{R}_c$ and rotation R.

7.1 Calculating the pose of a planar object

In the planar affine space, pose is recovered from a shape-vector $\mathbf{X}$ by first obtaining affine parameters $\mathbf{u}, M$ via (4.5) on page 78. Then $\mathbf{u}, M$ are used to obtain pose parameters. This is fairly straightforward under orthographic projection. For large fields of view however it is necessary to use weak perspective and then the pose-recovery computation must be elaborated to compensate for the obliqueness of projection in the periphery of the field of view.

Orthographic projection

First, a solution is presented for the pose of an object under orthographic projection, suitable for use when the field of view is significantly smaller than 1 radian. Given a planar affine vector $\mathbf{X}$, the algorithm computes object pose as the position vector $\mathbf{R}_c = (X_c, Y_c, Z_c)^T$ of the object's centre and the orientation as a rotation R relative to the reference pose. First the algorithm is given, then a short proof (optional).

1. Compute $\mathbf{u}$ and M from
$$\mathbf{X} = (u_1, u_2, M_{11} - 1, M_{22} - 1, M_{21}, M_{12})^T.$$

2. Compute the eigenvectors $\mathbf{v}_1, \mathbf{v}_2$ and eigenvalues λ_1, λ_2 (both must be positive and are ordered so that $\lambda_1 \geq \lambda_2$) of the matrix MM^T.

3.
$$Z_c = f/\sqrt{\lambda_1}$$

4.
$$\cos\theta = \left(\frac{Z_c}{f}\right)^2 \det M$$

5.
$$(\cos\phi, \sin\phi)^T = \mathbf{v}_1$$

6.
$$R(\psi) = S_y(1/\cos\theta)R(-\phi)\ M\ \frac{Z_c}{f}$$

$$\text{where }\ R(\psi) \equiv \begin{pmatrix} \cos\psi & -\sin\psi \\ \sin\psi & \cos\psi \end{pmatrix} \text{ and }\ S_y(\mu) \equiv \begin{pmatrix} 1 & 0 \\ 0 & \mu \end{pmatrix}.$$

7.
$$\text{Combined rotation: } R = R_z(\phi)R_x(\theta)R_z(\psi).$$

8.
$$\text{Translation: } \mathbf{R}_c = (X_c, Y_c, Z_c)^T \ \text{ where }\ X_c = u_1 Z_c/f, \ \ Y_c = u_2 Z_c/f.$$

Figure 7.1: Algorithm for recovery of the pose of a planar object.

The algorithm is summarised in figure 7.1 and a commentary follows. Given $\mathbf{u}, M$ from step 1 of the algorithm, we first work on M to recover the object's attitude as represented by the rotation matrix R. It is well known that a three-dimensional rotation can be decomposed in terms of three *Euler angles* θ, ϕ, ψ appearing in a sequence of three standard rotations. It is particularly convenient here to use a decomposition that is aligned with the image plane:

$$R(\phi, \theta, \psi) = R_z(\phi)R_x(\theta)R_z(\psi). \qquad (7.1)$$

This should be read from right to left as a z-rotation through an angle ψ followed by a rotation through angle θ about the x-axis and finally a z-rotation through angle ϕ, as illustrated in figure 7.2. Our algorithm recovers θ, ϕ and ψ so that the rotation matrix can be calculated as in (7.1). First Z_c, ϕ, θ and ψ are computed in steps 3–6. Note that a redundancy in ϕ arises from the indeterminacy of the sign of eigenvalue $\mathbf{v}_1$: for a given rotation R, ϕ can equally well be replaced by $\phi + 180^o$ (the substitution $\phi \rightarrow 180^o + \phi$, $\theta \rightarrow -\theta$, $\psi \rightarrow 180^o + \phi$ leaves R unchanged). Finally, the focal length f of the camera is known, so the image-plane components of translation $\mathbf{R}_c$ can be recovered directly in step 8. The result of applying this algorithm to a sequence of images of a moving hand are shown in figure 7.3.

Ambiguities

There are two kinds of ambiguity in the recovered parameter. The first is a straightforward indeterminacy in linear scale. If (X_c, Y_c, Z_c) was doubled and accompanied by a doubling in size of the object, there would be no visible effect on the image contour. The scale factor is fixed here by our convention in orthographic perspective that $Z_c = f$ in the standard view.

The second and more complex ambiguity is apparent in step (4) above: since $\cos\theta = \cos(-\theta)$, there must be two possible solutions $\pm\theta$. Unlike the indeterminacy of ϕ which represents simply an alternative representation of a given physical transformation, this is a genuine ambiguity, a reversal as illustrated in figure 7.4. In the 3D mouse application, for example, this could result in a "flip" of the object controlled by the mouse. One practical solution is to arrange for the camera to be oblique to the table so that the hand avoids the "singular" orientation $\theta = 0$, when image and object planes are parallel. Then at each time-step, the value $\pm\theta$ is chosen which give θ, ϕ and ψ closest to their values at the previous time-step.

Derivation of pose-recovery algorithm. The key point is that the effect of a rotation $R_x(\theta)$, on the image of a planar object lying on an xy plane, is to compress the image along

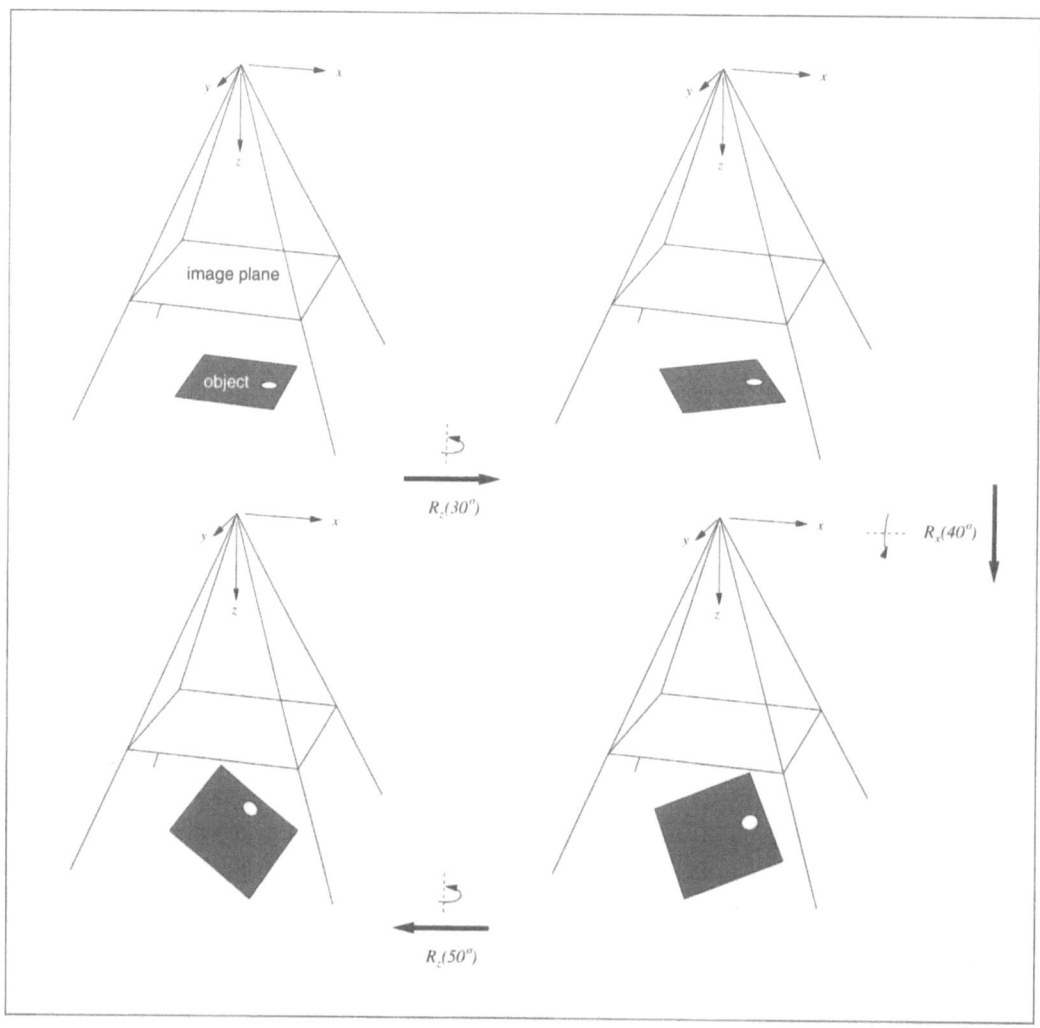

Figure 7.2: Euler angles. *The figure shows Euler angles to describe rotations in the camera coordinate frame. A general rotation R is expressed as a sequence (7.1) and the case $\phi = 50°$, $\theta = 40°$, $\psi = 30°$ is illustrated.*

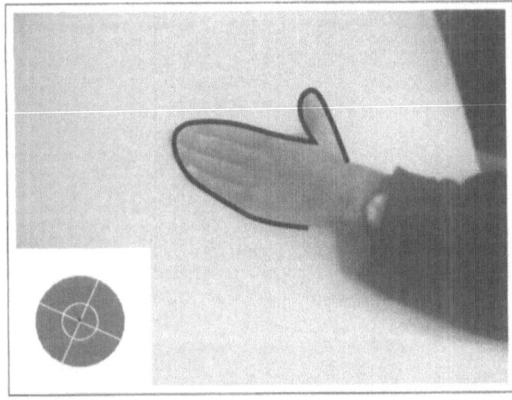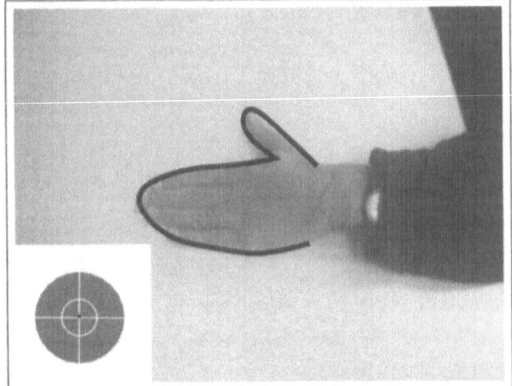

Figure 7.3: Computing pose. *The outline of a hand is tracked in a planar affine shape-space and the pose-recovery algorithm described above is applied. Computed pose is displayed using the "target" icon in the corner of each image.*

the y-axis, that is to apply a scaling transformation $S_y(\cos\theta)$. In that case, the affine image transformation due to the Euler angle transformations (7.1) and to distance scaling is:

$$M = \frac{f}{Z_c} R_z(\phi) S_y(\cos\theta) R_z(\psi).$$

The algorithm given above is simply a decomposition of this sequence, using the fact that

$$MM^T = \left(\frac{f}{Z_c}\right)^2 R_z(\phi) S_y(\cos^2\theta) R_z(-\phi)$$

which is in diagonal form.

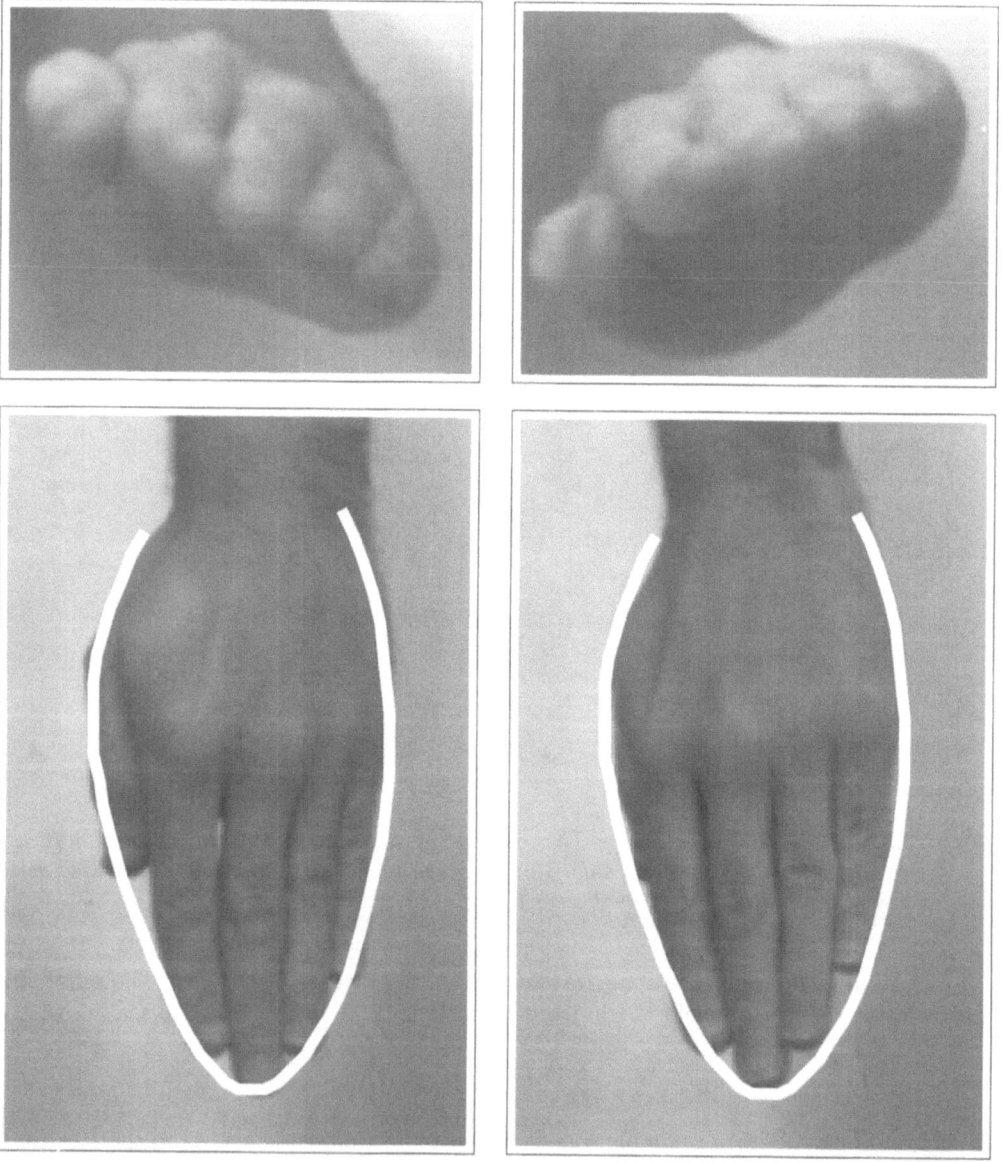

Figure 7.4: Ambiguity of pose recovery. *A hand in two alternative positions with equal and opposite slant (top row) gives rise to almost identical outline contours when viewed from above (bottom row).*

Weak Perspective projection

The pose-recovery method for orthographic projection can be simply extended for the case of weak perspective projection, when the field of view may be large but the object is still small. The most straightforward approach is to introduce a virtual orthographic image as in figure 7.5. The pose-recovery algorithm extended for weak perspective is

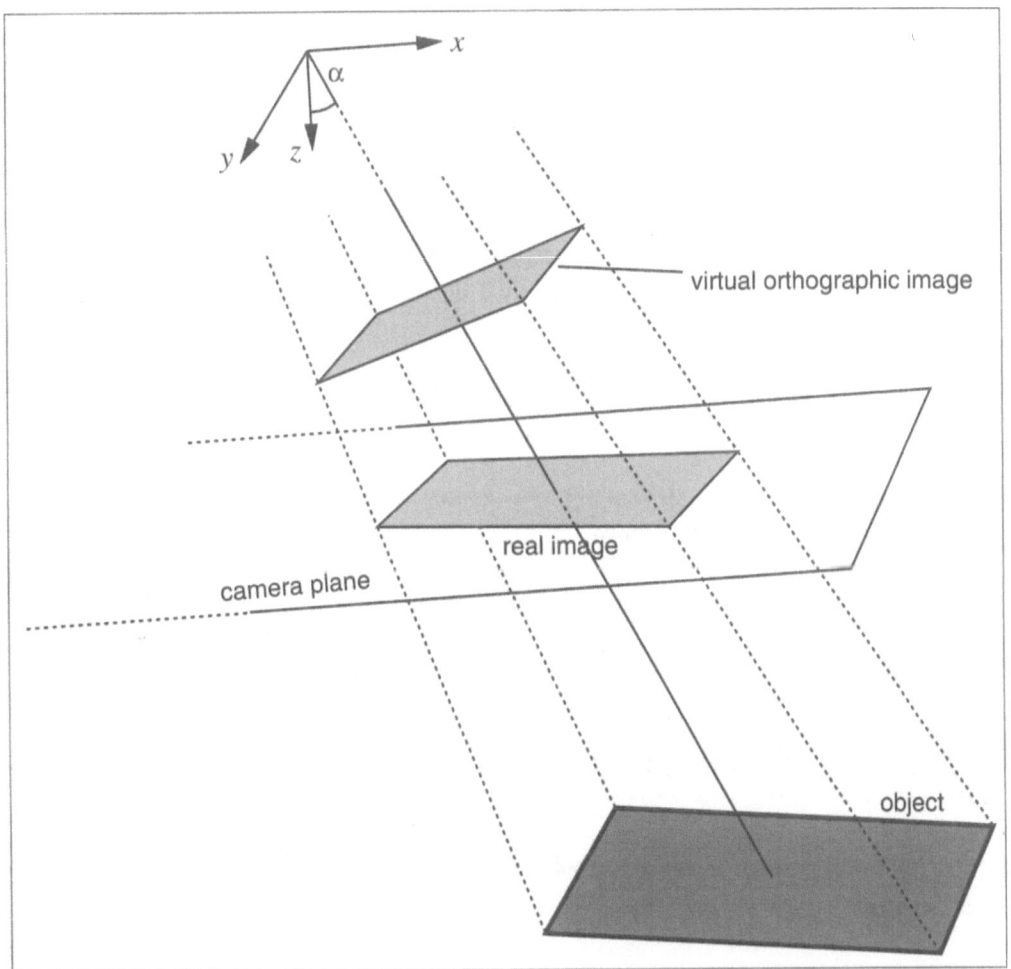

Figure 7.5: Pose correction for weak perspective. *Under weak perspective, pose recovery proceeds as in the orthographic case, on a virtual orthographic image as shown.*

1. Compute $\mathbf{u}$ and M from
$$\mathbf{X} = (u_1, u_2, M_{11} - 1, M_{22} - 1, M_{21}, M_{12})^T.$$

2. Compute polar angles α, β:
$$\alpha = \arctan(|\mathbf{u}|/f), \quad \beta = \arctan(\mathbf{u}).$$

3.
$$M' = S_x(\sec \alpha)R(-\beta)M \quad \text{where} \quad S_x(\mu) \equiv \begin{pmatrix} \mu & 0 \\ 0 & 1 \end{pmatrix}$$

($R(-\beta)$ is represented here as a 2×2 matrix.)

4. Apply steps 2–7 of the orthographic algorithm of figure 7.1 to M' (in place of M), giving R', Z'_c (in place of R, Z_c).

5.
$$R = R_z(\beta)R_y(\alpha) \ R'.$$

(These rotation matrices are now all 3×3)

6. Translation:
$$\mathbf{R}_c = (X_c, Y_c, Z_c)^T \quad \text{where} \quad Z_c = Z'_c \cos \alpha, \quad X_c = u_1 Z_c/f, \quad Y_c = u_2 Z_c/f.$$

Figure 7.6: Recovery of pose of a plane under weak perspective.

outlined in figure 7.6. First, express the image-translation vector $\mathbf{u}$ in terms of polar angles

$$\alpha, \ -\frac{\pi}{2} \leq \alpha \leq \frac{\pi}{2} \quad \text{and} \quad \beta, \ -\pi \leq \beta \leq \pi,$$

that is, so that

$$\mathbf{u} = f \tan \alpha \begin{pmatrix} \cos \beta \\ \sin \beta \end{pmatrix}. \tag{7.2}$$

The next step 3 is to replace the affine matrix M by the affine matrix M' on the virtual orthographic plane. Then the orthographic pose-recovery algorithm is applied to M' in place of M, to give a three-dimensional rotation matrix R' (subject, of course, to reversal ambiguity) and translation in depth Z'_c, both in a virtual orthographic frame. Finally, these parameters are transformed back into the camera coordinate frame in steps 5 and 6.

7.2   Pose recovery for three-dimensional objects

The planar affine space survives intact as a subspace in the three-dimensional affine case (see (4.17) on page 90), so the pose-recovery method for the planar case continues to be usable, in principle, for the three-dimensional problem. Coefficients of the first 6 elements of the configuration vector $\mathbf{X}$ could be used to obtain $\mathbf{u}, M$ as before and the planar pose-recovery algorithm would compute pose $\mathbf{R}_c, R$ correctly, subject to the usual ambiguities. However, the planar algorithm does not make optimal use of the information now available. The additional pair of shape parameters contain valuable *parallax* information which is complementary to the information available in the planar case. The planar algorithm is at its least accurate close to the standard pose. The extreme case is the singularity at the standard pose itself, which gives rise to the ambiguity in orientation discussed earlier. In contrast, the additional parallax information is at its most useful precisely at the singularity, improving accuracy of recovered pose, and removing ambiguity, as figure 7.7 shows.

Once $\mathbf{Q}_0^z$ has been estimated for a given object (see below), the recovery of pose from some outline $\mathbf{X}$ in an image of the object is relatively straightforward — the algorithm is given in figure 7.8. Step 1 computes planar affine parameters $\mathbf{u}, M$ and parallax parameter $\mathbf{v}$ from the shape-vector. Recovery of R, Z_c in steps 2 and 3 is based on solving (4.16) on page 88, expressing R in terms of its rows R_1, R_2, R_3, which must be unit vectors, and mutually orthogonal. The matrix $\hat{R}$ resulting from step 3 need not be precisely orthogonal so step 4 finds the closest orthogonal matrix by "singular value decomposition" (appendix A.1) of $\hat{R}$. Finally, translation is calculated as in the planar affine case.

Figure 7.9 illustrates that the pose of a three-dimensional object can be recovered effectively over a wide range of views, using parallax as in the algorithm above.

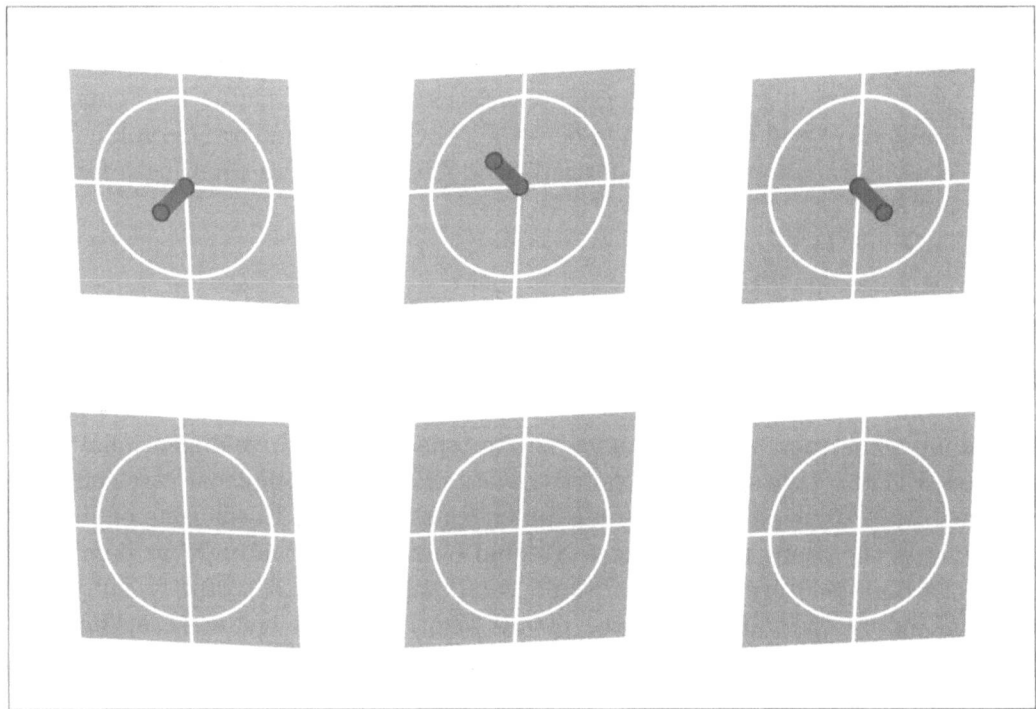

Figure 7.7: Parallax in pose recovery. *The poses of the top row of wheels, complete with axles, are easily discriminable. When parallax information is omitted by removing the axles, as on the bottom row, the orientations of the first two wheels are discriminable with difficulty and the last two actually appear identical because of the planar orientation ambiguity phenomenon.*

Estimating $\mathbf{Q}_0^z$

In principle, it would be possible to rotate the object, from its standard view, through 90° about the y-axis, and the new view would be

$$\begin{pmatrix} \mathbf{Q}_0^x \\ \\ \mathbf{Q}_0^z \end{pmatrix}$$

from which $\mathbf{Q}_0^z$ could be obtained. This would be possible with a bent wire object, but with a solid object such as the leaves above only a modest angle of rotation is possible without alterations in hidden line structure. Therefore, practically, a method is required for estimating $\mathbf{Q}_0^z$ from a rotation R through a modest angle θ about an

---

1. Compute $\mathbf{u}$, M and $\mathbf{v}$ from the shape-vector $\mathbf{X}$, using:
$$\mathbf{X} = (u_1, u_2, M_{11} - 1, M_{22} - 1, M_{21}, M_{12}, v_1, v_2).$$

2.
$$Z_c = 2f \left[|(M_{11}, M_{12}, v_1)| + |(M_{21}, M_{22}, v_2)| \right]^{-1}$$

3.
$$\hat{R}_3 = \hat{R}_1 \times \hat{R}_2 \quad \text{where} \quad \begin{pmatrix} \hat{R}_1 \\ \hat{R}_2 \end{pmatrix} = \frac{Z_c}{f} (M \mid \mathbf{v}).$$

4. Enforce orthogonality:
$$R = UV \quad \text{where} \quad \hat{R} = UDV$$

  (in which D should be approximately the identity I_3).

5.
  Translation: $\mathbf{R}_c = (X_c, Y_c, Z_c)^T$ where $X_c = u_1 Z_c / f$, $Y_c = u_2 Z_c / f$.

Figure 7.8: **Recovery of pose using parallax.**

axis lying approximately parallel to the image plane. It is assumed that the object does not translate significantly in depth (Z) during the rotation.

The rotated view $\mathbf{r}_1(s) = U(s)\mathbf{Q}_1$ is a transformed version of standard views:

$$\mathbf{r}_1(s) = \frac{f}{Z_c} \left[\mathbf{u}_1 + R_{2 \times 3} \begin{pmatrix} X_0(s) \\ Y_0(s) \\ Z_0(s) \end{pmatrix} \right],$$

in which $Z_0(s)$ is as yet unknown. The translation $f\mathbf{u}_1 / Z_c$ can be calculated simply as the centroid

$$\frac{f}{Z_c} \mathbf{u}_1 = \bar{\mathbf{r}}_1 = \begin{pmatrix} \mathbf{Q}_1^x \cdot \mathbf{1} \\ \mathbf{Q}_1^y \cdot \mathbf{1} \end{pmatrix}.$$

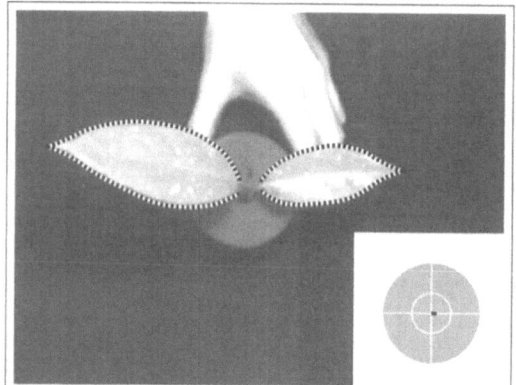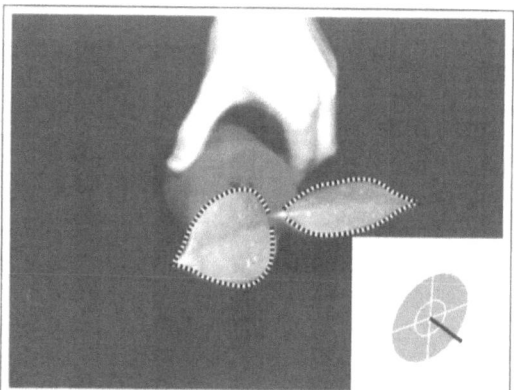

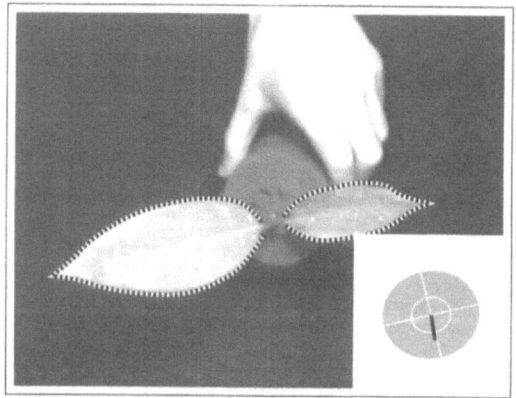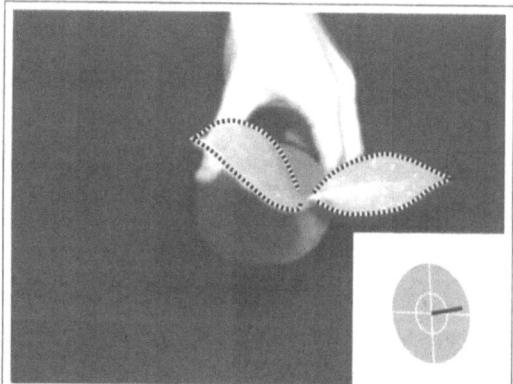

Figure 7.9: Pose recovery using the three-dimensional affine algorithm. *The pose can be accurately recovered over a wide range of transformations, including those near the planar affine degeneracy at the standard pose.*

Then $Z_0(s)$ can be estimated from

$$(\mathbf{a}^T\mathbf{a})Z_0(s) = \mathbf{a}^T\left[\left(\frac{Z_c}{f}\mathbf{r}_1(s) - \mathbf{u}_1\right) - R_{2\times2}\mathbf{r}_0(s)\right]$$

in which $\mathbf{a} = (R_{13}, R_{23})^T$. The resulting $Z_0(s)$ is a spline that can be expressed in terms of its control points as

$$\mathbf{Q}_0^z = \frac{1}{\mathbf{a}^T\mathbf{a}}(\mathbf{a}^T \otimes I)\left[\left(\frac{Z_c}{f}\mathbf{Q}_1 - \mathbf{u}_1 \otimes \mathbf{1}\right) - (R_{2\times2} \otimes I)\mathbf{Q}_0\right] \tag{7.3}$$

where I is the $N_B \times N_B$ identity matrix and $\otimes$ is the "Kronecker product" operation (see appendix A.1).

A more accurate method

If the rotation matrix R in (7.3) is not accurately known, $\mathbf{Q}_0^z$ will be imperfectly estimated. The quality of the estimate can be improved by using more views and a more sophisticated algorithm. Suppose the views are $(\mathbf{Q}_n^x, \mathbf{Q}_n^y)$, $n = 1, \ldots, N$, then the following simultaneous equations hold:

$$\begin{pmatrix} \mathbf{Q}_n^x \\ \mathbf{Q}_n^y \end{pmatrix} - \frac{f}{z_n} \left\{ \begin{pmatrix} x_n \mathbf{1} \\ y_n \mathbf{1} \end{pmatrix} + T_n \begin{pmatrix} \mathbf{Q}_0^x \\ \mathbf{Q}_0^y \\ \mathbf{Q}_0^z \end{pmatrix} \right\} = 0 \qquad (7.4)$$

where (x_n, y_n, z_n, T_n), $n = 1, \ldots, N$ and $\mathbf{Q}_0^z$ are unknown. Here (x_n, y_n, z_n) is the object translation in view n and T_n is the 2×3 rotation matrix. If f is only known approximately, it can also be treated as an unknown in the system. The unknowns can be estimated using any non-linear minimisation method, for example conjugate gradient descent. Convergence is quick, and likely to find the correct local minimum, if the unknowns are initialised approximately. An alternative linear algorithm which does not rely on careful initialisation, is the "image-stream factorisation" algorithm of Tomasi and Kanade. This works by approximating the image stream, laid out as a matrix

$$\begin{pmatrix} \mathbf{Q}_1^x & \mathbf{Q}_1^y & \mathbf{Q}_2^x & \mathbf{Q}_2^y & \cdots \end{pmatrix}$$

to have its theoretical rank (4 in this case), using singular value decomposition. Once this is done, $\mathbf{Q}_0^z$ could be calculated from the approximated image stream, subject to a certain fundamental affine indeterminacy.

7.3   Separation of rigid and non-rigid motion

In chapter 4 shape-spaces for combined rigid and non-rigid motion were constructed using key-frames. Such spaces are often too big for efficient or robust shape-fitting; they prove still to be useful for interpretation of shape, decomposing displacements into rigid and non-rigid components. This is done by projecting a fitted shape $\mathbf{Q}$ onto the shape-space and applying "Singular Value Decomposition" (SVD).

The decomposition algorithm is given in figure 7.10 (and justified below). The algorithm is illustrated in figure 7.11 for the problem of separating facial expression

---

Algorithm:

Given a fitted spline curve **Q**:

1. Express **Q** in the key-frame basis:
$$\left(u_1, u_2, \{\hat{Y}_i^j, \ i = 0, 1, \ldots, \ j = 1, 2, \ldots\}\right) = W^+\mathbf{Q},$$

   where W is as defined in (4.21) on page 93.

2. Treating $\hat{Y}_i^j$ as components of a rectangular matrix $\hat{Y}$, apply SVD:
$$\hat{Y} = UDV.$$

3. Approximate Y as a rank 1 matrix
$$Y = UD^*V$$

   where D^* is D with all but the largest singular value set to 0.

4. Recovered pose is given by the shape-vector
$$\mathbf{X} = (u_1, u_2, {Y_0}^1, {Y_0}^2, \ldots)^T.$$

5. Recovered non-rigid deformation is expressed as a contour in standard pose:
$$\mathbf{Q}^d = \mathbf{Q}_0 + \sum_i \lambda_i \mathbf{Q}_i \quad \text{where} \quad \lambda_i = \frac{Y_i^1}{Y_0^1}, \ i = 1, 2, \ldots$$

   and $\mathbf{Q}_i$ are key-frames as before.

Figure 7.10: Algorithm for decomposition of rigid and non-rigid displacement.

from head pose. Rigid transformations are modelled here as 3D affine, including parallax to accommodate the modest departure from co-planarity of tracked eyebrows, mouth and facial creases. Expression is represented in three-dimensional coordinates whose axes are defined by key-frames for smile, surprise and disgust, relative to a neutral expression.

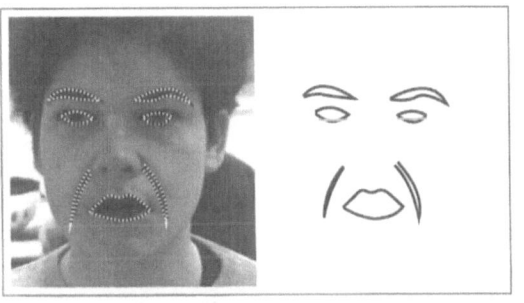 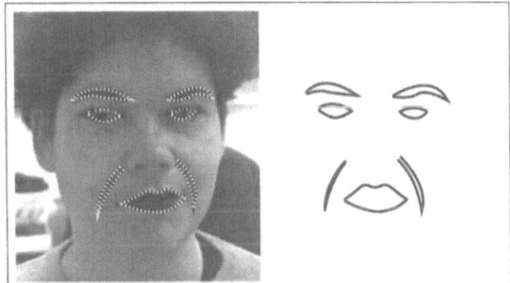

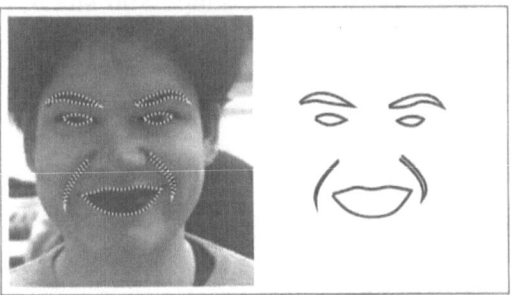 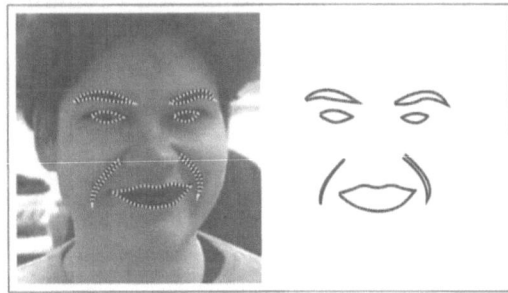

Figure 7.11: Pose-invariant transmission of facial expression. *Separation of non-rigid from rigid motion by SVD is used here to extract the facial expression of an actor. The extracted expression is displayed on this cat caricature in a fixed pose, and can be seen to be independent of the pose of the actor's head. (Figure courtesy of Benedicte Bascle.)*

Derivation of decomposition algorithm. Shape parameters Y_i^j, $i = 0, \ldots, N_k$, $j = 1, \ldots, N_r$ are highly interdependent, in principle, because, given the structure of shape-space in (4.21) on page 93, Y is a product of two vectors:

$$Y_i^j = \lambda_i X_{j+2},$$

where $\lambda_0 = 1$ is the weight of the template $\mathbf{Q}_0$ and λ_i is the weight of key-frame $\mathbf{Q}_i$ for $i = 1, \ldots, N_k$. (The index $_{j+2}$ is an offset to allow for two image-translation parameters.) Thus Y should have rank 1, and this is enforced by approximating $\hat{Y}$ to rank 1 in the SVD step. Note that one should expect D to contain one dominant singular value, indicating that Y is a good approximation to Y^*.

Bibliographic notes

Recovery of the pose of an object from its shape-space vector is based on an eigenvalue method for recovery of the pose of a textured surface in (Blake and Marinos, 1990). The "image-stream factorisation" algorithm (Tomasi and Kanade, 1991) uses singular value decomposition (SVD) (Barnett, 1990) of the image stream for motion analysis. Here, an adaptation of the algorithm was suggested for calibration of pose recovery using parallax. The algorithm for decomposition of rigid and non-rigid displacement is yet another application of SVD to bilinear decomposition problems in vision, others being structure and motion (Tomasi and Kanade, 1991), and shape and shading (Freeman and Tenenbaum, 1997).

Part II

Probabilistic Modelling

Chapter 8

Probabilistic models of shape

The purpose of this second part of the book is to put Active Contours into a probabilistic setting. As chapter 2 claimed, the probabilistic framework is essential for dealing with *classes* of shapes and motions. It is valuable even with deformable templates, in static problems, to describe classes of shapes. Then probabilistic modelling is extended to dynamic problems, to mesh with the powerful Kalman filtering formalism, in which cumulative temporal uncertainty about shape is counterbalanced by the inflow of measurements from an image sequence.

This chapter concentrates on the application of probabilistic models to static problems. The ideas discussed so far about fitting curves by regularisation are to be re-interpreted probabilistically. The deterministic approach of chapter 6 aimed to generate a unique estimate $\hat{\mathbf{X}}$ of curve shape from data $\mathbf{r}_f(s)$, moderated via regularisation towards a template $\overline{\mathbf{X}}$. Now, in a more general probabilistic setting, $\hat{\mathbf{X}}$ is merely one property, typically the mean or mode, of an entire probability distribution. The solution to the fitting problem is therefore no longer just a single value, but a whole family of possible curves. In that case, the variance of the distribution is a measure of how accurate the fitted curve is and can be used to generate a range of plausible fitted curves. One application for this is to sweep out the search region for image-processing operations. Another application, developed in later chapters on tracking, is to achieve a fusion of shape information accumulated over time and the latest visual measurements.

The distribution for curve shape $\mathbf{X}$ obtained from probabilistic fitting is expressed as a *posterior density* $p(\mathbf{X}|\mathbf{r}_f)$. This is the conditional probability density for the curve shape $\mathbf{X}$ *given* the observed data $\mathbf{r}_f(s)$. According to Bayes' formula for densities (see appendix A.3) the posterior density can be obtained as a product of a *prior density* $p_0(\mathbf{X})$ and an *observation density* $p(\mathbf{r}_f|\mathbf{X})$:

$$p(\mathbf{X}|\mathbf{r}_f) \propto p(\mathbf{r}_f|\mathbf{X})p_0(\mathbf{X}). \tag{8.1}$$

Note that although $p(\mathbf{r}_f|\mathbf{X})$ is a probability density over $\mathbf{r}_f$, in this formula $\mathbf{r}_f$ is considered fixed and it is the variation of $p(\mathbf{r}_f|\mathbf{X})$ with $\mathbf{X}$ — the "likelihood" of $\mathbf{X}$ — that is of interest. The prior density is the probabilistic mechanism for regularisation. For example, the norm-squared regulariser $\alpha\|\mathbf{X} - \overline{\mathbf{X}}\|^2$ used in chapter 6 becomes a Gaussian density function $p(\mathbf{X})$ whose mean is the curve $\overline{\mathbf{X}}$. The other term $\|\mathbf{r} - \mathbf{r}_f\|^2$ in the regularisation problem, conveying the influence of the data, also becomes a Gaussian density whose value is high when the hypothesised curve $\mathbf{r}$ fits the data closely. Then Bayes' rule simply combines the competing influences of the prior and the observations into a single density, also Gaussian.

Note that Bayesian principles of image interpretation reach far beyond Gaussian modelling. However, the Gaussian is the simplest case and has attractive properties that facilitate efficient computation. Consideration of more general distributions is left until much later in the book, in chapter 12.

8.1 Probability distributions over curves

The Gaussian prior probability density for curves in shape-space $\mathcal{S}$ consistent with the general quadratic regulariser $(\mathbf{X} - \overline{\mathbf{X}})^T \overline{S}(\mathbf{X} - \overline{\mathbf{X}})$ is:

$$p_0(\mathbf{X}) \propto \exp -\frac{1}{2}(\mathbf{X} - \overline{\mathbf{X}})^T \overline{S}(\mathbf{X} - \overline{\mathbf{X}}), \tag{8.2}$$

where the matrix $\overline{S}$ is the "information" matrix introduced in chapter 6. Note that probability density decreases as the curve $\mathbf{X}$ deviates further from the mean $\overline{\mathbf{X}}$ of the distribution, as expected. Probabilistically, $\overline{S}$ has a particular interpretation: its inverse $P_0 = \overline{S}^{-1}$, if it exists, is a *covariance matrix* (appendix A.3), a multi-dimensional measure of the variability of curve shape across a distribution. It can be used to compute the positional variability of a given point $\mathbf{r}(s)$ on a curve as a 2×2 covariance matrix

$$P_{\mathbf{r}}(s) = U(s)W P_0 W^T U(s)^T, \tag{8.3}$$

in a two-dimensional Gaussian distribution $\mathcal{N}(\bar{\mathbf{r}}(s), P_{\mathbf{r}}(s))$. In general, this distribution can be depicted as an ellipse whose axes are eigenvectors of $P_{\mathbf{r}}(s)$ and whose semi-axes, representing positional variances, are the eigenvalues. The spatial variance $P_{\mathbf{r}}(s)$ is represented by an uncertainty ellipse giving a mean-square displacement

$$\rho_0^2(s) = \mathrm{tr}(P_{\mathbf{r}}(s)) \tag{8.4}$$

where $\mathrm{tr}(\cdot)$ denotes the trace of a matrix. The mean-square displacement along the entire curve can then be computed easily (the proof follows below) as

$$\bar{\rho}_0^2 = \mathrm{tr}(P_0\mathcal{H}). \tag{8.5}$$

An isotropic prior, uniform in s, with mean-square displacement along the curve $\bar{\rho}_0^2$ is projected onto shape-space $\mathcal{S}$ as

$$p_0(\mathbf{X}) \propto \exp -\frac{N_X}{2\bar{\rho}_0^2}\|\mathbf{X} - \overline{\mathbf{X}}\|^2.$$

This can be seen by considering its information matrix in shape-space, which is

$$\overline{S} = \frac{N_X}{\bar{\rho}_0^2}\mathcal{H}. \tag{8.6}$$

From (8.5) it is apparent that the mean-square displacement along the curve is

$$\mathrm{tr}\left(\frac{\bar{\rho}_0^2}{N_X}\mathcal{H}^{-1}\mathcal{H}\right) = \frac{\bar{\rho}_0^2}{N_X}\mathrm{tr}(I_{N_X}) = \bar{\rho}_0^2,$$

as required.

Spline space is an interesting special case in which the covariance at a given point $\mathbf{r}(s)$ is $P_{\mathbf{r}}(s) \propto I_2$, a multiple of the identity matrix, so that positional error at each point is isotropically distributed (see below for the derivation). The average displacement of $\mathbf{r}(s)$ from its mean $\bar{\mathbf{r}}(s)$, is $\rho_0(s)$ and in fact $\rho_0(s) \approx \bar{\rho}_0$ for all s. This suggests circular confidence regions for $\mathbf{r}(s)$ (figure 8.1). (The average pointwise displacement $\rho_0(s)$ varies a little with s, relative to $\bar{\rho}_0$ — by around $\pm 10\%$, for instance, on a regular, closed, quadratic B-spline.)

In chapter 6, there was some discussion of the way in which image-processing operations are applied along curve normals. In the interests of efficiency, processing is limited to a segment of each normal within a search region. The probabilistic

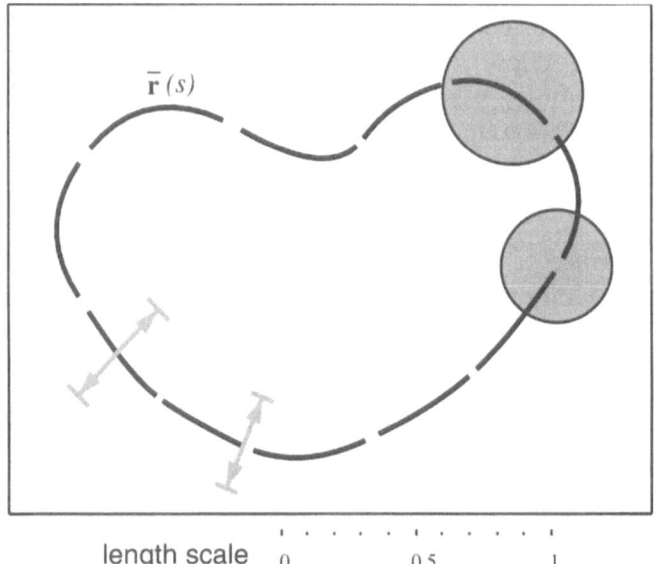

Figure 8.1: Probability distribution for curves. *An example of a Gaussian probability distribution in spline space with mean shape $\bar{\mathbf{r}}(s)$ and uniform, isotropic covariance P_0. The distribution of the position of a given point $\mathbf{r}(s)$ is Gaussian, with circular confidence intervals of radius $1.73\rho_0(s)$ (95% confidence). The normal displacement has a Gaussian distribution with standard deviation $\rho_n(s)$ and $\rho_n(s) = \rho_0(s)/\sqrt{2}$ in the special case of this example. The search segments shown (arrowed) are intervals of length $\pm 2\rho_n(s)$ for approximately 95% statistical confidence. (Average displacement $\bar{\rho}_0 = 0.14$ here.)*

framework provides a rationale, using the prior distribution, for fixing the search segment along the normal $\bar{\mathbf{n}}(s)$ at $\bar{\mathbf{r}}(s)$. The normal component of displacement of the curve is Gaussian with mean 0 and standard deviation $\rho_n(s)$ where

$$\rho_n^2(s) = \bar{\mathbf{n}}(s)^T P_{\mathbf{r}}(s)\bar{\mathbf{n}}(s).$$

or, directly in terms of shape-space covariance P_0,

$$\rho_n^2(s) = \mathbf{h}(s)^T P_0 \mathbf{h}(s) \tag{8.7}$$

where $\mathbf{h}(s)$ was defined in (6.15) on page 124 in chapter 6. In the special case of the norm-squared prior ($S \propto \mathcal{H}$) in spline space, the isotropy of $P_{\mathbf{r}}(s)$ means that $\rho_n(s) = \rho_0(s)/\sqrt{2}$.

A reasonable search segment is $\bar{\mathbf{r}}(s) \pm 2\rho_n(s)\bar{\mathbf{n}}(s)$, corresponding to a confidence interval (for the one-dimensional Gaussian distribution) at a level of approximately 95%. Such search segments are illustrated in figure 8.1. This idea is elaborated later for validation gates, taking account not only of the prior but also the observation density.

Derivation of average radius and isotropy results

The formula (8.5) for the root-mean-square displacement for the covariance ellipse is derived first. Pointwise, the mean-square displacement is $\rho_0^2(s)$ whose average value along the curve, from (8.4), is

$$\frac{1}{L}\int_0^L \mathrm{tr}\,(P_{\mathbf{r}}(s))\ ds$$

$$\underset{(8.3)}{=} \frac{1}{L}\int_0^L \mathrm{tr}\,\left(U(s)WP_0W^TU(s)^T\right)\ ds$$

$$= \mathrm{tr}\,\left(P_0W^T\left[\frac{1}{L}\int_0^L U(s)^TU(s)\,ds\right]W\right)$$

$$\underset{(3.23)}{=} \mathrm{tr}\,\left(P_0W^T\mathcal{U}W\right)$$

$$\underset{(4.7)}{=} \mathrm{tr}\,\left(P_0\mathcal{H}\right),$$

as required.

The special case of the norm-squared prior (8.6) in spline space $\mathcal{S} = \mathcal{S}_Q$ was considered in which $W = I_{N_Q}$ and the pointwise covariance is

$$P_{\mathbf{r}}(s) = U(s)P_0U(s)^T.$$

Since $P_0 \propto \mathcal{U}^{-1}$,

$$P_{\mathbf{r}}(s) \propto (\mathbf{B}(s)^T\mathcal{B}^{-1}\mathbf{B}(s))I_2$$

— a multiple of the identity matrix as claimed.

Random sampling

A graphic illustration of a statistical family of curves can be made by sampling randomly from its distribution. Random curve sampling is a practical technique in its own right, partly as a debugging aid to check whether a particular probabilistic prior model of shape is appropriate to a given application. More importantly it forms the basis of some powerful algorithms for recognising patterns, used when the posterior density for shape given data is too complex to be represented exactly and is represented instead by a set of samples.

Given an N_X-dimensional Gaussian distribution $\mathcal{N}(\overline{\mathbf{X}}, P)$ in shape-space with mean $\overline{\mathbf{X}}$ and covariance matrix P, a random variate $\mathbf{X}$ from the distribution can be generated by taking a vector $\mathbf{w}$ of N_X independent "standard" normal variables each distributed as $\mathcal{N}(0, 1)$ and transforming $\mathbf{w}$ linearly:

$$\mathbf{X} = B\mathbf{w} + \overline{\mathbf{X}} \tag{8.8}$$

where

$$B = \sqrt{P}, \tag{8.9}$$

a matrix square root with the property that $BB^T = P$ (Note that this does not uniquely specify B, but any admissible B serves equally well, and generates the desired distribution; see also appendix A.1). This technique is used to illustrate prior distributions of curves in spline space $\mathcal{S}_Q$ (figure 8.2) and in various shape-spaces (figure 8.3). As before, the distributions are given by norm-squared densities over shape-space.

8.2   Posterior distribution

The aim of the chapter is to express curve fitting in terms of the posterior distribution for shape $\mathbf{X}$ given observed data $\mathbf{r}_f$. Realistically, $\mathbf{r}_f$ must be obtained by processing image data, and that is addressed in the next section. For now, the posterior is illustrated with artificial data, using the random sampling method. As at the start of chapter 6, the artificial image feature is modelled as a spline curve with control-vector $\mathbf{Q}_f$. What was characterised there, in (6.3) on page 117, as fitting by regularisation, becomes the construction by Bayes' formula (8.1) of the posterior distribution $p(\mathbf{X}|\mathbf{Q}_f)$, given a Gaussian prior $p_0(\mathbf{X})$ and an observation density $p(\mathbf{Q}_f|\mathbf{X})$. The measurement term $\|\mathbf{Q} - \mathbf{Q}_f\|^2$ in the regularisation problem is interpreted as an observation density

$$p(\mathbf{Q}_f|\mathbf{X}) \propto \exp -\frac{N_X}{2\overline{\rho}_f^2}\|\mathbf{Q} - \mathbf{Q}_f\|^2 \tag{8.10}$$

where a variance constant $\overline{\rho}_f^2/N_X$ has been included here to represent explicitly the uncertainty of measurements. The distance constant $\overline{\rho}_f$ can be interpreted as the average displacement, at a given point on the fitted curve, due solely to observation error and without any influence from the prior. Finally, multiplying prior and observation densities, in accordance with Bayes' formula, the posterior distribution proves to be a Gaussian

$$\mathbf{X}|\mathbf{Q}_f \sim \mathcal{N}(\hat{\mathbf{X}}, P). \tag{8.11}$$

Its mean $\hat{\mathbf{X}}$ is precisely the solution (6.7) to the earlier regularisation problem, adjusted now to allow for the observation variance constant:

$$\hat{\mathbf{X}} = S^{-1}\left(\overline{S}\,\overline{\mathbf{X}} + \frac{N_X}{\overline{\rho}_f^2}\mathcal{H}\mathbf{X}_f\right) \tag{8.12}$$

where

$$S = \overline{S} + \frac{N_X}{\overline{\rho}_f^2}\mathcal{H}. \tag{8.13}$$

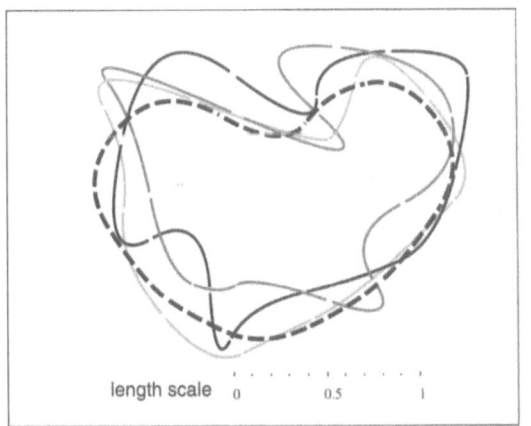 

Figure 8.2: Sampling from curve families. *Samples are drawn at random (left) from a uniform, isotropic, Gaussian distribution in the spline space $\mathcal{S}_Q$ such that $\|\mathbf{r}(s) - \overline{\mathbf{r}}(s)\|$ has a root-mean-square value of 0.2 length units. (Mean shape $\overline{\mathbf{r}}$ is shown dashed). Parallel curves depicting confidence intervals for normal displacement (spatially averaged), at 95% significance, contain the random curves as expected (right).*

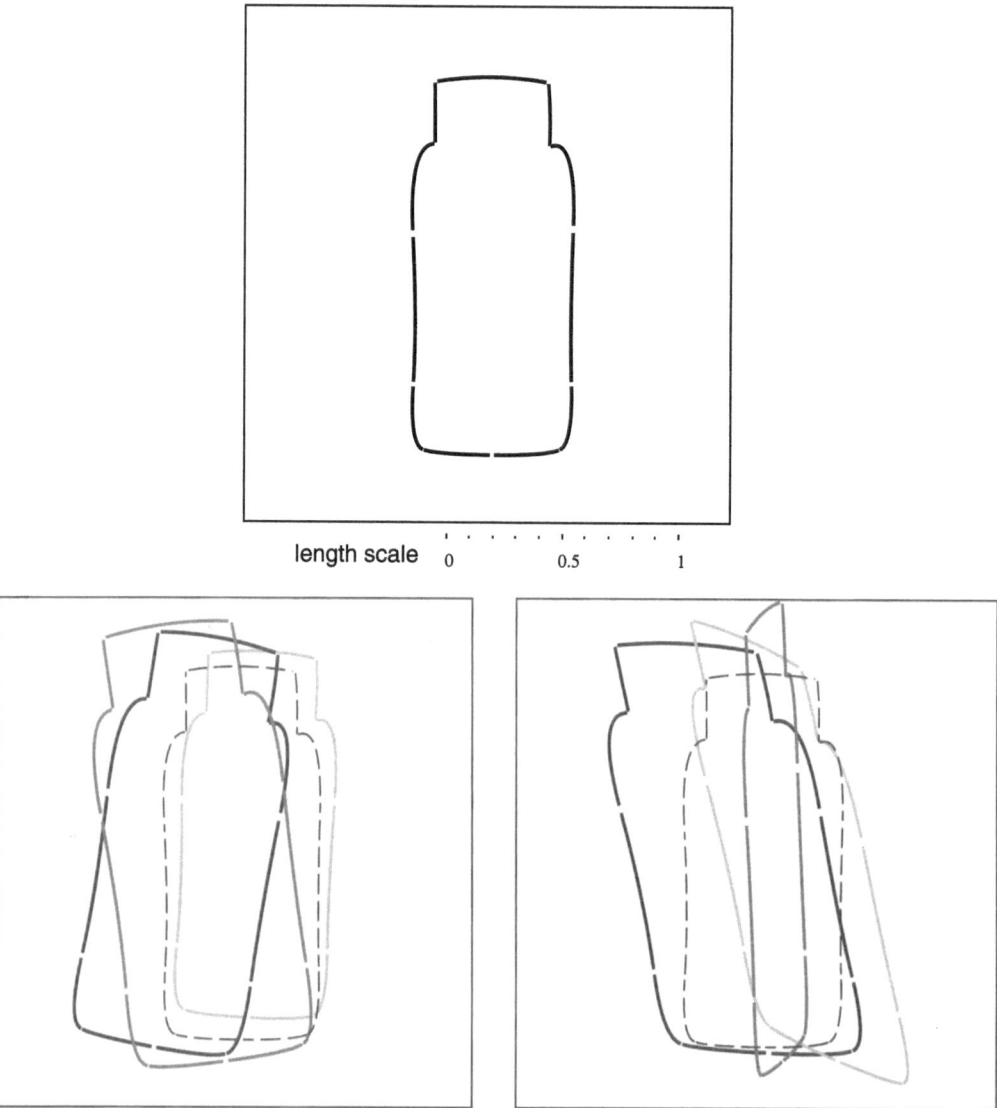

length scale 0   0.5   1

Figure 8.3: Random sampling in shape-space *The families of bottle-outlines shown have mean shape (top) and a uniform, isotropic, Gaussian distribution in shape-space with root-mean-square displacement of 0.3 length units. Euclidean similarities (left); affine space (right).*

The estimate $\hat{\mathbf{X}}$ is also known as a "Maximum A Posteriori" (MAP) estimate because it is the value of $\mathbf{X}$ at which the posterior probability density $p(\mathbf{X}|\mathbf{Q}_f)$ achieves its peak value. The covariance of the posterior distribution is $P = S^{-1}$, and S, the total information matrix for the posterior, is just the sum of information in the prior and observation distributions. (The derivation of the posterior distribution is essentially a replay of the derivation on page 118 of the solution to the regularisation problem.)

The posterior distribution for the curve-fitting problem of figure 6.2 on page 119 can be illustrated by random sampling. Again, regularisation is made invariant to transformations in a subspace $\mathcal{S}_s$. As in chapter 6, this is represented probabilistically by modifying the information matrix (8.6) for a uniform, isotropic, Gaussian prior using a projection E^d outside the invariant subspace:

$$\overline{S} = \frac{N_X}{\overline{\rho}_0^2} E^{d^T} \mathcal{H} E^d. \tag{8.14}$$

This has the effect of producing a prior distribution that is invariant to Euclidean similarities so that

$$p_0(\mathbf{X} + \mathbf{X}^s) = p_0(\mathbf{X}) \quad \text{for any } \mathbf{X}^s \in \mathcal{S}_s.$$

The posterior distribution, illustrated by random sampling in figure 8.4, clearly shows the likely range of variability of possible solutions, something that regularisation alone could not convey. This is also reflected mathematically by the fact that there is an extra constant in the system. As before there is a regularisation constant α which is now a ratio of the coefficients of the information matrices for the observation and for the prior:

$$\alpha = \overline{\rho}_f^2 / \overline{\rho}_0^2$$

and which determines the mean $\hat{\mathbf{X}}$ of the posterior distribution. In addition there is now also the spatial variance parameter $\overline{\rho}_0^2$ for the prior which also affects the variability of fitted curves as modelled by the posterior distribution.

Finally, exploiting the additivity of information in (8.13), a mean spatial variance $\overline{\rho}^2$ for the posterior can be computed from

$$\frac{1}{\overline{\rho}^2} = \frac{1}{\overline{\rho}_0^2} + \frac{1}{\overline{\rho}_f^2}$$

and is exact if there is no invariant subspace ($\mathcal{S}_s = 0$), and otherwise is a lower bound on mean variance. A confidence region based on $\overline{\rho}$ is shown in figure 8.4.

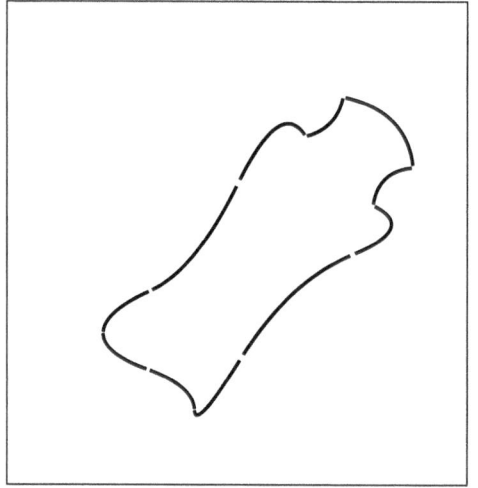

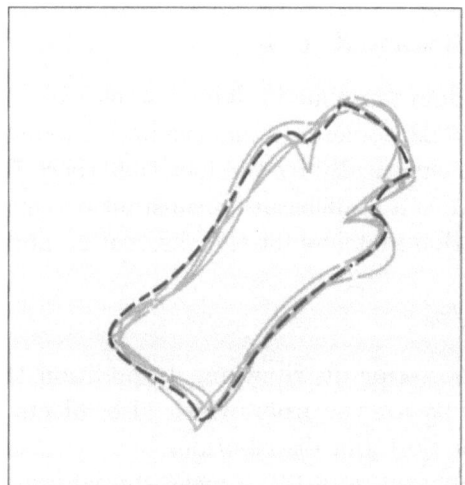
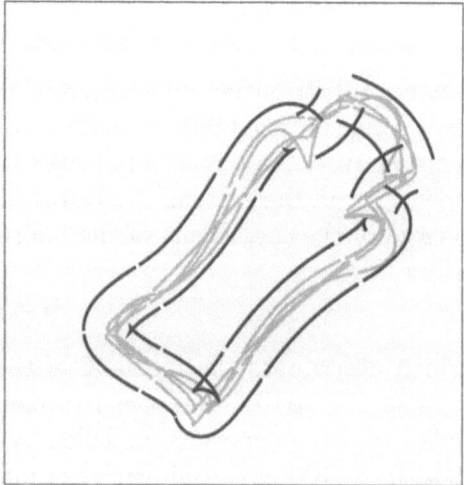

Figure 8.4: Sampling from the posterior *Data (top) is given, together with a prior over spline space with the Euclidean similarities as an invariant subspace (to allow free rotation and translation). Curves sampled from the posterior distribution sweep out a distribution (left), whose mean (dashed line) is simply the regularised estimate obtained in chapter 6 (figure 6.2 on page 119), with $\alpha = 0.5$. The sampled curves (right) lie comfortably inside a 95% confidence interval.*

Note that the matrix $\overline{S}$ in (8.14) has become singular, so that the covariance matrix P_0 does not exist and the prior density p_0 cannot formally be normalised. In fact p_0 can still be treated as a consistent Gaussian distribution. It consists of a valid Gaussian density restricted to shape-space $\mathcal{S} \ominus \mathcal{S}_s$ excluding the invariant subspace $\mathcal{S}_s$, together with a uniform density over $\mathcal{S}_s$. It is the uniform density that cannot be normalised and gives rise to unbounded covariance. We could treat it formally as a sequence of Gaussians of increasing variance, understanding that any expression incorporating p_0 will be evaluated in the limit, where one exists. In fact this is achieved for our purposes simply by applying additivity of information to $\overline{S}$ just in the same way as for a non-singular $\overline{S}$. The result is a *posterior* Gaussian whose covariance generally exists and which can be normalised.

8.3   Probabilistic modelling of image features

The example above of a posterior distribution is a contrived one, for tutorial purposes, driven by ideal data in the form of a spline curve $\mathbf{Q}_f$. In chapter 6, methods were developed for fitting to real data $\mathbf{r}_f$ in the form of a curve sampled from an image. The importance of using only the normal component of displacement was explained. Now the probabilistic approach needs to be developed to encompass normal displacements in real image data.

As in the previous section, the data-term in the regularisation problem can be interpreted probabilistically, but now based on normal displacement to give an observation density:

$$p(\mathbf{r}_f|\mathbf{X}) \propto \exp -\frac{N_X}{2\overline{\rho}_f^2}\|\mathbf{r} - \mathbf{r}_f\|_{\overline{\mathbf{n}}}^2 \quad \text{with} \quad \mathbf{r}(s) = U(s)(W\mathbf{X} + \mathbf{Q}_0).$$
(8.15)

It can be shown (see below) that $\overline{\rho}_f^2$ is the contribution to mean-square value of the normal displacement $(\mathbf{r} - \overline{\mathbf{r}}) \cdot \mathbf{n}$, in the fitted curve, due solely to measurement error.

Now the discrete form (6.13) from chapter 6 for $\|\mathbf{r} - \mathbf{r}_f\|_{\overline{\mathbf{n}}}^2$ can be used, giving approximately:

$$p(\mathbf{r}_f|\mathbf{X}) \propto \exp -\frac{1}{2\sigma^2}\sum_{i=1}^{N}[(\mathbf{r}_f(s_i) - \mathbf{r}(s_i)) \cdot \overline{\mathbf{n}}(s_i)]^2 .$$
(8.16)

where

$$\sigma = \overline{\rho}_f\sqrt{\frac{N}{N_X}}.$$
(8.17)

One interpretation of this observation density is in terms of the variability of individual measurements. Suppose the marginal density of a single measurement

$$[\mathbf{r}_f(s_i) - \mathbf{r}(s_i)] \cdot \overline{\mathbf{n}}(s_i)$$

were a Gaussian $\mathcal{N}(0, \sigma^2)$ variable. Then the joint density (8.16) could be interpreted as treating the data $\mathbf{r}_f$ as a series of sampled image features

$$\mathbf{r}_f(s_i) \sim \mathcal{N}((\mathbf{r}(s_i), \sigma^2) \tag{8.18}$$

with mutually *independent* Gaussian distributions. To some extent such independence is plausible but it depends rather on what is the chief source of the variability in image-feature measurements. If the variability were thought to derive from noise, for example electrical or optical noise at individual pixels, it is probably defensible to regard noise sources at neighbouring pixels as independent; even then, noise due to lighting flicker (especially fluorescent lighting) or power supply ripple would be highly correlated across pixels. In practice, however, the variability in feature position due to such noise is negligible. The dominant sources of variability reflect grosser discrepancies than mere device noise from the camera.

- Imperfections in the shape-space model mean that the image outline $\mathbf{r}_f$ will not lie entirely in the shape-space $\mathcal{S}$. The difference

$$\mathbf{e}(s) = \mathbf{r}_f(s) - U(s)(WX_f + \mathbf{Q}_0)$$

  between the outline and its projection onto shape-space acts as spatially correlated noise.

- Background texture and occasional obscuration by passing objects will cause features to be picked up by image processing that do not belong to the modelled object. Such errors may be gross and, again, spatially correlated.

The result is that statistical independence between neighbouring $\mathbf{r}_f(s_i)|\mathbf{X}$ is plausible only if features are not sampled too densely, and this is a limitation of the observation model (8.15). It is therefore important to note that the interpretation of (8.16) in terms of statistically independent measurements $\mathbf{r}_f(s_i)$ is by no means the only consistent interpretation (see discussion at the end of this section).

Algorithm to construct the posterior

Construction of the posterior density $p(\mathbf{X}|\mathbf{r}_f)$ from image measurements $\mathbf{r}_f(s_i)$, $i = 1,\dots,N$ uses precisely the recursive fitting algorithm from chapter 6 (figure 6.7 on page 127), but now interpreted probabilistically. The algorithm computes an "aggregated observation" $\mathbf{Z}$ and the associated statistical information for estimation of $\mathbf{X}$ is S, also output by the algorithm. In fact

$$\mathbf{Z}|\mathbf{X} \sim \mathcal{N}(S\mathbf{X}, S),$$

so that $\mathbf{Z}$ is an unbiased estimator for $S\mathbf{X}$ (rather than for $\mathbf{X}$ directly). Its covariance is in fact S *(sic)*, following from the fact that the statistical information in $S^{-1}\mathbf{Z}$ is S.

The posterior distribution is a Gaussian

$$\mathbf{X}|\mathbf{Z} \sim \mathcal{N}(\hat{\mathbf{X}}, P) \quad \text{where} \quad P = (\overline{S} + S)^{-1}. \tag{8.19}$$

An example of a posterior constructed in this way from an image is given in figure 8.5.

Independence of measurements

The functional form of the density $p(\mathbf{r}_f|\mathbf{X})$ does not in fact uniquely determine the conditional distribution for observations until a parameterisation is specified for $\mathbf{r}_f(s)$, the image feature. For example $\mathbf{r}_f(s)$ could be modelled as a spline with parameters $\mathbf{Q}_f$ of dimension N_f. Then the conditional density can be used to compute probabilities via integrals of the form

$$\int p(\mathbf{r}_f|\mathbf{X}) \, d\mathbf{Q}_f.$$

To examine the question of independence of the $\mathbf{r}_f(s_i)$, the conditional density must be transformed from the $\mathbf{Q}_f$ parameterisation to a new parameter set $(\mathbf{r}_f(s_1),\dots,\mathbf{r}_f(s_N))$. This could be done (without loss of generality) by expressing the new parameters in terms of the first N components in the vector $\mathbf{Q}_f$, and integrating over the remaining $N_f - N$ components. This is possible *only if* $N < N_f$. Otherwise there exists no density in the new variables, let alone an independent one. This is intuitively reasonable. When the underlying parameterisation $\mathbf{Q}_f$ of the image curve has relatively few degrees of freedom (N_f is small), this represents an image curve with strong spatial correlation, such as a spline with few control points. In that case, successive measurements made on the curve cannot be mutually independent. To summarise, the form of (8.16) does not force the deduction that individual image measurements $\mathbf{r}_f(s_i)$ are independent. This conclusion is important to the operation of the validation gate, discussed below.

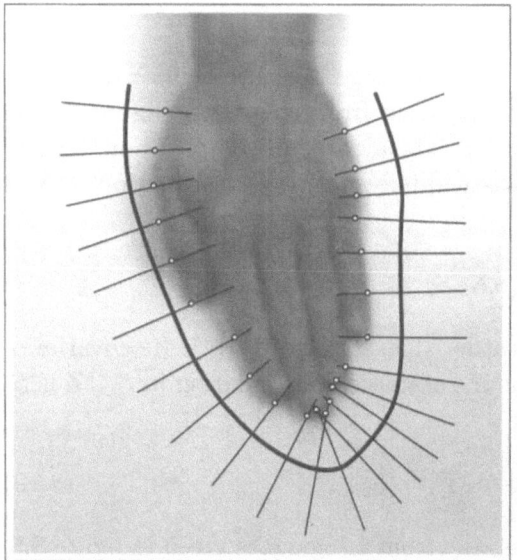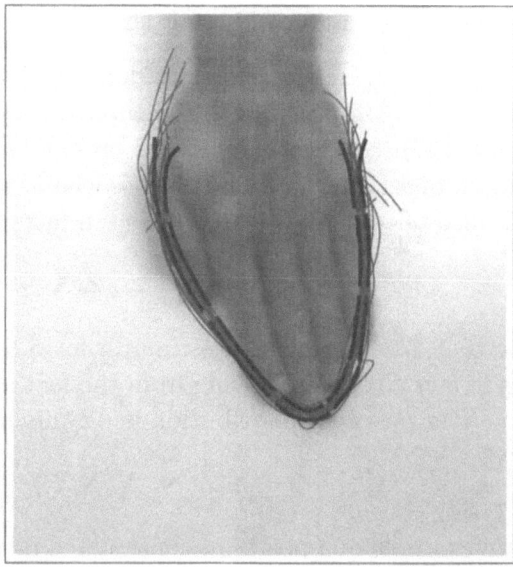

Figure 8.5: Constructing a posterior distribution from an image. *The image data (left) is shown overlaid with a broad search region determined by the chosen prior. When combined with image observations, a relatively tight posterior distribution results, illustrated here by randomly sampled curves and 95% confidence region (right).*

Mean-square normal displacement

It was claimed above that $\bar{\rho}_f^2$ is the mean-square value of the normal displacement $(\mathbf{r} - \mathbf{r}_f) \cdot \mathbf{n}$. This could be proved directly by integration, as was done earlier for $\bar{\rho}_0$ in the norm-squared prior. A concise alternative is to appeal to the well-known property of Boltzmann distributions in statistical mechanics for random systems whose energy $U(\mathbf{X})$ is a quadratic function of parameters $\mathbf{X}$ and is distributed as $p(\mathbf{X}) \propto \exp{-U}$. The mean energy of such a system is simply $\overline{U} = N_X/2$, where N_X is the number of degrees of freedom of the parameter set.

8.4   Validation gate

The validation gate was introduced in chapter 6 (section 6.3) as a mechanism for detecting outliers, features that are misplaced or missing when, for instance, part of an object is obscured. The idea was to test the innovation ν_i to ensure that it was not too large; the difficulty was to decide how large is too large. Now in the probabilistic framework, there is a basis for making this test as a statistical hypothesis test on the innovation, whose Gaussian distribution is known. The innovation (6.14) can be expanded as

$$\nu_i = [\mathbf{r}(s_i) - \bar{\mathbf{r}}(s_i)] \cdot \bar{\mathbf{n}}(s_i) \ + \ [\mathbf{r}_f(s_i) - \mathbf{r}(s_i)] \cdot \bar{\mathbf{n}}(s_i),$$

in which the first term has a $\mathcal{N}(0, \rho_n^2(s))$, distribution, from (8.7). The second term is a little more difficult to deal with. In the case that the measurements $\mathbf{r}_f(s_i)$ are considered independent, it is Gaussian $\mathcal{N}(0, \sigma^2)$ from (8.18). At the other extreme, in the limit of high spatial interdependence of measurements, its variance should tend towards 0. This means that

$$\nu_i \ \sim \ \mathcal{N}(0, \rho^2(s)) \ \text{ where } \ \rho_n^2(s) \le \rho^2(s) \le \rho_n^2(s) + \sigma^2. \tag{8.20}$$

A conservative assumption of high spatial dependence would suggest

$$\rho(s) = \rho_n(s). \tag{8.21}$$

The distribution of the innovation suggests a test that if $|\nu_i| > \kappa \rho(s)$ then the measurement $\mathbf{r}_f(s)$ is invalidated. Taking the default value $\kappa = 2$, at approximately 95% statistical significance the measurement is regarded as an outlier. In practice this is implemented by using a search segment (figure 8.1) of length $\rho(s)$ so that when $|\nu_i| > 2\rho(s)$, no feature will be found. The recursive fitting algorithm of chapter 6 can be modified to take account of outliers. The iterative step in the original algorithm (figure 6.7 on page 127) is rewritten to ignore outliers, as in figure 8.6.

Note that the modified algorithm requires that $\overline{S}$ have full rank and that excludes the use of an invariant subspace in the prior which always makes $\overline{S}$ singular. This is a real and reasonable limitation. The idea of the invariant subspace is that it allows certain shape variables to be unconstrained. In that case, the absolute position of a given point $\mathbf{r}(s)$ on the curve is also unconstrained by the prior (even though constraints may apply to the relative positions of pairs of points). Validity checks are only effective when the constraints imposed by the prior on individual points are reasonably tight. This requires the prior on the shape $\mathbf{X}$ as a whole to be reasonably tight — none of the eigenvalues of $\overline{S}$ should be too close to 0. The effect of outlier rejection on fitting is illustrated in figure 8.7.

Iterative fitting step

Iterate, for $i = 1, \ldots, N$:

$$\nu_i = (\mathbf{r}_f(s_i) - \overline{\mathbf{r}}(s_i)) \cdot \overline{\mathbf{n}}(s_i);$$

$$\mathbf{h}(s_i)^T = \overline{\mathbf{n}}(s_i)^T U(s_i) W;$$

$$\rho_i = \sqrt{\mathbf{h}(s_i)^T \overline{P} \mathbf{h}(s_i)};$$

If $(|\nu_i| < \kappa \rho_i)$ **then**

$$S_i = S_{i-1} + \frac{1}{\sigma_i^2} \mathbf{h}(s_i) \mathbf{h}(s_i)^T;$$

$$\mathbf{Y}_i = \mathbf{Y}_{i-1} + \frac{1}{\sigma_i^2} \mathbf{h}(s_i) \nu_i;$$

Else

$$S_i = S_{i-1}; \ Y_i = Y_{i-1};$$

Figure 8.6: Curve fitting with outliers. *The recursive curve-fitting algorithm in figure 6.7 on page 127 can be modified to ignore outliers. Its iterative step is replaced by the one shown here. The search-line length factor is taken to be $\kappa = 2$ by default. (Note that $\overline{P} = \overline{S}^{-1}$ needs to have been computed at the start of the algorithm, and that $\overline{S}$ must therefore be of full rank.)*

8.5 Learning the prior

The importance of a prior distribution is that it draws interpretations of data towards the more plausible shapes in a class of curves. The value of such stabilising behaviour is only as good as the prior model itself, embodied by the coefficients in the information matrix $\overline{S}$. So far, we have proposed priors based on uniform, isotropic error ($\overline{S} \propto \mathcal{H}$) and a modification that allows for an invariant subspace. These are reasonable choices if little is known precisely about the class of curves. If, however, a more specific prior could be obtained from actual snapshots of a curve in a series of representative configurations, stabilisation should be very much more effective, and this does indeed

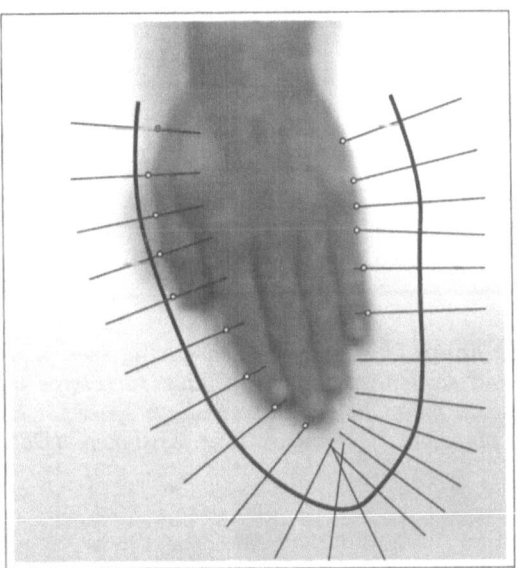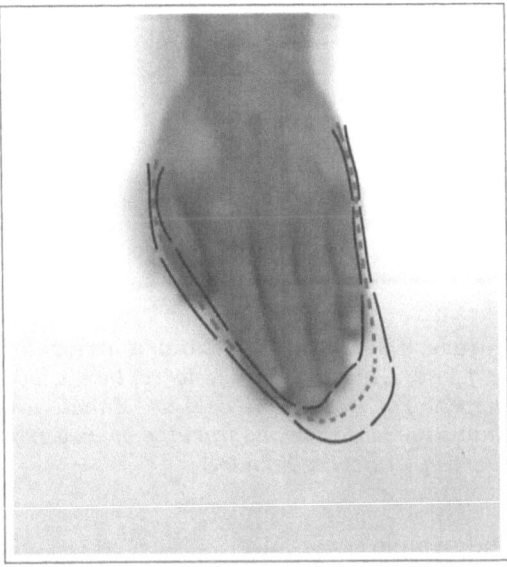

Figure 8.7: Validation failure causes a bulge in the posterior confidence region. *Validation fails around the fingertips (left) resulting in a bulge in the 95% confidence region (right) such that it still contains the true edge of the hand.*

prove to be the case.

The first step towards learning is to acquire a training set, a set of curves $\{\mathbf{X}_k, \ k = 1, \ldots, M\}$, where $\mathbf{X}_k$ are vectors in a shape-space $\mathcal{S} = \mathcal{L}(W, \mathbf{Q}_0)$ which may either be the spline space $\mathcal{S}_Q$ or a subspace of $\mathcal{S}_Q$. The set is supposed to capture the outline being modelled in many characteristic poses. For current purposes, the order in which poses appear in the sequence is immaterial, though order will matter in later chapters when object dynamics are being modelled. One could think of each $\mathbf{X}_k$ as a contour drawn by hand on a given image but in practice the temporal tracking technology developed later allows these outlines to be acquired automatically.

If the prior distribution is assumed to be Gaussian $\mathcal{N}(\overline{\mathbf{X}}, \overline{P})$, then "maximum likelihood" estimators for the parameters of the distribution are given by the sample mean

$$\overline{\mathbf{X}} = \frac{1}{M} \sum_{k=1}^{M} \mathbf{X}_k \qquad (8.22)$$

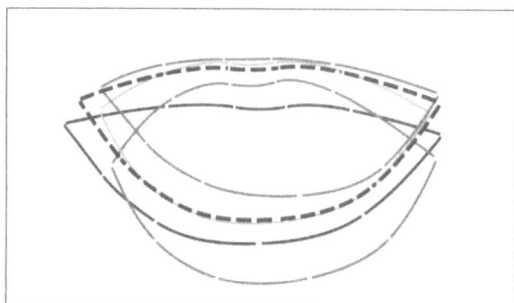

Figure 8.8: Sampling from a prior for lip shape *Lips are tracked during speech, as in figure 1.10 on page 14, to capture a 60 second sequence comprising 3000 successive lip shapes. The sequence is used to estimate a Gaussian prior distribution in shape-space for lip shape; random sampling from the prior generates plausible lip configurations, as shown. (Data courtesy of Robert Kaucic.)*

and sample covariance

$$\overline{P} = \frac{1}{M} \sum_{k=1}^{M} (\mathbf{X}_k - \overline{\mathbf{X}})(\mathbf{X}_k - \overline{\mathbf{X}})^T. \tag{8.23}$$

For example, in figure 8.8 a Gaussian prior distribution has been learned from a long sequence of lip motion during speech. The resulting prior is simulated by random sampling and the sampled curves are plausible lip configurations.

The matrix $\overline{P}$ is a $N_X \times N_X$ matrix and should be invertible so that $\overline{S} = \overline{P}^{-1}$ can be calculated as required for the recursive curve-fitting algorithm. However, from (8.23),

$$\mathrm{rank}(\overline{P}) \leq M - 1$$

so that $\overline{P}$ certainly cannot be inverted unless $M > N_X$, and in practice one would expect a data set several times that minimum size. Alternatively, if a sufficiently large data set is not available and the estimated $\overline{P}$ is singular then one approach is to restrict the shape-space sufficiently that the reduced covariance matrix is no longer singular, and this is discussed next.

8.6   Principal Components Analysis (PCA)

Even if the estimated prior covariance matrix $\overline{P}$ is not technically singular, it is frequently close to singularity even when the training sequence is long $(M > N_X)$. This happens when the typical motions of the object under study are largely accounted for by a few independent modes of motion. If $\overline{P}$ is found to have a large condition number, larger say than 100, this suggests that, on the basis of the available data, the shape-space $\mathcal{S}$ in which data was collected is unnecessarily large. Efficiency of the matching algorithms described earlier, and of later tracking algorithms, would be considerably enhanced if it could be replaced by a smaller shape-space. The covariance matrix $\overline{P}$ can be used to determine a smaller shape-space $\mathcal{S}' = \mathcal{L}(W', \mathbf{Q}'_0) \subset \mathcal{S}$, a subspace of $\mathcal{S}$ with dimension N'_X, that spans, at least approximately, all of the shapes in a training sequence $\mathbf{X}_1, \dots, \mathbf{X}_M$. This idea was first introduced by Cootes and Taylor, in the special case of polygonal contour models, which they dubbed the "Point Distribution Model" or PDM. The PCA method is explained starting with "classical PCA" which is well-known and easily accessible in textbooks, via two refinements, to "L_2-norm PCA in shape-space," the recommended method for dealing with curves. In the special case that $M \leq N_X$ and the covariance matrix $\overline{P}$ is actually singular, the simplest way to build a reduced shape-space is to take appropriate linear combinations of the frames in the training sequence — so-called key-frames, as described in chapter 4.

Classical PCA

The classical approach to PCA would allow the following version of the approximation problem to be solved. Given a long $(M > N_X)$ training sequence, solve:

$$\min_{W', \mathbf{Q}'_0, \mathbf{X}'_1, \dots, \mathbf{X}'_{N'_X}} \left(\sum_{k=1}^{M} |\mathbf{Q}_k - \mathbf{Q}'_k|^2 \right) \tag{8.24}$$

where

$$\mathbf{Q}'_k = W'\mathbf{X}'_k + \mathbf{Q}'_0 \quad \text{and} \quad \mathbf{Q}_k = W\mathbf{X}_k + \mathbf{Q}_0.$$

The distance measure $|\cdot|$ is the Euclidean norm, that is, $|\mathbf{Q}|^2 \equiv \mathbf{Q}^T\mathbf{Q}$. The well known solution to this problem gives $\mathbf{Q}'_0 = \overline{\mathbf{Q}}$, the mean of the training sequence, and W' is a matrix whose columns are the first N'_X of the orthonormal eigenvectors of the covariance matrix

$$\Sigma = \frac{1}{M} \sum_{k=1}^{M} (\mathbf{Q}_k - \overline{\mathbf{Q}})(\mathbf{Q}_k - \overline{\mathbf{Q}})^T,$$

ordering the eigenvectors in descending order of their (necessarily positive) eigenvalues. The intuitive interpretation is that the eigenvalues represent variance in the training set in the mutually orthogonal directions of the eigenvectors; the first N'_X eigenvectors form a basis for the subspace of dimension N'_X that "explains" as much as possible of the variance in the training set. Note that since the data $\mathbf{Q}_k$ all live within a shape-space of dimension N_X there will be at most N_X non-zero eigenvalues of Σ.

L_2-norm PCA

In previous chapters, it has been argued that the induced L_2-norm is the natural norm over spline space. In that case, the classical problem above is not entirely applicable. Instead, the approximation problem should be posed as follows:

$$\min_{W',\mathbf{Q}'_0,\mathbf{X}'_1,\dots,\mathbf{X}'_{N'_X}} \left(\sum_{k=1}^{M} \|\mathbf{Q}_k - \mathbf{Q}'_k\|^2 \right), \tag{8.25}$$

in which the L_2-norm $\| \cdot \|$ has been substituted where Euclidean norm $| \cdot |$ appeared before. The solution to this problem, which can be derived from the solution to the classical problem (see below), is that $\mathbf{Q}'_0 = \overline{\mathbf{Q}}$ as before and W' is a matrix whose columns are the first N'_X eigenvectors of the matrix $\Sigma \mathcal{U}$.

L_2-norm PCA over shape-space

Finally L_2-norm PCA can be re-expressed in terms of training-set covariance $\overline{P}$ in shape-space, without recourse to the (considerably larger) covariance matrix Σ in spline space. This gives the recommended algorithm in figure 8.9. Results from the application of the algorithm to the lip-motion sequence are shown in figure 8.10. PCA analysis for full facial expression is illustrated in figure 8.11. In neither case are individual PCA components recognisable as particular expressions; rather they are mixtures of expressions. It is when they are taken as a set that they are meaningful, as a basis for the repertoire of commonly occurring deformations.

Remember that Euclidean norm $|\mathbf{X}|$ in shape-space has no clear geometrical meaning. As a result classical PCA applied to the shape-space training set $\mathbf{X}_1,\dots,\mathbf{X}_M$ would certainly not be meaningful.

Residual PCA

A constructive description of shape-space, in which the shape-matrix is composed from transformations of a template or from key-frames, is rather desirable because

Problem: given training data $\mathbf{X}_1, \ldots, \mathbf{X}_M$ over shape-space $\mathcal{S}$, find a sub-space $\mathcal{S}' = \mathcal{L}(W', \mathbf{Q}'_0)$ of dimension N'_X to minimise

$$\sum_{k=1}^{M} \|\mathbf{Q}_k - \mathbf{Q}'_k\|^2$$

where

$$\mathbf{Q}_k = W\mathbf{X}_k + \mathbf{Q}_0 \quad \text{and} \quad \mathbf{Q}'_k = W'\mathbf{X}'_k + \mathbf{Q}'_0.$$

Algorithm

1. Construct the training-set mean

$$\overline{\mathbf{X}} = \frac{1}{M} \sum_{k=1}^{M} \mathbf{X}_k.$$

2. Construct the training-set covariance

$$\overline{P} = \frac{1}{M} \sum_{k=1}^{M} (\mathbf{X}_k - \overline{\mathbf{X}})(\mathbf{X}_k - \overline{\mathbf{X}})^T.$$

3. Find eigenvectors $\mathbf{v}_1, \ldots, \mathbf{v}_{N_X}$ of $\overline{P}\mathcal{H}$, in descending order of eigenvalue.

4. Construct $W'' = \left(\mathbf{v}_1, \ldots, \mathbf{v}_{N'_X}\right)$.

5. The parameters $\mathbf{Q}'_0$ and W' of the shape-subspace are then:
$$\begin{aligned} \mathbf{Q}'_0 &= W\overline{\mathbf{X}} + \mathbf{Q}_0 \\ W' &= WW''. \end{aligned}$$

Figure 8.9: **Algorithm for L_2 PCA in shape-space.**

each of the components of the shape-vector $\mathbf{X}$ has a clear interpretation. For example, the shape-space (4.20) composed of key-frames under Euclidean similarity is a constructive shape-space for which a given $\mathbf{X}$ represents an explicit combination of rigid transformation and facial expression. In contrast, deriving a shape-matrix from

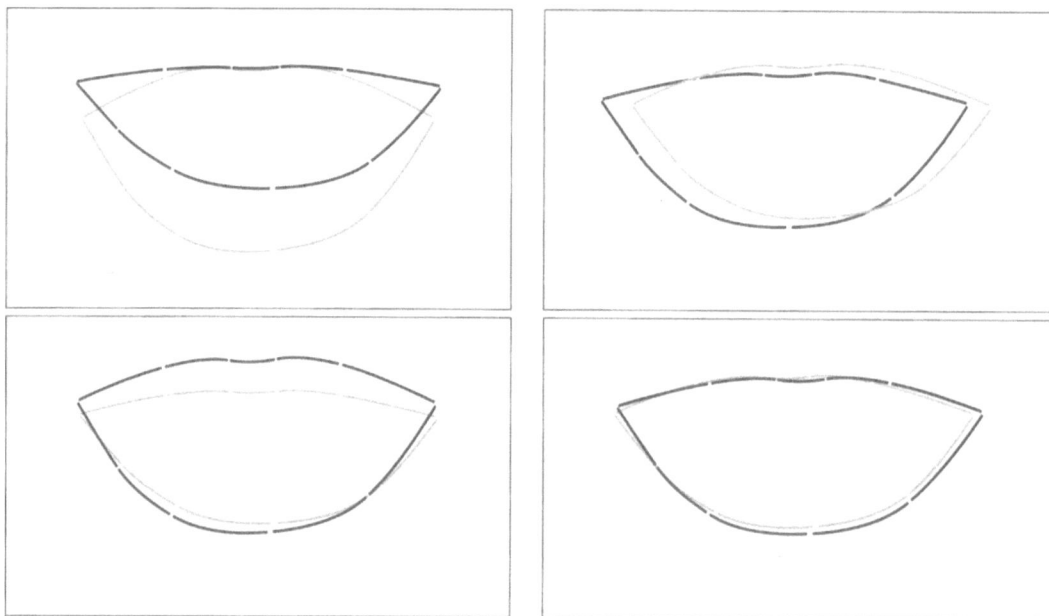

Figure 8.10: *The lip-motion sequence of figure 8.8 on page 176 is used here in PCA. The first four eigenvectors, illustrated here, capture over 95% of the variance in the data set. (Eigenvectors are displayed here as displacements either side of the mean, with magnitude equal to their standard deviation over the sequence.) (Figures by courtesy of Robert Kaucic.)*

PCA loses the clear interpretation, but has the advantage that no prior insight into likely transformations or deformations is required. It is possible, to some extent, to enjoy the best of both worlds. Residual PCA operates on a constructive shape-space that does not totally cover a certain data set, and fills in missing components by PCA. Then the constructive subspace retains its interpretation and only the residual components, covered by PCA, cannot be directly interpreted.

Given training data $\mathbf{X}_1, \ldots, \mathbf{X}_M$, collected from a shape-space $\mathcal{S}$, and a constructive subspace $\mathcal{S}_c = \mathcal{L}(W^c, \mathbf{Q}_0^c)$, it is desired to augment $\mathcal{S}_c$, increasing its dimension by N_X', to cover the training-data set more completely. Residual PCA achieves this by computing, after steps 1,2 of the algorithm of figure 8.9, the mean and variance of the residue of the training set:

$$\overline{\mathbf{X}}^r = (I - E^c)\overline{\mathbf{X}} \tag{8.26}$$
$$\overline{P}^r = (I - E^c)\overline{P}(I - E^c)^T \tag{8.27}$$

Figure 8.11: *Facial expression as in figure 1.2 on page 6 is analysed by PCA. The first two eigenvectors are illustrated here. (Figures by courtesy of Benedicte Bascle.)*

where $E^c = W(W^c)^+$ is the matrix for projection onto the subspace $\mathcal{S}_c$. Step 3 of the algorithm computes the first N'_X eigenvectors $\mathbf{v}_1, \dots, \mathbf{v}_{N'_X}$ now of $\overline{P}^r$, rather than $\overline{P}$ as in the original algorithm. Steps 4 and 5 proceed as before, computing the template $\mathbf{Q}'_0$ and shape-matrix W' which now form a shape-space for the *residual* variation in the training set that was not covered by $\mathcal{S}_c$. The augmented shape-space is then

$$\mathcal{S}_a = \mathcal{L}(W^a, \mathbf{Q}_0^a)$$

where

$$\mathbf{Q}_0^a = E^c \mathbf{Q}_0^c + W' \overline{\mathbf{X}}^r$$

and the shape-matrix

$$W^a = (W^c | W^r),$$

a concatenation of columns from the shape-matrices for the constructive and residual spaces.

As an illustration, the Residual PCA algorithm is applied to the lip-motion sequence used earlier. Instead of applying PCA directly to the motion sequence as in figure 8.13, a constructive shape-space $\mathcal{S}_c$, a two-dimensional space of rigid translations, is used. Translation accounts for about 58% of the motion of the training set

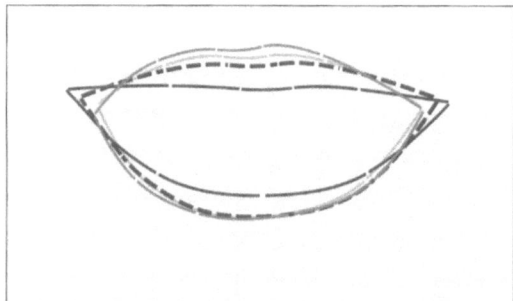

Figure 8.12: Sampling from a prior for lip shape, excluding translation *Random sampling illustrates how a learned prior generates plausible lip configurations, as in figure 8.8, but now with the rigid translation due to head motion excluded, leaving just the lip motions associated with speech.*

(horizontal 24%, vertical 34%) and since head motion is likely to be somewhat independent of speech, it is natural to try to discount it. The covariance $\overline{P}^r$ of the residue of the sequence with translation removed can be used in its own right to construct a prior, as in figure 8.13. Then the first two components from residual PCA, illustrated in the figure, can be expected to bear the bulk of the visual information that is related to speech.

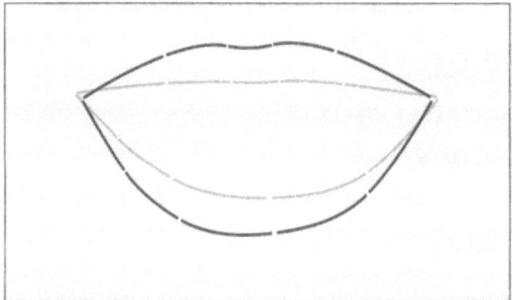 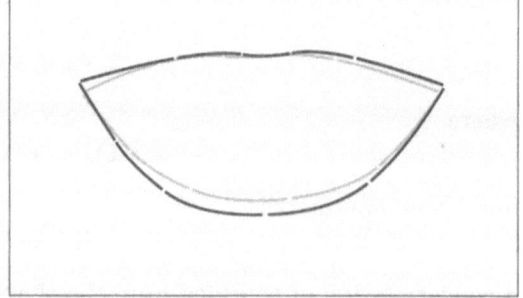

Figure 8.13: The lip-motion sequence is analysed by residual PCA. *All rigid translation is first excluded from the motion sequence by projection. Translation and the further 2 eigenvectors shown here account for over 95% of the variance in the data set.*

Derivation of PCA algorithms

First L_2-norm PCA is derived from classical PCA. Noting that $\|\mathbf{Q}\| = |\mathcal{U}^{1/2}\mathbf{Q}|$, the problem (8.25) can be recast as a minimisation, in the form of the classical problem (8.24), of

$$\sum_{k=1}^{M} |\mathcal{U}^{1/2}\mathbf{Q}_k - \mathcal{U}^{1/2}\mathbf{Q}'_k|^2$$

with

$$\mathcal{U}^{1/2}\mathbf{Q}'_k = \mathcal{U}^{1/2}W'\mathbf{X}'_k + \mathcal{U}^{1/2}\mathbf{Q}'_0,$$

Its solution is therefore that

$$\mathcal{U}^{1/2}\mathbf{Q}'_0 = \mathcal{U}^{1/2}\overline{\mathbf{Q}}$$

and that $\mathcal{U}^{1/2}W'$ is made up of columns which are eigenvectors of the sample variance

$$\text{Var}(\mathcal{U}^{1/2}\mathbf{Q}_k) = \mathcal{U}^{1/2}\Sigma\mathcal{U}^{1/2}.$$

This means that $\mathbf{Q}'_0 = \overline{\mathbf{Q}}$. Furthermore, a column $\mathcal{U}^{1/2}w$ of $\mathcal{U}^{1/2}W'$ satisfies

$$\mathcal{U}^{1/2}\Sigma\mathcal{U}^{1/2}\left(\mathcal{U}^{1/2}w\right) = \lambda\left(\mathcal{U}^{1/2}w\right)$$

which simplifies to

$$\Sigma\mathcal{U}w = \lambda w,$$

each column of W' is an eigenvector of $\Sigma\mathcal{U}$, as claimed. Note that equivalently w is a *generalised* eigenvector of the real, symmetric matrices $\mathcal{U}\Sigma\mathcal{U}$ and $\mathcal{U}$ (i.e. solutions of $\mathcal{U}\Sigma\mathcal{U}\mathbf{v} = \lambda\mathcal{U}\mathbf{v}$) and hence the eigenvalues are still positive, so that their descending order remains well-defined.

The shape-space version of the algorithm (figure 8.9) is obtained from the L_2-norm algorithm above by substituting

$$\mathbf{Q}_k = W\mathbf{X}_k + \mathbf{Q}_0 \quad \text{and} \quad \mathbf{Q}'_k = W'\mathbf{X}'_k + \mathbf{Q}'_0$$

so that

$$\overline{\mathbf{Q}} \rightarrow W\overline{\mathbf{X}} + \mathbf{Q}_0,$$

as in the algorithm, and

$$\Sigma\mathcal{U} \rightarrow W\overline{P}W^T\mathcal{U}$$

whose eigenvectors give W'. Writing $W' = WW''$, W'' must therefore be composed of eigenvectors of $\overline{P}W^T\mathcal{U}W$, but $W^T\mathcal{U}W = \mathcal{H}$ and this gives the required form for W'.

Bibliographic notes

The chapter is based on probabilistic analysis — for a review of basic probability, there is an excellent introductory book (Papoulis, 1990).

The use of a statistically estimated $\overline{P}$ in a Gaussian prior is equivalent to replacing the norm $\|\cdot\|$ in the regulariser of chapter 6 with a "Mahalanobis metric" (Rao, 1973) that emphasises improbable distortions, so that they are penalised more than probable ones.

Confidence regions for shapes were delimited by parallel curves. These are a well-known geometrical construct; one pitfall is that a smooth base curve can have offset curves that are not smooth but contain cusps (Bruce and Giblin, 1984). They are also known in CAD as "offset" curves (Faux and Pratt, 1979).

Principal Components Analysis (PCA) is a standard statistical technique (Rao, 1973). The idea of constructing a Gaussian curve model of reduced dimension, using PCA, was first developed for the case of polygonal outlines and termed the "Point Distribution Model" or PDM (Cootes et al., 1993), and subsequently extended to splines (Baumberg and Hogg, 1994). A related idea based on image intensities rather than curves is the eigen-image (Turk and Pentland, 1991) which can be used directly in tracking e.g. (Bregler and Omohundro, 1994; Black and Jepson, 1996). PCA is sometimes applied jointly to outline shape and image intensity distributions (Lanitis et al., 1995; Beymer and Poggio, 1995; Vetter and Poggio, 1996), to impressive effect. In standard PCA, it is common to pre-process the data by applying a diagonal scaling transformation (Ripley, 1996). This proved to be too restrictive for building curve models, for which general linear transformations were needed. In particular the idea of performing PCA in the L_2 norm, using the metric matrix $\mathcal{H}$ was developed in (Baumberg and Hogg, 1995a), and this is equivalent to scaling the data by $\mathcal{H}^{1/2}$. Generalised eigenvalues (Golub and van Loan, 1989) are used to combine scaling and PCA efficiently.

Shape priors discussed here have had a Gaussian form in shape-space. In the case of a norm-squared density over spline space ($S \propto \mathcal{U}$), the prior is actually a Gaussian Markov Random Field (MRF), of second-order in the case of quadratic splines. MRFs have been used widely for modelling prior distributions for curves (Grenander et al., 1991; Ripley and Sutherland, 1990; Storvik, 1994).

Chapter 9

Dynamical models

The remainder of the book aims to establish effective procedures for tracking curves in *sequences* of images. As with single images, the importance of powerful prior models of shape holds good, but now prior models can be extended to capitalise on the *coherence* of typical motions through a sequence. Crudely this could mean a repeated application of the regularised curve-fitting of chapter 6, in which the fitted curve in the $k-1$th frame of a sequence is used as an initial estimate of curve position and shape for the kth frame. In the probabilistic context of chapter 8 this would involve applying, to each frame, a Gaussian prior distribution with fixed covariance but whose mean was simply the estimated shape from the previous frame. This immediately suggests a more subtle approach. Rather than fixing the form of the prior via one constant covariance for all frames, it seems more natural to take the *posterior* from frame $k-1$ as the prior for frame k. In that way, it would not be merely an estimated shape that would pass from time-step to time-step but an entire probability distribution.

This idea is still too crude as it stands, for two reasons. First, it lacks a mechanism for extrapolation of motion between successive time-steps, and this is reasonable only for tracked objects which are moving slowly (figure 9.1). Secondly, as the posterior at one time-step is handed on to be the prior at the next, statistical information from measurements steadily accumulates. Accumulation proceeds without bound so that the statistical information $S(t_k)$ in the posterior continues to increase in successive frames and the corresponding covariance $P(t_k)$ decreases towards zero. This is counterintuitive — if the prior at time t_k has a vanishing covariance, its regularising effect will become too strong and image measurements will have an ever decreasing influence on estimated shape.

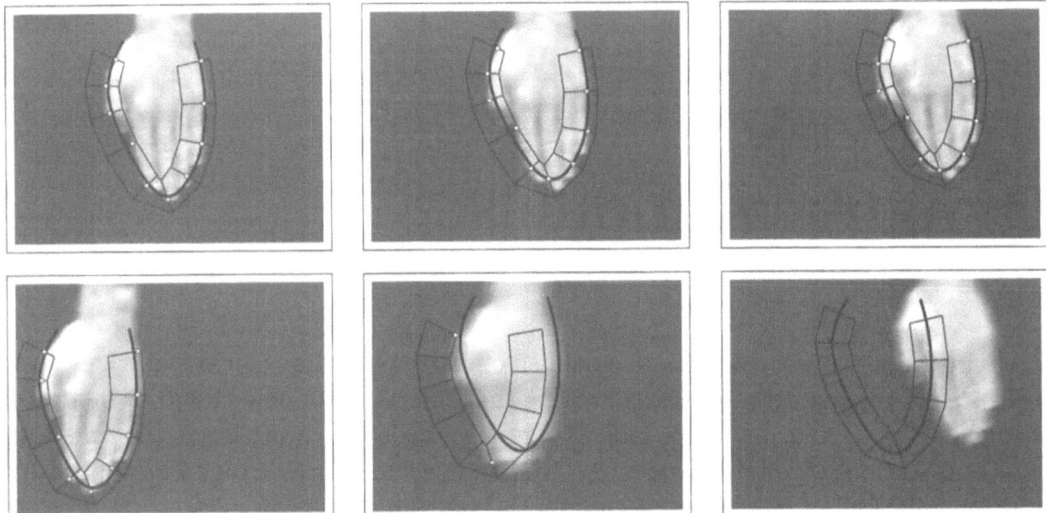

Figure 9.1: The need to exploit coherence of motion. *Example of hand tracking, showing that using the fitted curve at time t_{k-1} as the initial frame at time t_k is of limited use — it tracks slow motions (top) but not faster ones (bottom).*

An adequate statistical framework for motion tracking must therefore provide not only a prior for the first frame, similar to the prior in the static contour fitting problem but, far more importantly, a prior for possible motions, in the broad sense of rigid motion plus deformation of shape. This dynamical prior distribution should apply between all pairs $k-1, k$ of successive frames. It must have a deterministic part, giving the expected displacement between successive frames, and this addresses the first objection above about extrapolation of motion. Less obviously perhaps, it also needs a stochastic component, an injection of randomness, causing a steady "leakage" of information to counteract the otherwise unlimited accumulation of information that led to the second objection above.

In the previous chapter, the probabilistic framework for the fitting of curves to static images was expounded in three parts. First, the encoding of prior knowledge of curve shape had the form of a Gaussian prior probability distribution in shape-space. Secondly, the posterior probability distribution of curve shape given image data was computed by the recursive algorithm. Thirdly, rather than conjuring up a prior distribution which is merely plausible, a specific prior was actually learned from a training-data set. Now it is time to extend these three ideas, in this and the next two chapters respectively, to deal with curves in moving images. First, this chapter deals with prior models for curve shape and motion.

9.1 Some simple dynamical prior distributions

Consider the problem of building an appropriate dynamical model for the position of a hand-mouse engaged in an interactive graphics task. This is the problem that was illustrated in figure 1.16 on page 20 in which a hand whose motion is monitored visually controls a simulated object rendered in three dimensions. A typical trace in the xy plane of a finger drawing letters is shown in figure 9.2. If the entire trajectory were

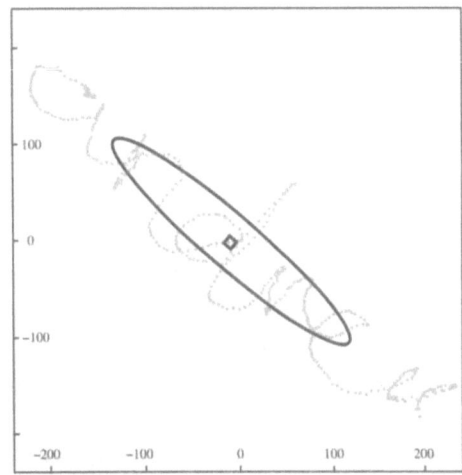

Figure 9.2: The moving finger writes. *The finger trajectory (left) which has a duration of about 10 seconds executes a broad sweep over the plane. If the trajectory is treated as a training set, the learned Gaussian prior is broad, as the covariance ellipse (right) shows. Clearly though, successive positions (individual dots represent samples captured every 20 ms) are much more tightly constrained.*

treated as a training set, the methods of the previous chapter could be applied to learn a Gaussian prior distribution for finger position. The learned prior is broad, spanning a sizable portion of the image area, and places little constraint on the measured position at any given instant. Nonetheless, it is quite clear from the figure that successive positions are tightly constrained. Although the prior covariance ellipse spans about

300×50 pixels, successive sampled positions are seldom more than 5 pixels apart!

For sequences of images, then, a global prior $p_0(\mathbf{X})$ is not enough. What is needed is a conditional distribution $p(\mathbf{X}(t_k)|\mathbf{X}(t_{k-1}))$ giving the distributions of possibilities for the kth shape $\mathbf{X}(t_k)$ *given* the $k-1$th shape $\mathbf{X}(t_{k-1})$. This amounts to a "first-order Markov chain" model in shape-space in which, although in principle $\mathbf{X}(t_k)$ may be correlated with all of $\mathbf{X}(t_1) \ldots \mathbf{X}(t_{k-1})$, only correlation with the immediate predecessor is explicitly acknowledged — the "Markov" assumption is made that

$$p(\mathbf{X}(t_k)|\mathbf{X}(t_1)\ldots\mathbf{X}(t_{k-1})) = p(\mathbf{X}(t_k)|\mathbf{X}(t_{k-1})).$$

This is an assumption that greatly simplifies the probabilistic model. (It will be argued later that it is actually an oversimplification and that a second-order Markov model, taking two predecessors into account, is needed.)

Continuing the Gaussian theme of earlier chapters, a simple, isotropic, first-order Gaussian Markov process in shape-space is

$$p(\mathbf{X}(t_k)|\mathbf{X}(t_{k-1})) \propto \exp -\frac{1}{2b^2}\|\mathbf{X}(t_k) - \mathbf{X}(t_{k-1})\|^2. \qquad (9.1)$$

For example, in a shape-space $\mathcal{S}$ of translations in the plane, this process represents simply a two-dimensional, random walk of the centroid of a rigid shape, in which the average (root-mean-square) step length is $b\sqrt{2}$ and is isotropic — steps are equally likely to occur in all directions (figure 9.3 (top)). This particular form of random walk is known as "Brownian motion" and is well-known as a physical model of the thermodynamics of microscopic particles. Such a distribution looks too random to be generally useful for modelling the motion of real, massive objects. A predominantly deterministic motion with an added random component is more plausible (figure 9.3 (left)). A modest elaboration of the model above achieves this by including a constant offset $\mathbf{D}$:

$$p(\mathbf{X}(t_k)|\mathbf{X}(t_{k-1})) \propto \exp -\frac{1}{2b^2}\|\mathbf{X}(t_k) - \mathbf{X}(t_{k-1}) - \mathbf{D}\|^2. \qquad (9.2)$$

In the case of planar translation, $\mathbf{D}$ is a vector representing the mean displacement in each time-step. The endpoints of random walks consisting of N steps, starting from a fixed origin, are distributed isotropically as a Gaussian in the plane (figure 9.3 (right)). Its distribution is the sum of N Gaussians, which is the single Gaussian whose covariance ellipse (a circle), and any confidence interval derived from it, grows steadily, in

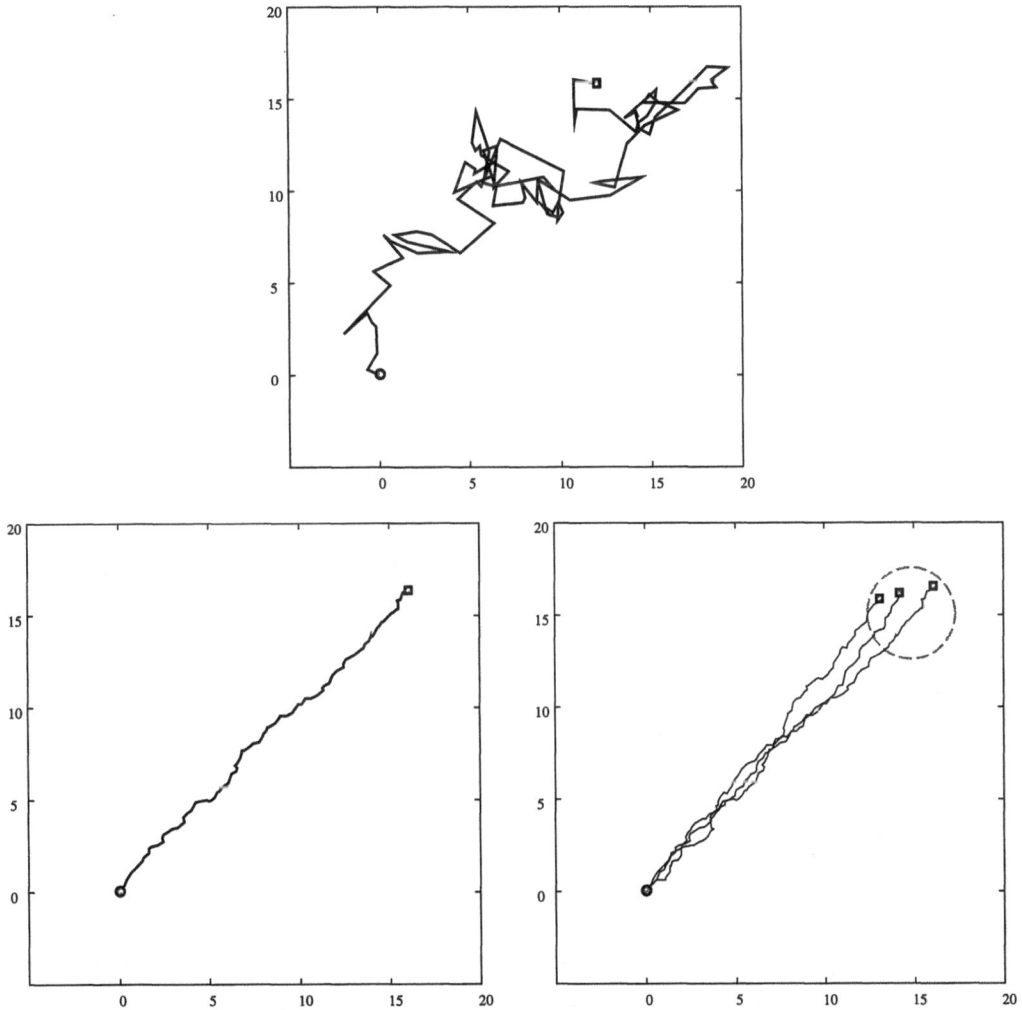

Figure 9.3: Brownian motion in the plane. *Plots show successive positions of the centroid of some rigid shape undergoing Brownian motion. (Top) an isotropic random walk (100 steps, b = 1). A random walk (left) with a superimposed drift velocity (100 steps, b = 0.1, drift (0.15,0.15) per time-step). Several samples (right) of random walk with drift; the endpoints of the random walks (marked with squares) have an isotropic Gaussian distribution in the plane (95% confidence ellipse shown dotted).*

fact in proportion to $\sqrt{N}$. This is precisely the steady leakage of information which was claimed earlier to be an essential component of a dynamical model.

Let us accept for the time being the random walk with drift as a plausible model of the translational motion of an object (though later in the chapter it will be argued that its trajectories are insufficiently smooth as a model of the motion of real physical bodies). Now consider instead deformations of the object, governed by the components of, say, a 4-dimensional subspace $\mathcal{S}_d$ containing all the affine components except translation. The random walk with drift is certainly not an appropriate model for deformation because, as we saw in the examples above, its variance grows unboundedly over time. In the case of the bottle example of the previous chapter this would imply a bottle distorting progressively, without limit. Somehow the small deformations that occur in a continuous fashion from frame to frame need to be forced to stay with an overall envelope. What is needed is a Markov process that looks like a random walk on a short time-scale but sweeps out some bounded distribution, presumably a Gaussian one, over long times. It turns out that a minor modification of discrete Brownian motion (9.1):

$$p(\mathbf{X}(t_k)|\mathbf{X}(t_{k-1})) \propto \exp -\frac{1}{2b^2}\|(\mathbf{X}(t_k) - \overline{\mathbf{X}}) - a(\mathbf{X}(t_{k-1}) - \overline{\mathbf{X}})\|^2 \qquad (9.3)$$

with $0 \leq a \leq 1$ gives a *constrained* Brownian process that has exactly the desired properties (figure 9.4). Initially it appears to behave like the unconstrained process but as the long-time limit is approached it becomes more and more evident that it lies within a limiting Gaussian envelope. Exactly what that distribution is and how it depends on a and b is clarified later.

Now, such a distribution, visualised so far only in terms of points in the plane, can be illustrated over a realistic shape-space. Take the bottle example of the last chapter, over an affine space $\mathcal{S}$, partitioned into 2-dimensional translational and 4-dimensional deformation subspaces $\mathcal{S}_s$ and $\mathcal{S}_d$ respectively. A plausible dynamical model for a translating bottle whose shape undergoes small distortions about a mean shape $\overline{\mathbf{X}}$ due to projective effects on the changing viewpoint might have two components. One is a constrained Brownian process in the deformation space $\mathcal{S}_d$. The other is an unconstrained Brownian process in the translational space $\mathcal{S}_s$. As figure 9.3 (left) showed, adding drift to Brownian motion models directional motion, but unfortunately only in a preassigned, fixed direction. If the direction of motion is not known *a priori* then the only model we have so far for translation (see later for better models) is pure Brownian motion. The combined model is expressed in terms of the following

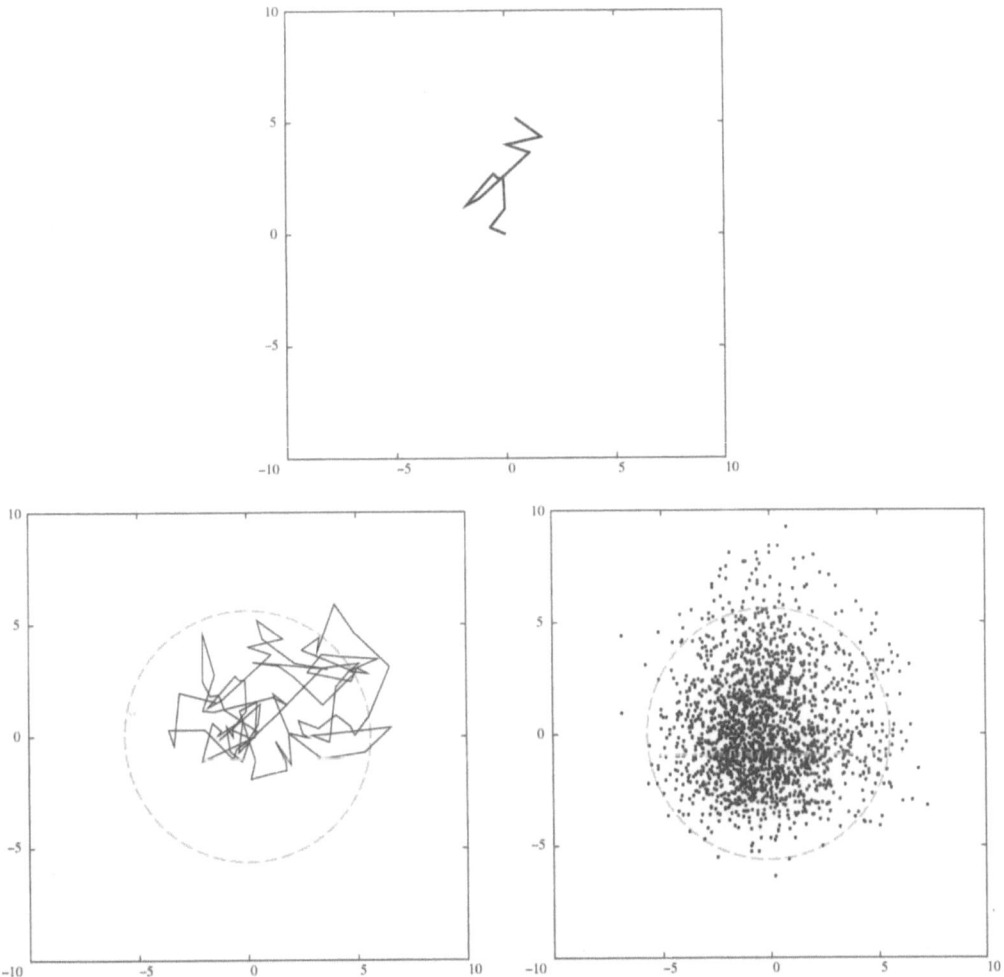

Figure 9.4: Constrained Brownian motion *with $b = 1$ and $a = 0.9$. On a small scale, the process appears like unconstrained Brownian motion with $b = 1$ so that (top) the first 12 steps are similar to the first 12 steps of figure 9.3 (top) but after 100 steps (left) things look quite different — the process is constrained by an overall Gaussian envelope whose 95% confidence ellipse is marked. After 2000 steps (right) the steady-state Gaussian envelope is quite evident (positions only shown, for clarity).*

first-order Markov conditional density:

$$p(\mathbf{X}(t_k)|\mathbf{X}(t_{k-1})) \quad \propto \quad \exp -\frac{1}{2}\left\| \frac{1}{b^s}E^s\left(\mathbf{X}(t_k) - \mathbf{X}(t_{k-1})\right) \right.$$

$$\left. + \frac{1}{b^d}E^d\left((\mathbf{X}(t_k) - \overline{\mathbf{X}}) - a^d(\mathbf{X}(t_{k-1}) - \overline{\mathbf{X}})\right) \right\|^2 . \tag{9.4}$$

(Subspace projection matrices E^s and E^d were defined two chapters back, in (6.5) on page 117.) A sample path from a simulation of such a dynamical model is illustrated in figure 9.5. It is perhaps not altogether clear from the Markov density (9.4) exactly how simulation can be done: in fact, expressing the density instead as an "Auto-regressive" process amounts to an explicit prescription for simulation and this is explained next.

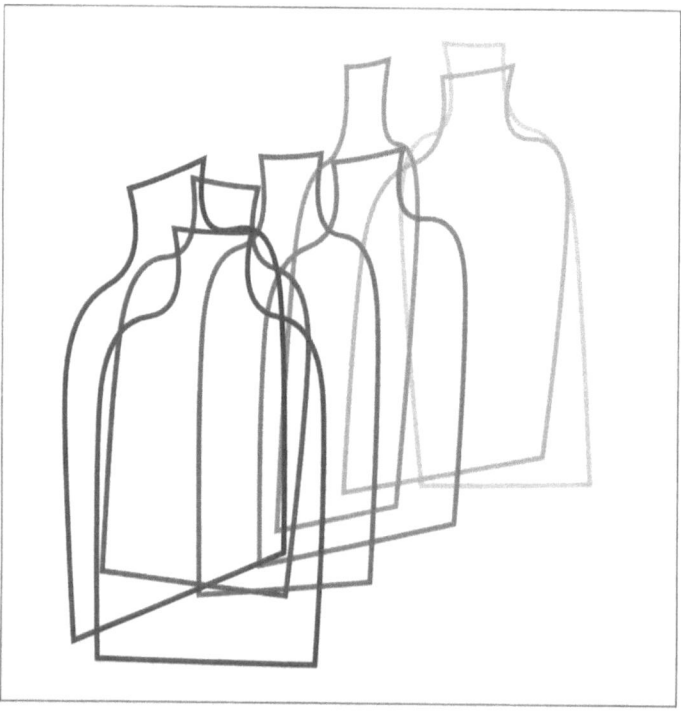

Figure 9.5: First-order stochastic motion. *The figure shows a sample path from a distribution (9.4) which translates as an unconstrained Brownian process ($b^s = 30$ pixels) and deforms as a constrained Brownian process ($a^d = 0.5$, $b^d = 15$ pixels). The sequence shown displays every 10th frame.*

9.2 First-order Auto-regressive processes

We have seen some specific examples of useful Markov processes and generally they can be expressed in the form

$$p(\mathbf{X}(t_k)|\mathbf{X}(t_{k-1})) \propto \exp{-\frac{1}{2}\|B^{-1}(\mathbf{X}(t_k) - A\mathbf{X}(t_{k-1}) - \mathbf{D})\|^2}. \qquad (9.5)$$

Just as Gaussian priors $\mathbf{X} \sim \mathcal{N}(\overline{\mathbf{X}}, \overline{P})$ in the previous chapter could be expressed as $\mathbf{X} = \overline{\mathbf{X}} + B\mathbf{w}$ which was amenable to simulation, so also the Markov process can be expressed in a generative form:

$$\mathbf{X}(t_k) = A\mathbf{X}(t_{k-1}) + \mathbf{D} + B\mathbf{w}_k, \qquad (9.6)$$

specifying a "sample path" for the process, in which each $\mathbf{w}_k$ is a vector of N_X independent random $\mathcal{N}(0,1)$ variables and $\mathbf{w}_k$, $\mathbf{w}_{k'}$ are independent for $k \neq k'$. A form which is not quite as general but will prove convenient is

$$\mathbf{X}(t_k) - \overline{\mathbf{X}} = A(\mathbf{X}(t_{k-1}) - \overline{\mathbf{X}}) + B\mathbf{w}_k \qquad (9.7)$$

and this is the standard form for a first-order "Auto-regressive" (AR) process. The constants in the two forms are related by

$$(I - A)\overline{\mathbf{X}} = \mathbf{D}.$$

The AR standard form has the advantage that the parameter $\overline{\mathbf{X}}$ has a clear interpretation: it is the steady-state limit of the mean value of $\mathbf{X}(t_{k-1})$. As $k \to \infty$, and provided the process is stable (a condition for stability is $\|A\|_2 < 1$ — see appendix B.1), the distribution $\mathbf{X}(t_{k-1})$ approaches a steady state $\mathcal{N}(\overline{\mathbf{X}}, P_\infty)$ for some limiting covariance P_∞. The steady-state distribution is exactly the overall envelope distribution for constrained Brownian motion.

In fact the AR form of the process can be used to derive the distribution $\mathbf{X}(t_k) \sim \mathcal{N}(\hat{\mathbf{X}}(t_k), P(t_k))$ at all times t_k. Since, by definition, $\hat{\mathbf{X}}(t_k) = \mathcal{E}[\mathbf{X}(t_k)]$ and $P(t_k) = \mathcal{V}[\mathbf{X}(t_k)]$, taking the expectation and variance of (9.7) gives, respectively, the "mean-state" equation and

$$\hat{\mathbf{X}}(t_k) - \overline{\mathbf{X}} = A(\hat{\mathbf{X}}(t_{k-1}) - \overline{\mathbf{X}}) \qquad (9.8)$$

or, in the alternative form,

$$\hat{\mathbf{X}}(t_k) = A\hat{\mathbf{X}}(t_{k-1}) + \mathbf{D}, \qquad (9.9)$$

and the "covariance" or Riccati equation

$$P(t_k) = AP(t_{k-1})A^T + BB^T. \tag{9.10}$$

It is clear by inspection that $\mathbf{X}(t_k) = \mathbf{X}(t_{k-1}) = \overline{\mathbf{X}}$ satisfies (9.8) so that the mean of the steady-state distribution must be $\overline{\mathbf{X}}$. The steady-state covariance $P_\infty \equiv \lim_{k\to\infty} P(t_k)$ must be a fixed point of (9.10):

$$P_\infty = AP_\infty A^T + BB^T \tag{9.11}$$

and although this equation can be solved exactly (by diagonalising A) the most straightforward method, if not the most efficient, is simply to iterate (9.10) to convergence.

Examples of AR processes

A variety of Markov processes can be succinctly expressed in the AR formalism.

1. **Unconstrained Brownian motion**

   Setting $A = I$, the AR process (9.7) simplifies to

   $$\mathbf{X}(t_k) = \mathbf{X}(t_{k-1}) + B\mathbf{w}_k$$

   which is independent of $\overline{\mathbf{X}}$. The fact that $\overline{\mathbf{X}}$, the steady-state mean, should not need to be specified is reasonable given the process has no steady state. In fact the mean value is constant throughout: $\hat{\mathbf{X}}(t_k) = \mathbf{X}(t_0)$, and, from (9.10), the covariance evolves as

   $$P(t_k) = P(t_{k-1}) + BB^T$$

   so that, given an exact initial value for $\mathbf{X}(t_0)$, $P(t_k) = kBB^T$ which is unbounded, with no limiting value as $k \to \infty$. The covariance ellipse has a mean-square radius of $\mathrm{tr}(P)$ and therefore grows in proportion to $\sqrt{k}$. There remains the choice of B which is effectively a scaling transformation for the process. For example, in the plane, $B = I_2$ produces a statistically isotropic random walk, whereas

   $$B = \begin{pmatrix} 2 & 0 \\ 0 & 1 \end{pmatrix}$$

   is a random walk which has been stretched out by a factor of 2 in the horizontal direction. In any shape-space, a norm-squared Brownian process (9.1) has the special form $B = b_0 \mathcal{H}^{-1/2}$ so that $P(t_k) = k \, b_0^2 \, \mathcal{H}^{-1}$, and $\mathbf{X}(t_k)$ is distributed as a norm-squared Gaussian in which average curve displacement is proportional to $\sqrt{k}$.

2. **Unconstrained Brownian motion with drift**

   In fact the addition of drift is outside the scope of the standard form (9.7) but can be expressed in the alternative form (9.6) with $A = I$ and $\mathbf{D}$ being the drift per unit time-step. Still there is no steady state but, from (9.9) and (9.10), the distribution $\mathbf{X}(t_k) \sim \mathcal{N}(\hat{\mathbf{X}}(t_k), P(t_k))$ is given by

   $$\hat{\mathbf{X}}(t_k) = k\mathbf{D} + \hat{\mathbf{X}}(t_0) \quad \text{and} \quad P(t_k) = kBB^T,$$

   so that the covariance ellipse grows in proportion to $\sqrt{k}$ as before but also drifts at a constant rate.

3. **Constrained Brownian motion**

   The constrained Brownian motion model of (9.3), with its norm-squared Gaussian density, can be expressed in the standard form of the AR process with coefficients $A = aI$, $B = b\mathcal{H}^{-1/2}$. If $a^2 = 1 - \epsilon$ with $0 < \epsilon \ll 1$ then, on a small time-scale, the process is almost indistinguishable from the case $\epsilon = 0$ of unconstrained Brownian motion, but in the longer term the distinction which was clear graphically in figure 9.4 is apparent also from the covariance equation. Since $\|A\|_2 = a < 1$, the covariance reaches a steady state, at a fixed point (9.11)

   $$P_\infty = \frac{1}{\epsilon}BB^T = \frac{b^2}{\epsilon}\mathcal{H}^{-1}$$

   and the limiting Gaussian envelope for the process is then $\mathcal{N}(\overline{\mathbf{X}}, P_\infty)$. The results of the previous chapter imply that the curve has an average displacement from the template of $\overline{\rho} = b\sqrt{N_X/\epsilon}$ in the steady state.

   In the simple example of figure 9.4, in which $a = 0.9$, $b = 1$ and the space is translational, we get $\overline{\rho} = \sqrt{2/(1 - 0.9^2)} = 3.24$ units. The 95% confidence circle has a radius approximately 1.73 times as great, approximately 5.6 units, as illustrated in the figure.

4. **Motion in affine space**

   A Markov process suitable for an affine shape-space was set out above (9.4) and simulated in figure 9.5. It can be expressed as an AR process in standard form (9.7) with $\overline{\mathbf{X}}$ as the mean shape for the object and the other coefficients can be shown to be

$$A = E^s + a^d E^d,$$

simply applying a different multiplier in each subspace, and similarly

$$B = \left(b^s E^s + b^d E^d \right) \mathcal{H}^{-\frac{1}{2}},$$

partitioned across the two subspaces.

This model does not reach a steady state because the Brownian motion in the translational subspace causes unbounded variance $P(t_k)$. However, the deformation component $E^d \mathbf{X}(t_k)$ is a constrained Brownian process which reaches a steady state as above.

9.3   Limitations of first-order dynamical models

First-order AR processes seem to meet some of the requirements for dynamical modelling. They are stochastic and can model entire families of motions. They deal with changes of position and shape and so can impose strong incremental constraints when global constraints are weak. They allow a global distribution to be chosen in addition to and independently of the local process. They appear to be the simplest available models that achieve these properties. Certainly, a first-order model is better than no dynamical model at all, as far as effective tracking is concerned. They do have limitations however.

First, unconstrained Brownian motion with drift, although it seems to model noisy directional motion at a constant average velocity, has an average direction that is fixed over time and must be known in advance. However, a good model for the motion of the writing finger would be a prior distribution that favours smooth motion of approximately constant velocity, but in an arbitrary direction, and allowing that velocity to change slowly in magnitude and direction. Another example is tracked vehicle motion (figure 9.6) in which slow, gradual changes of velocity are typical. More rapid changes in a given component of velocity are likely to occur during vehicle manoeuvres. Again, a model is called for that allows for substantial changes in velocity components over time, albeit mostly slow changes, and this is beyond the scope of the first-order model.

A second requirement is to be able to model oscillations that occur in many dynamical systems of interest. For example the tracked motion of the beating heart

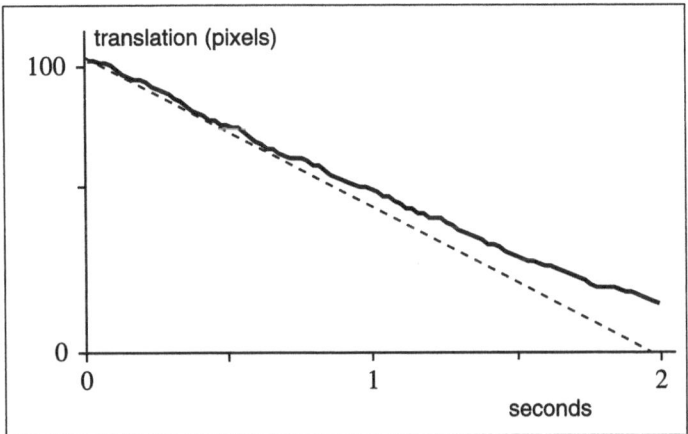

Figure 9.6: Tracking moving vehicles. *One translational component of tracked image motion for cars on a highway (figure 1.3 on page 7) is plotted here. Velocity is approximately constant over short time-scales. Here image velocity decreases gradually as the vehicle recedes.*

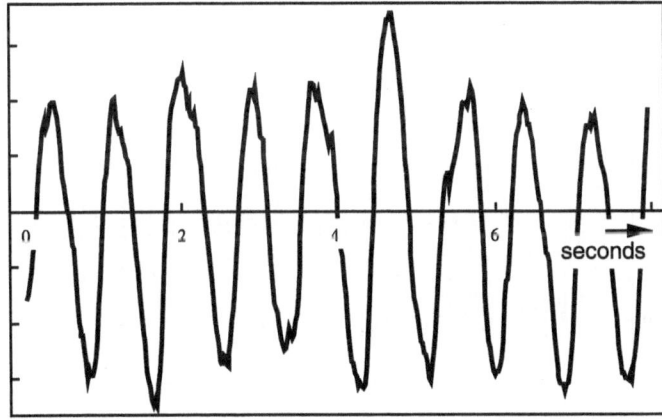

Figure 9.7: Tracked motion of a beating heart. *The first principal component of the motion of a beating heart, tracked in an ultrasound image sequence (figure 1.12 on page 16) is, of course, highly periodic. (Data courtesy of Gary Jacob and Alison Noble.)*

in the introductory chapter of the book is, not surprisingly, strongly oscillatory, as figure 9.7 shows. Any first-order AR process $\mathbf{X}(t_k)$ is a regularly sampled version of an underlying continuous-time process $\mathbf{X}(t)$, sampled at regular intervals $t_k = k\tau$ (see appendix B.1). The underlying continuous process has a characteristic time-course or "impulse response" that follows an exponential decay of the form $\exp -\beta t$, without any oscillatory component. Such a process would appear, therefore, to be unable to model oscillatory signals adequately.

Clearer evidence of the inadequacy of first-order models comes from examining the "spectral" characteristics of motion. In an AR process, the decaying or resonant response is driven by Gaussian random noise. Thus a first-order AR process, despite its decaying impulse response, does not actually decrease in magnitude because it is continuously excited by noise. The result is a complex superposition of exponentials which is succinctly characterised by its "power spectrum," the distribution of signal power as a function of frequency. For a first-order system, the power density is greatest at low frequency, falling off steadily as frequency increases.

Consider the example of lip motion during speech. It does not approach perfect periodicity as did the motion of the heart. Its motion does appear to have distinct periodic elements (figure 9.8) but over a spread of frequencies. A spectral analysis

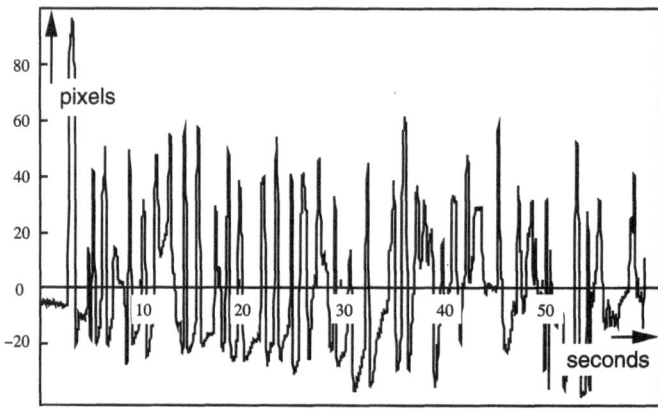

Figure 9.8: Lip motion during speech is oscillatory *This plots the opening/shutting motion of frontally viewed lips during a 60 second sequence of continuous speech. The density of peaks and troughs suggests oscillation at frequencies around 0.5–1 Hz.*

of the signal (figure 9.9) shows the "power spectrum" of the speech signal — its distribution of power as a function of frequency. This is done by computing the Discrete Fourier Transform (DFT) of the signal and plotting the square of its complex amplitude. The resulting power spectrum shows a background power distribution across all frequencies, decreasing in magnitude as frequency increases. Superimposed

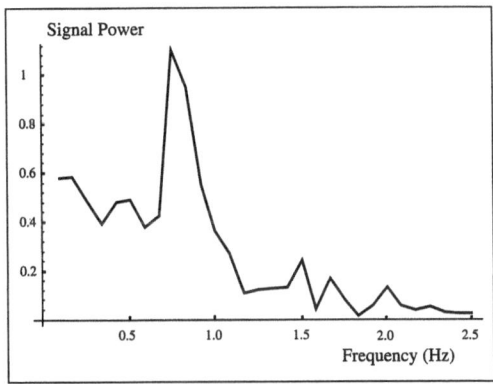 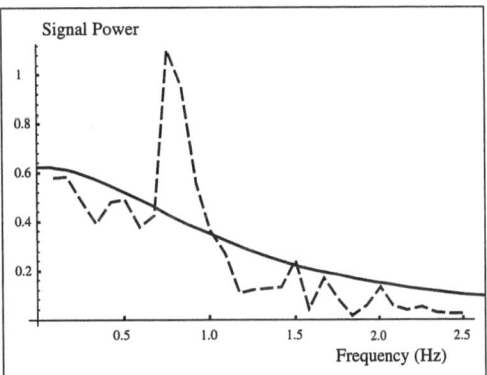

Figure 9.9: Spectral analysis of lip motion suggests second-order dynamics *Signal power in the lip-motion signal of figure 9.8 peaks (left) at frequency of about 0.8 Hz, reflecting the average spacing of peaks and troughs. A first-order AR process can be chosen with a spectrum (right) that models the background power level effectively but does not capture the peak.*

on this is a clear "resonant" peak at a frequency which appears to correspond well with the spacing of peaks and troughs in the original signal. Now it is known that for a first-order AR process in 1 dimension

$$x(t_k) = ax(t_{k-1}) + bw_k$$

the corresponding power spectrum has the form (see appendix B.1):

$$S_{xx}(f) \propto \frac{1}{1 + \gamma f^2} \tag{9.12}$$

where γ is a constant that depends on the parameters a and b of the AR process. This spectrum always has maximum power at zero frequency $f = 0$ and hence cannot have a resonant peak at some frequency $f > 0$. Choosing a value of γ by hand to fit the lip-motion spectrum, it is possible to explain the background power distribution (figure 9.9) but the peak cannot be modelled. When the model is used to generate a sample (figure 9.10), the simulated signal is a poor replica of the original data. Transitions in the simulated signal are too rapid, and this is consistent with the excessive signal power at high frequencies in the first-order model. The simulated signal does not appear to contain an underlying oscillation and this is consistent with the absence of a resonant peak in the first-order fitted spectrum.

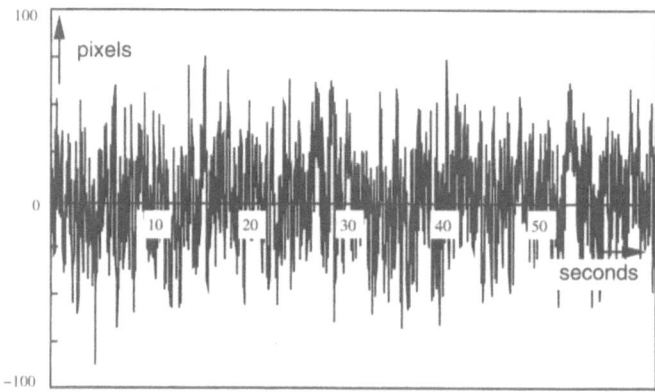

Figure 9.10: Simulated first-order lip motion *Parameters for a first-order AR process are chosen to correspond to the fitted power spectrum in figure 9.9 (left). A sample path for the process is illustrated here and should be compared with the original signal in figure 9.8. The simulated signal comprises far more rapid transitions than the original data it is supposed to model.*

9.4   Second-order dynamical models

Randomly excited harmonic motion

The simplest auto-regressive processes that meet the additional requirements, both for oscillatory and translational motion, are "second-order" processes (see below). They are typically resonant with a characteristic time-course in the form of a damped oscillation $\exp -\beta t \cos 2\pi f t$. The power spectrum of a second-order AR process (see appendix B.2) is:

$$S_{xx}(f) \propto \frac{1}{\pi^2(f_0^2 - f^2)^2 + \beta_0^2 f^2} \tag{9.13}$$

where β_0 is a "damping" constant with the dimensions of inverse time. It is clear that this spectral density reaches a maximum at approximately $f = f_0$ (provided $\beta_0 \ll f_0$), and is therefore capable of representing resonant or frequency-tuned behaviour. The width of the resonant peak, determined jointly by parameters f_0 and β_0, is $\Delta f_0 \approx f_0^2/\beta_0$. Choosing f_0 and β_0 by hand to match the centre and width of the resonant peak in the lip-motion data of figure 9.8 gives a much improved fit, as figure 9.11 shows. Now the peak in the spectral power distribution is represented, at least approximately, although the background power density may have been approximated better by the

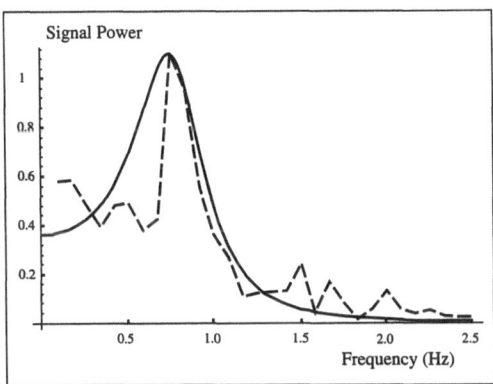

Figure 9.11: Second-order power spectrum model *A second-order model can represent the spectral peak in the lip-motion data of figure 9.8.*

first-order model, especially at high frequency — power density in the second-order model tails off rather too fast. As a final piece of empirical evidence in favour of second-order models, the parameters f_0 and β_0 can be used to specify an AR process representing a second-order Markov chain (see next section) which can be sampled by driving it with randomly generated noise $\mathbf{w}_k$, as was done earlier with first-order AR processes. The result (figure 9.12) appears quite plausibly to be a signal of a similar type to the original, real lip-motion data.

Physical realisability

A further point of principle that favours second-order models over first-order ones comes from considering a temporal process $x(t)$ with power spectrum $S_{xx}(f)$ as a component of the physical motion $\mathbf{X}(t)$ of a body with mass. In that case, the corresponding velocity $v(t) = \dot{x}(t)$ has a power spectrum proportional to $f^2 S_{xx}(f)$ and the mean kinetic energy associated with that motion is proportional to

$$\mathcal{E}[v(t)^2] \propto \int_{-\infty}^{\infty} f^2 S_{xx}(f)\, df$$

and this integral must give a finite result for the model to be realisable as a physical process. For the first-order power spectrum $S_{xx}(f)$ in (9.12) the integral diverges, ruling out first-order models as unrealisable (other than with ideal, massless bodies). For the second-order spectrum (9.13) however, the integral converges and this is true of higher order AR models also.

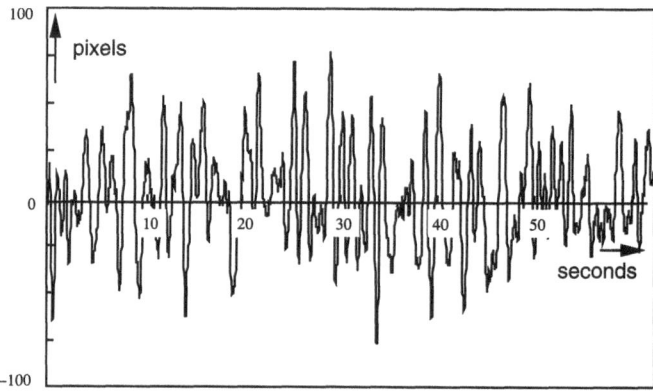

Figure 9.12: Simulated second-order lip motion *Parameters for a second-order AR process are chosen to correspond to the fitted power spectrum in figure 9.11. A sample path for the process is illustrated here and should be compared with the original signal in figure 9.8 — the distribution of spacings between peaks appears to be similar.*

It was noted above that although the resonant peak in the lip-motion spectrum was captured by a second-order model, the tail of the spectrum was actually modelled better by the first-order model. This might suggest that a weighted sum of the first- and second-order processes would fit better than either first- or second-order alone. This may be so but the mixed model would share the unrealisability of the pure first-order model; this is because the first-order spectrum dominates at large f, causing divergence of the kinetic energy integral. Mixtures of second- and higher-order models would satisfy realisability constraints. (In the language of "poles" and "zeros", popularly used to describe spectra, a pure nth order AR model has n poles and no zeros — an "all-pole" spectrum. The mixed first- and second-order model has 2 poles and 1 zero and is unrealisable. Generally physical realisability, in the mechanical sense, demands at least two more poles than zeros.) Furthermore, learning dynamics (see chapter 11) is rather more difficult for the mixed model than for the pure second-order model.

If physical realisability demands second-order or higher, conventional dynamical modelling suggests that second-order may be sufficient. For a system with just one degree of freedom, the situation is relatively straightforward. If it is assumed that potential energy and frictional power dissipation are each quadratic functions $\frac{1}{2}kx^2$ and $\frac{1}{2}\nu\dot{x}^2$ respectively then the equation of motion is not only second-order but also linear:

$$m\ddot{x} = -\nu\dot{x} - kx + f$$

where f is an externally applied force. Note that the quadratic assumptions cover the cases of potential energy due to elasticity (Hooke's law) or gravity and dissipation due to viscous forces at relatively low speeds. If f comes from some class of possible temporal force functions the result will be a class of motions $x(t)$. The simplest random model for f is as a "white noise" signal $f(t) = bw(t)$, where $w(t)$ is a Wiener process, the continuous counterpart to the uncorrelated discrete noise signal $\mathbf{w}_k$ in AR processes. In that case, the spectrum of $x(t)$ is second-order and all-pole.

Multidimensionally, in shape-space, motion $\mathbf{X}(t)$ is a little more difficult to describe. A general formalism ("Lagrangian dynamics" — see bibliographic notes) is available to convert a physical description of mass distribution, potential energy and dissipation into equations of motion. As before, the equations are of second-order, and linear if potential energy and power dissipation are quadratic:

$$\ddot{\mathbf{X}} = F_0 \mathbf{X} + F_1 \dot{\mathbf{X}} + G_0 \mathbf{w}, \qquad (9.14)$$

where $\mathbf{w}(t)$ is a vector of N_X independent Wiener processes and F_0, F_1 and G_0 are $N_X \times N_X$ matrices. Physical realisability means that F_0 and F_1 must be negative definite, with real eigenvalues and eigenvectors, but are not generally symmetric. These conditions guarantee stability of $\mathbf{X}(t)$, as might be expected for a physical process.

Lastly, note that $\mathbf{X}(t)$, being a shape-space variable, refers to the image of a physical object, rather than to the object itself. However, it was argued in chapter 4 that projection from world to image is (approximately) a linear mapping. That ensures that the general form of a linear model in world coordinates is preserved in the image plane, and also in shape-space given that its parameter $\mathbf{X}$ is defined to be linear.

Signal phase

So far, the argument about model order has concentrated on the power spectrum of a signal. This represents the amplitude of the signal spectrum but neglects its phase. Phase is also important but is less amenable to graphical arguments about goodness of fit. In fact, attempting to fit also the phase distribution of the lip-motion signal inevitably causes the power spectrum fit to deteriorate somewhat. This is evident later, in chapter 11, where an automatic fitting algorithm is developed. It fits models directly to signals, rather than to their power spectra, and therefore is bound to take both amplitude and phase into account.

9.5   Second-order AR processes in shape-space

A second-order AR process in shape-space is a natural extension of the first-order process (9.7). It has the form

$$\mathbf{X}(t_k) - \overline{\mathbf{X}} = A_2(\mathbf{X}(t_{k-2}) - \overline{\mathbf{X}}) + A_1(\mathbf{X}(t_{k-1}) - \overline{\mathbf{X}}) + B_0 \mathbf{w}_k \qquad (9.15)$$

in which A_2, A_1 and B_0 are all $N_X \times N_X$ matrices. In a second-order process, the shape-vector at a given time depends on two previous time-steps, rather than just one as in the first-order process. It can be expressed more compactly by defining a "state-vector"

$$\mathcal{X}(t_k) = \begin{pmatrix} \mathbf{X}(t_{k-1}) \\ \mathbf{X}(t_k) \end{pmatrix} \tag{9.16}$$

and then writing

$$\mathcal{X}(t_k) - \overline{\mathcal{X}} = A(\mathcal{X}(t_{k-1}) - \overline{\mathcal{X}}) + B\mathbf{w}_k \tag{9.17}$$

where

$$A = \begin{pmatrix} 0 & I \\ A_2 & A_1 \end{pmatrix}, \quad \overline{\mathcal{X}} = \begin{pmatrix} \overline{\mathbf{X}} \\ \overline{\mathbf{X}} \end{pmatrix} \quad \text{and} \quad B = \begin{pmatrix} 0 \\ B_0 \end{pmatrix}. \tag{9.18}$$

It is straightforward to show that (9.17) is precisely equivalent to (9.15). Replacing the shape-vector $\mathbf{X}$ by a state-vector $\mathcal{X}$ of double size allows the notation for a first-order process to continue to be used, albeit with some restrictions (9.18) on the form of the coefficient matrices A and B.

The second-order state $\mathcal{X}$ has mean and covariance

$$\hat{\mathcal{X}}(t_k) = \mathcal{E}[\mathcal{X}(t_k)] \quad \text{and} \quad \mathcal{P}(t_k) = \mathcal{V}[\mathcal{X}(t_k)] \tag{9.19}$$

where $\mathcal{P}(t_k)$ is a $2N_X \times 2N_X$ matrix that is naturally decomposed into submatrices:

$$\mathcal{P}(t_k) = \begin{pmatrix} P''(t_k) & P'(t_k)^T \\ P'(t_k) & P(t_k) \end{pmatrix} \tag{9.20}$$

in which

$$\begin{aligned} P''(t_k) &= \mathcal{E}[(\mathbf{X}(t_{k-1}) - \hat{\mathbf{X}}(t_{k-1}))(\mathbf{X}(t_{k-1}) - \hat{\mathbf{X}}(t_{k-1}))^T] \\ P'(t_k) &= \mathcal{E}[(\mathbf{X}(t_k) - \hat{\mathbf{X}}(t_k))(\mathbf{X}(t_{k-1}) - \hat{\mathbf{X}}(t_{k-1}))^T] \\ P(t_k) &= \mathcal{E}[(\mathbf{X}(t_k) - \hat{\mathbf{X}}(t_k))(\mathbf{X}(t_k) - \hat{\mathbf{X}}(t_k))^T]. \end{aligned} \tag{9.21}$$

The most interesting of these is $P(t_k)$ which represents covariance in shape-space at time t_k. The other submatrices need to be "carried" by $\mathcal{P}$ simply in order to allow second-order propagation of $P(t_k)$. (Note that, as a consequence of second-order propagation, $P''(t_k) = P(t_{k-1})$.) Propagation is governed by mean-state and covariance equations for the second-order model, by analogy with the first-order case (9.8) and (9.10):

$$\hat{\mathcal{X}}(t_k) - \overline{\mathcal{X}} = A(\hat{\mathcal{X}}(t_{k-1}) - \overline{\mathcal{X}}) \qquad (9.22)$$

and

$$\mathcal{P}(t_k) = A\mathcal{P}(t_{k-1})A^T + BB^T. \qquad (9.23)$$

A steady-state $\mathcal{P}_\infty$ can be computed as before, by iterating (9.23) to convergence. Its lower-right submatrix P_∞ then represents the covariance of the Gaussian envelope in shape-space for the second-order Brownian process.

The AR process (9.16) is defined over discrete time, but can in fact be regarded as a regular sampling of an underlying continuous-time process. In the next section this discrete–continuous correspondence maps damped harmonic motion, described in continuous time, into discrete time. It serves as a "synthetic" predictive dynamical model, used in tracking in the next chapter. The correspondence is also useful in the reverse direction, to decompose and interpret an AR process in physical terms. This will be important in chapter 11 where models are learned from training sequences and learned coefficients A and B are interpreted as collections of damped harmonic oscillators, each associated with a vibrational mode — effectively a one-dimensional shape-subspace. Relevant details of the discrete–continuous correspondence are set out, for completeness, in appendix B.2.

9.6 Setting dynamical parameters

The best dynamical models for tracking, in the sense of being most appropriately tuned to expected motions, are obtained by learning and this is discussed fully in chapter 11. Until such learned models are available, we have to be content to use default models, synthesised by hand to match general, intuitive expectations about the motions to be observed. These expectations are addressed partly by the static constraints embodied in the choice of shape-space, and this has been dealt with in earlier chapters. Dynamical characteristics must also be specified in order to define fully an operational tracker (see the next chapter) which could be regarded as an end-product in its own right. Alternatively, the tracker may be applied just once, to capture a training sequence which is then used to learn a more refined shape-space via PCA (chapter 8) and more refined dynamics (chapter 11).

This section offers a systematic approach to synthesising dynamical models. The synthetic models are based on harmonic oscillators driven by spatially homogeneous noise. Shape-space is decomposed into natural subspaces and a stochastic harmonic oscillator is set up in each subspace.

Stochastic, harmonic motion in one dimension

First, to describe the harmonic oscillator which is the building block for synthesised models, an oscillation with damping rate β and frequency of oscillation f is expressed discretely by

$$A = \begin{pmatrix} 0 & 1 \\ a_2 & a_1 \end{pmatrix} \quad \text{and} \quad B = \begin{pmatrix} 0 \\ b_0 \end{pmatrix} \tag{9.24}$$

where

$$a_2 = -\exp(-2\beta\tau) \quad \text{and} \quad a_1 = 2\exp(-\beta\tau)\cos(2\pi f\tau). \tag{9.25}$$

(The derivation of these relations is given below.) A practically important special case, is *critical damping* in which $f = 0$, particularly suitable for motions which are not expected to be oscillatory; that leaves just $1/\beta$ to be specified, as a characteristic time-scale for the motion. Note that $f = 0$ implies the constraint on discrete parameters that $-a_1^2 = 4a_2$. Provided $\beta > 0$, the process is stable and has a steady-state distribution with finite spatial variance. To obtain a desired steady-state variance $\bar{\rho}^2$, choose

$$b_0 = \bar{\rho}\sqrt{1 - a_2^2 - a_1^2 - 2\frac{a_2 a_1^2}{1 - a_2}}. \tag{9.26}$$

(This formula comes directly from the steady-state solution of the covariance equation (9.23).) A more intuitive relationship is obtained by taking the "continuous time" limit that $\beta\tau \ll 1$, (an assumption that holds in most cases of practical interest):

$$b_0 = \bar{\rho}\left[2(\beta\tau)^{1/2}\sin(2\pi f\tau)\right]. \tag{9.27}$$

Constant-velocity model

A further sub-case of harmonic motion, useful particularly for translational motion, is the *constant-velocity* model in which $\beta = f = 0$, so that $a_2 = -1$ and $a_1 = 2$. As expected, for constant-velocity parameters the formulae (9.25) and (9.26) give $b_0 = 0$ or, equivalently, for any $b_0 > 0$ the variance $\bar{\rho}^2$ is unbounded. Given that there is no steady state, $\bar{\rho}$ cannot be used to characterise the stochastic component of the model. Instead, a rate of growth parameter γ_0 is specified. It can be shown that, asymptotically, the root-mean-square search-region diameter grows semicubically:

$$\bar{\rho}(t) = \gamma_0 t^{3/2} \quad \text{where} \quad b_0 = \gamma_0 \tau^{3/2}. \tag{9.28}$$

Harmonic motion in shape-space

A straightforward way to construct a dynamic model for a shape-space S_X is to extend the one-dimensional harmonic motion just described, to the entire space. This is done by choosing

$$A_2 = a_2 I_{N_X}, \quad A_1 = a_1 I_{N_X} \quad \text{and} \quad B_0 = \frac{b_0}{\sqrt{N_X}} \mathcal{H}^{-\frac{1}{2}}$$

where a_2, a_1 and b_0 are chosen as above. The resulting A-matrix then represents "degenerate" modes, N_X independent motions, with common frequency and damping constants, that span the shape-space. The driving Brownian noise has the usual norm-squared density, giving spatial uniformity and isotropy but subject to the constraints of the shape-space. The steady-state covariance $\mathcal{P}_\infty$ can be shown from (9.23) to be

$$P_\infty = \frac{\bar{\rho}^2}{N_X} \mathcal{H}^{-1}, \tag{9.29}$$

where $\bar{\rho}$ and b_0 are related as in (9.26). This represents a Gaussian envelope with mean $\mathbf{\bar{X}}$ and root-mean-square displacement $\bar{\rho}$ along the curve.

Partitioned harmonic motion across shape-subspaces

Finally, it is usually desirable to partition shape-space into several subspaces and apply different harmonic models to each. Earlier, affine space was partitioned into translation and deformation components using projection matrices E^s and E^d, and a first-order model (9.4) on page 192, defined across the partitioned space. This was chosen to allow relatively free translation but tightly constrained deformation of shape. A partitioned second-order model can similarly be specified by defining A and B matrices as follows:

$$A_2 = a_0^s E^s + a_0^d E^d, \quad A_1 = a_1^s E^s + a_1^d E^d \tag{9.30}$$

$$B_0 = \left(\frac{b_0^s}{\sqrt{N_s}} E^s + \frac{b_0^d}{\sqrt{N_d}} E^d \right) \mathcal{H}^{-\frac{1}{2}},$$

where N_s and N_d are the dimensions of the shape-subspaces. Parameters a_0^d, a_1^d and b_0^d are set for appropriate frequency f^d, damping rate β^d and average displacement $\overline{\rho}^d$ for deformation, using (9.26) above, and similarly for translation. Any desired number of partitions of shape-space can be chosen, and partitioned dynamics defined additively as above. Variances of the Gaussian envelopes are additive so that, in the two-component case above,

$$\overline{\rho}^2 = \sqrt{(\overline{\rho}^s)^2 + (\overline{\rho}^d)^2} \tag{9.31}$$

is the root-mean-square displacement in model as a whole.

Examples

Examples of dynamical models set up by hand are given here. The first is for vehicles on a motorway, as in the traffic-monitoring application described in chapter 1. Sample trajectories are displayed in figure 9.13. Translational dynamics are constant-velocity, with characteristic semicubical growth of the region of positional uncertainty, reaching 35 pixels (RMS) after one second. Affine deformation is set to vary slowly, over a 10-second time-scale, to allow the shrinkage of the template under perspective scaling, as vehicles recede into the distance. Ideally this shrinkage should be coupled with the translational motion in the dynamical model itself, rather than simply in the initial conditions, and limited to scaling rather than allowing all affine deformations. That is rather more elaborate than can reasonably be programmed by hand. What can be done is to use the model above as a predictor in a tracker that is just capable of gathering a training set. The training set can then be used in the learning algorithm to be described in chapter 11 to build automatically a model that incorporates the appropriate couplings.

A second example of a hand-programmed dynamical model is designed to represent the motion of a dancing girl, as in figure 9.14. This time the model is explicitly oscillatory, with independent dynamics for horizontal and vertical motion. As with

Figure 9.13: Dynamical model for traffic. *Three simulated trajectories for an ARP set up to represent the traffic pattern, each initialised with a velocity in shape-space corresponding to translation along the road, coupled with shrinkage consistent with perspective scaling. Translational parameters are $\beta = f = 0$ (constant-velocity) with $\gamma_0 = 35 \, \text{pixel.s}^{-3/2}$, allowing the paths to veer over longer times. Affine deformations are critically damped ($f = 0$) with a long time-constant ($1/\beta = 10 \, s$), and largely deterministic ($\bar{\rho} = 2 \, pixels$), allowing the outline to shrink steadily. Contours are plotted here at intervals of $200 \, \text{ms}$.*

the example of traffic above, a dynamical model such as this is sufficient for use as a predictor in a tracker that can capture a training set, which is then used to learn automatically an auto-regressive process model that is more specifically tuned to the observed motions.

Derivation of simple harmonic oscillator coefficients a_2, a_1, in terms of f, β as in (9.25):

Figure 9.14: Dynamical model for a dancer. *Three simulated trajectories, each of 0.84 s duration, are shown of an ARP model in a shape-space of translations. Horizontal oscillation is slow ($\beta = 0.2\,\text{s}^{-1}$, $f = 0.2\,\text{Hz}$) and broadly distributed ($\bar{\rho} = 300$ pixels), representing the motion of the dancer to and fro, across the room. Vertical oscillation is faster ($\beta = 0.5\,\text{s}^{-1}$, $f = 1\,\text{Hz}$) and more tightly distributed ($\bar{\rho} = 100$ pixels), representing the bobbing motion of the head.*

given that a_2, a_1 are to be chosen to correspond to the damped exponential $\exp -\beta t \exp \pm 2\pi f t$, over a time-step τ, the eigenvalues of A in (9.24) must be $\exp -\beta \tau \exp \pm 2\pi f \tau$, whose product and sum, respectively the determinant and trace of A, give

$$a_2 = -\det A = -\exp(-2\beta\tau) \quad \text{and} \quad a_1 = \operatorname{tr} A = 2\exp(-\beta\tau)\cos(2\pi f\tau),$$

as required.

Bibliographic notes

Dynamical models exploit "coherence of motion" — a term coined by Yuille and Grzywacz (Yuille and Grzywacz, 1988) to refer to the spatial continuity of motion across an image, and used here to refer both to spatial but especially also to temporal continuity. Dynamical models were based on the Markov process. This is a development of the Markov Chain (see for instance (Rabiner and Bing-Hwang, 1993)) in a continuous-valued form, and is a core tool in control theory (Astrom and Wittenmark, 1984). Markov processes can be expressed either in continuous or discrete time and the relation between the two is explained in (Gelb, 1974). Continuous-time stochastic processes form the basis of the arguments about physical realisability of models in the chapter, and a detailed treatment of continuous-time stochastic processes is given in (Astrom, 1970). Furthermore continuous-time analysis establishes the semicubical growth of the search region under constant-velocity dynamics (Blake et al., 1993).

Markov processes can also be viewed in spectral terms as the output of certain linear systems driven by noise (Papoulis, 1991), and often displaying resonances in its spectral power distribution. The use of resonant modes in visual motion modelling was first developed by Pentland and Horowitz (1991) and applied to spline contour models by Baumberg and Hogg (1994). In both cases the dynamical models are purely deterministic, essentially the deterministic component of the stochastic models described in the chapter. Such deterministic dynamical models describe distributed, massive systems (Landau and Lifshitz, 1972), and can be derived using "Lagrangian analysis." Second order differential equations, expressed in terms of positions, velocities and accelerations, are obtained from the functions of position and velocity that specify the potential energy and power dissipation of a system. A useful introduction is given in (Terzopoulos and Szeliski, 1992).

Chapter 10

Dynamic contour tracking

In the previous chapter, dynamical models were characterised by a second-order state density $p(\mathcal{X}(t))$, evolving temporally, and representing the *prior* distribution for the state $\mathcal{X}$ at each time t. In this chapter, both the prior dynamical model and visual measurements are to be taken into account. The result is a fusion of information, both prior and observational, as was set out in chapter 8 for single images, but done now for image sequences, to track motion.

The natural mechanism for temporal fusion, when distributions are Gaussian, is the Kalman filter. It computes the evolution of the Gaussian density for the state of the tracked object, as figure 10.1 illustrates. A simple practical example consists of a planar affine shape-space, based on a hand-shaped template, with constant-velocity dynamics, which is capable of tracking motion of an outstretched hand (figure 10.2). All results in this chapter show tracking tasks that can be performed in real time on a desktop workstation.

10.1 Temporal fusion by Kalman filter

Observation history

The Kalman filter maintains a Gaussian distribution of the state

$$\mathcal{X}(t_k) \sim \mathcal{N}(\hat{\mathcal{X}}(t_k), \mathcal{P}(t_k))$$

given both the prior dynamical model for $\mathcal{X}(t_k)$ *and* the measurement history $\underline{\mathbf{Z}}(t_k)$:

$$\underline{\mathbf{Z}}(t_k) = (\mathbf{Z}(t_1), \dots, \mathbf{Z}(t_k)) \tag{10.1}$$

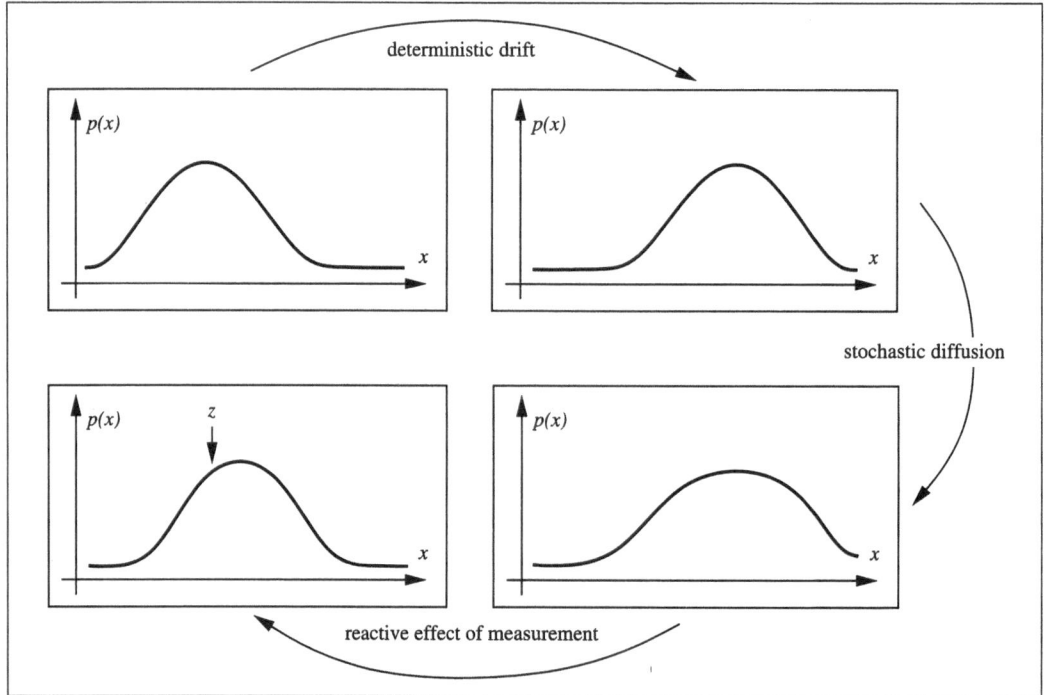

Figure 10.1: Kalman filter as density propagation. *In the case of Gaussian prior, process and observation densities, and assuming linear dynamics, the state density propagates as a Gaussian represented completely by its evolving (multi-variate) mean and variance.*

where $\mathbf{Z}(t_k)$ is the version at time $t = t_k$ of the aggregated observation $\mathbf{Z}$ that is produced by the recursive image-curve fitting algorithm of chapter 6 (figure 6.7 on page 127). As in that static algorithm, so also now in the dynamic case, the statistical information associated with $\mathbf{Z}(t_k)$ for estimating $\mathbf{X}$ is $S(t_k)$ (S in the fitting algorithm) and $\mathbf{Z}(t_k)$ is an unbiased estimate of $S(t_k)\mathbf{X}(t_k)$ (rather than of $\mathbf{X}(t_k)$ itself).

In the previous chapter, $\hat{\mathcal{X}}(t_k)$ and $\mathcal{P}(t_k)$ were defined (in (9.19) on page 204) to be the *prior* mean and covariance of the state. Now $\hat{\mathcal{X}}$ and $\mathcal{P}$ are redefined as the *posterior* mean and variance

$$\hat{\mathcal{X}}(t_k) = \mathcal{E}[\mathcal{X}(t_k)|\underline{\mathbf{Z}}(t_k)] \quad \text{and} \quad \mathcal{P}(t_k) = \mathcal{V}[\mathcal{X}(t_k)|\underline{\mathbf{Z}}(t_k)], \tag{10.2}$$

conditioned on the measurement history $\underline{\mathbf{Z}}(t_k)$. The principle behind the evolution of $\hat{\mathcal{X}}(t_k)$, together with $\mathcal{P}(t_k)$, is straightforward. The estimate at time t_{k-1} is propagated

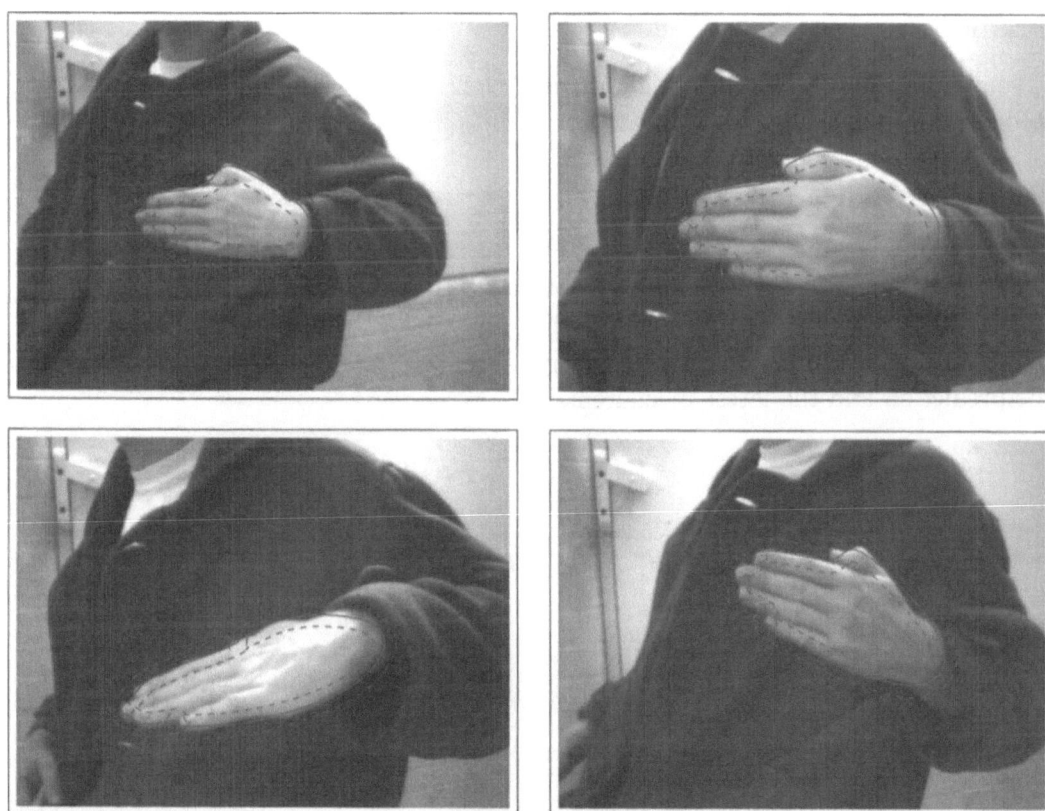

Figure 10.2: Hand tracking in planar affine shape-space. *A planar affine shape-space with constant-velocity dynamics is sufficient to track movements of an outstretched hand at modest speed.*

to time t_k in the two standard steps of the Kalman filter: prediction, and assimilation of observations. This engages all of the probabilistic machinery developed so far in the book: prediction is based on the dynamical models of the previous chapter; assimilation of observations is based on information weighting as in chapters 6 and 8.

Prediction

A single time-step of the dynamical model ((9.22) on page 205) is applied to $\hat{\mathcal{X}}(t_{k-1})$ to obtain a predicted state $\tilde{\mathcal{X}}(t_k)$:

$$\tilde{\mathcal{X}}(t_k) - \overline{\mathcal{X}} = A(\hat{\mathcal{X}}(t_{k-1}) - \overline{\mathcal{X}}) \tag{10.3}$$
$$\tilde{\mathcal{P}}(t_k) = A\mathcal{P}(t_{k-1})A^T + BB^T.$$

Assimilation by information weighting

In its most direct form, assimilation computes an information weighted mean between the prediction and the latest measurement:

$$\mathcal{P}(t_k) = \left(\tilde{\mathcal{P}}^{-1}(t_k) + H^T S(t_k)H\right)^{-1} \tag{10.4}$$
$$\text{and}$$
$$\hat{\mathcal{X}}(t_k) = \tilde{\mathcal{X}}(t_k) + \mathcal{P}(t_k)H^T \mathbf{Z}(t_k)$$

where the matrix

$$H = \left(\begin{array}{cc} 0 & I \end{array}\right) \tag{10.5}$$

maps $\mathcal{X}(t_k) \rightarrow \mathbf{X}(t_k)$ from state-space into shape-space.

It is clear that in the first of these equations, information is summed and that in the second, information weighting is applied. However, this assimilation procedure is somewhat inefficient because of the need to invert large ($2N_X \times 2N_X$) matrices. Fortunately the Kalman filtering canon offers a more efficient alternative in which only a single $N_X \times N_X$ matrix inversion is required at each time-step.

Assimilation by Kalman gain

An exactly equivalent algorithm for assimilation, free of inversions of the full state covariance $\mathcal{P}$, is based on a "Kalman gain" matrix $\mathcal{K}$, as follows:

$$\mathcal{K}(t_k) = \tilde{\mathcal{P}}(t_k)H^T \left(S(t_k)H\tilde{\mathcal{P}}(t_k)H^T + I\right)^{-1} \tag{10.6}$$
$$\hat{\mathcal{X}}(t_k) = \tilde{\mathcal{X}}(t_k) + \mathcal{K}(t_k)\mathbf{Z}(t_k)$$
$$\mathcal{P}(t_k) = (I - \mathcal{K}(t_k)S(t_k)H)\,\tilde{\mathcal{P}}(t_k).$$

Note that the only matrix inversion occurs in the expression for the Kalman gain and involves an $N_X \times N_X$ matrix, as promised. A particular form has been used here for the Kalman gain such that if the aggregated measurement $\mathbf{Z}(t_k)$ is deficient so that $S(t_k)$ has less than full rank, the assimilation step is still well-defined and free of singularities. In fact, even if measurement fails altogether so that $S(t_k) = 0$ and $\mathbf{Z}(t_k) = \mathbf{0}$, the assimilation step simply becomes $\hat{\mathcal{X}}(t_k) = \tilde{\mathcal{X}}(t_k)$, accepting the prediction without modification.

Block decomposition

Finally, the algorithm above can be expressed most efficiently, using submatrix decomposition of the state-space covariance $\mathcal{P}$ into submatrices P, P' and P'', as earlier in (9.20) on page 204. In addition, the Kalman gain is decomposed into two $N_X \times N_X$ submatrices:

$$\mathcal{K} = \begin{pmatrix} K' \\ K \end{pmatrix}. \tag{10.7}$$

This allows the special form of the dynamical matrix A and the symmetry of $\mathcal{P}$ to be exploited. The full algorithm is given in figure 10.3. It consists of one predict–observe–assimilate cycle for each time-step. The predicted shape $\tilde{\mathbf{X}}(t_k)$ is used as the basis for obtaining the aggregated measurement vector $\mathbf{Z}(t_k)$. It replaces, in steps 2 and 4 of the fitting algorithm of figure 6.7 on page 127, the shape-vector $\overline{\mathbf{X}}$ which defined the estimated curve $\overline{\mathbf{r}}(s)$ from which image features ν_i are measured. Finally, the estimated curve shape $\mathbf{X}(t_k)$ is computed, as required. Other variables $P(t_k)$, $P'(t_k)$, $P''(t_k)$ and $\hat{\mathbf{X}}'(t_k)$, computed in the assimilation step, are carried forward for use in prediction at the following time-step.

Initial conditions

The algorithm of figure 10.3 sets out the formation of the estimate at time t_k from the one at time t_{k-1}. All that remains to complete the algorithm is to specify initial conditions at time t_0. There are various natural ways to do this corresponding to various possible assumptions about the initial state of the object to be tracked. In general, values of $\hat{\mathbf{X}}(t_0)$ and $\mathcal{P}(t_0)$ are set to reflect the prior distribution for object state. The state distribution specifies the distributions of configurations at times

Algorithm for propagation over one time-step

Predict

$$
\begin{aligned}
\tilde{P}''(t_k) &= P(t_{k-1}) \\
\tilde{P}'(t_k) &= A_2 P'^{T}(t_{k-1}) + A_1 P(t_{k-1}) \\
\tilde{P}(t_k) &= A_2 P''(t_{k-1}) A_2^T + A_1 P'(t_{k-1}) A_2^T \\
&\quad + A_2 P'^{T}(t_{k-1}) A_1^T + A_1 P(t_{k-1}) A_1^T + B_0 B_0^T
\end{aligned}
$$

$$
\begin{aligned}
\tilde{\mathbf{X}}'(t_k) &= \hat{\mathbf{X}}(t_{k-1}) \\
\tilde{\mathbf{X}}(t_k) &= A_2 \hat{\mathbf{X}}'(t_{k-1}) + A_1 \hat{\mathbf{X}}(t_{k-1}) + (I - A_2 - A_1)\overline{\mathbf{X}}.
\end{aligned}
$$

Measure

Apply the algorithm of figure 6.7 on page 127, to the image at time t_k using $\tilde{\mathbf{X}}(t_k)$ as estimated contour from which normals are cast. Use $\tilde{P}(t_k)$ as the value of $\overline{P}$ to compute validation-gate width (figure 8.6 on page 174). Obtain the aggregated observation vector $\mathbf{Z}$ and information matrix S, denoted here as $\mathbf{Z}(t_k)$ and $S(t_k)$.

Assimilate

$$
\begin{aligned}
K'(t_k) &= \tilde{P}'(t_k)\left[S(t_k)\tilde{P}(t_k) + I\right]^{-1} \\
K(t_k) &= \tilde{P}(t_k)\left[S(t_k)\tilde{P}(t_k) + I\right]^{-1}
\end{aligned}
$$

$$
\begin{aligned}
P''(t_k) &= \tilde{P}''(t_k) - K'(t_k)S(t_k)\tilde{P}'(t_k) \\
P'(t_k) &= \tilde{P}'(t_k) - K(t_k)S(t_k)\tilde{P}'(t_k) \\
P(t_k) &= \tilde{P}(t_k) - K(t_k)S(t_k)\tilde{P}(t_k)
\end{aligned}
$$

$$
\begin{aligned}
\hat{\mathbf{X}}'(t_k) &= \tilde{\mathbf{X}}'(t_k) + K'(t_k)\mathbf{Z}(t_k) \\
\hat{\mathbf{X}}(t_k) &= \tilde{\mathbf{X}}(t_k) + K(t_k)\mathbf{Z}(t_k)
\end{aligned}
$$

Figure 10.3: Second-order Kalman filter for image sequences.

t_0, t_{-1}, that is $\mathbf{X}(t_0)$ and $\mathbf{X}'(t_0)$ (recall $\mathbf{X}'(t_0) \equiv \mathbf{X}(t_{-1})$). Alternatively, it can be thought of as specifying a prior distribution for initial position $\mathbf{X}(t_0)$ and velocity $\mathbf{V}(t_0) = \dot{\mathbf{X}}(t_0)$.

Known, static, initial position. This is an appropriate assumption for interactive applications such as the hand-mouse of figure 1.16 on page 20 in which the hand might typically be inserted into a marked outline with shape $\mathbf{X}_0$, and held still while tracking is switched on. In that case, suitable initial conditions are given by the *closed-loop* steady state of the filter. This can be obtained most simply by running the tracking algorithm with constant "artificial" measurements $S(t_k) = S$, $\mathbf{Z}(t_k) = S\mathbf{X}_0$ in which S takes the value for a full set of successful image measurements in the curve-fitting algorithm. Under these circumstances, values of $P(t_\infty)$, $P'(t_\infty)$, $P''(t_\infty)$, and of $\hat{\mathbf{X}}(t_\infty), \mathbf{X}'(t_\infty)$ should settle to a steady state. This serves as a suitable initial condition once tracking is switched on by suspending the artificial measurements and allowing real image measurements to flow in.

Unknown initial position. If initial position is unknown, then the only available prior information is what is embodied in the dynamical model itself. Then the assumption is that, at switch-on, the object may be anywhere within the Gaussian envelope implied by the model. It amounts to setting initial conditions to the *open-loop* steady state, determined by running the dynamical model in the absence of measurement. This is obtained by running the tracker on artificial measurements, this time with $S(t_k) = 0$, $\mathbf{Z}(t_k) = \mathbf{0}$, until tracking is switched on as before. Note that the method can be used only if the AR process has a steady state and that excludes "constant velocity" models (see previous chapter, page 206).

Known initial velocity. In many applications a prior estimate of the initial velocity of the object is available, at least over a subspace of the shape-space. For example, observing plants on the ground from a moving tractor (figure 1.6 on page 10) or objects moving on a conveyor belt. In that case the estimated velocity is incorporated in a velocity offset vector $\mathbf{V}_0$ in shape-space and the initial state becomes:

$$\hat{\mathbf{X}}(t_0) = \mathbf{X}_0, \quad \hat{\mathbf{X}}'(t_0) = \mathbf{X}_0 - \mathbf{V}_0\tau.$$

with $\mathcal{P}$ initialised to the closed-loop steady-state value as before.

Alternative Kalman Filter

An alternative version of the Kalman filter is possible with an assimilation step that is equivalent to (10.6) but with a modified presentation. It is based on the alternative curve-fitting algorithm of chapter 6 (figure 6.10 on page 133), rather than the original recursive algorithm. The detailed implementation of the algorithm is omitted here. The comparison between the two algorithms, here in the dynamic case, is similar to that for the static case: for small N_X the original algorithm is more efficient, but as N_X increases the alternative algorithm becomes relatively more efficient.

Computational cost

The computational cost of the tracking algorithm is in fact $O(N_X^3)$ so that increasing the dimension N_X of shape-space carries a heavy cost penalty. The cost is comprised of two components. Prediction and assimilation (figure 10.3) incur a cost of $O(N_X^3)$ in the multiplication and inversion of the various $N_X \times N_X$ matrices. The generation of the aggregated measurements $\mathbf{Z}(t_k)$ also costs (argued in chapter 6) approximately $O(N_X^3)$.

10.2   Tracking performance

At last, all three components of a fully dynamic, active contour tracker are in place: visual measurement, the dynamic model and temporal filtering. The validation-gate mechanism of chapter 8, used in the recursive fitting algorithm (figure 8.6 on page 174) to acquire the aggregated observations, is crucial here. It relates the width of the search region to the position covariance P, now a time-varying quantity $\tilde{P}(t_k)$. Combining the three components in the algorithm of figure 10.3, and using a validation gate, produces a tracker that is robust in a number of respects.

Initialisation

One sense in which tracking is robust is that it is capable, in the initialisation phase, of "locking on" to an object: uncertainty in object position and shape is tolerated initially but rapidly resolved as the continuity of tracked motion is recognised. This is reflected in the collapsing search region in figure 10.4.

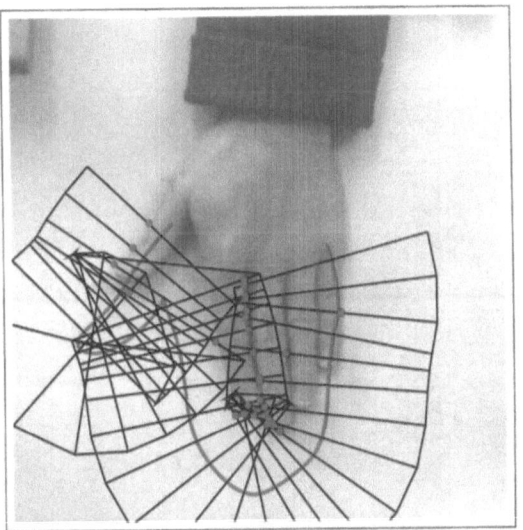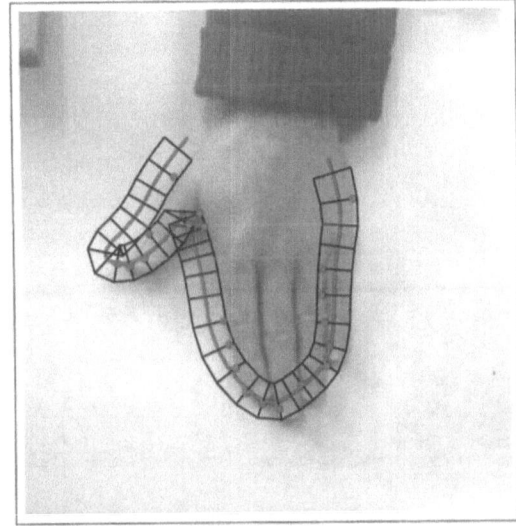

Figure 10.4: Locking on from the open-loop steady state. *The search region begins at its largest, reflecting initial uncertainty of location. It rapidly collapses as observations decrease the uncertainty.*

Adaptive search

If image observations fail along the contour, the search region expands over the affected segments until observations succeed again, as figure 10.5 shows. The extreme case is when observations fail along the entire length of the contour, in which case the search region regresses towards its open-loop steady state. Note the characteristically rapid recovery of tracking, in the figure, between $t = 0.16\,\text{s}$ and $t = 0.20\,\text{s}$. This is explained by the increased positional variance at $t = 0.16\,\text{s}$ ($\tilde{P}(t_k)$ in figure 10.3), leading to a correspondingly increased Kalman gain ($K(t_k)$). Once back in the steady state, covariance decreases and Kalman gain also decreases so that the relative influence of the predictive dynamical model becomes stronger again. To summarise, there is an inverse relationship between spatial scale and temporal scale. In the special case of constant-velocity dynamics, it is an inverse square law; the time-scale for smoothing varies as the inverse square of the spatial extent of the search region. This inverse relationship can be observed even within one contour, as figure 10.6 shows.

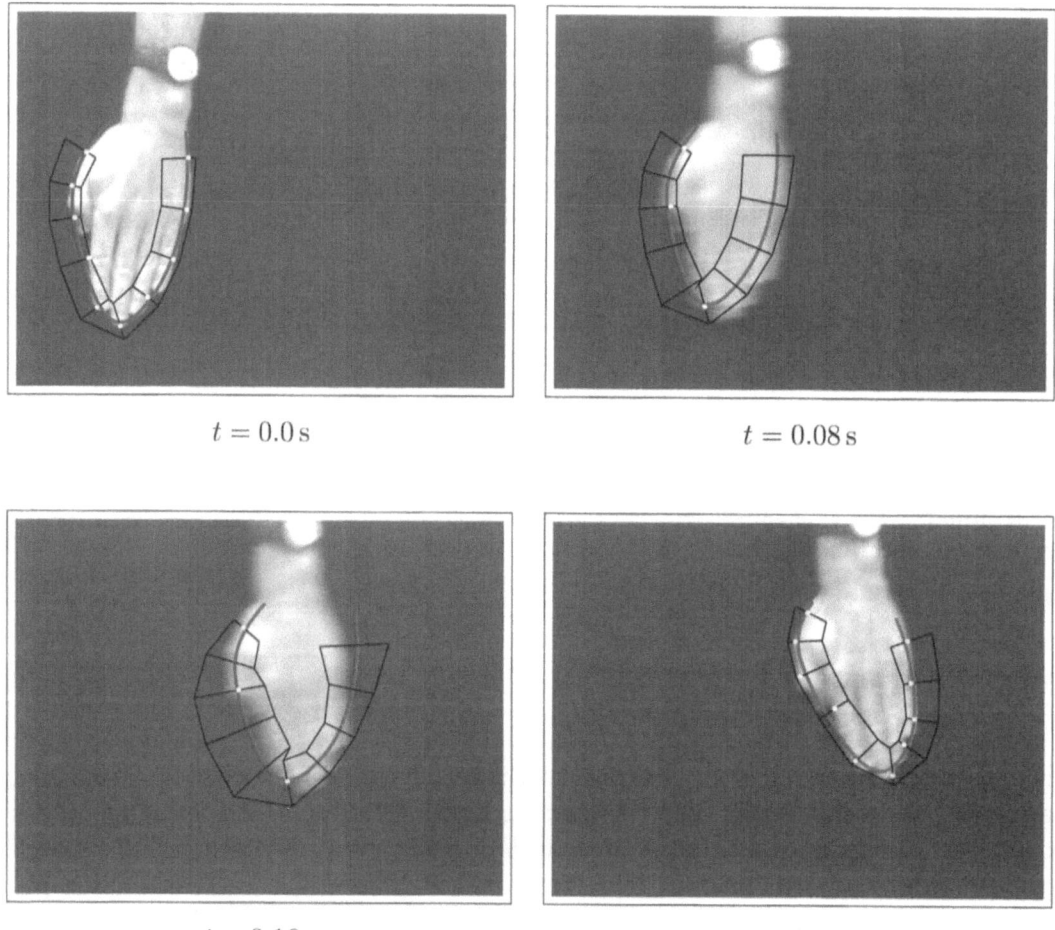

$$t = 0.0\,\text{s} \qquad\qquad t = 0.08\,\text{s}$$

$$t = 0.16\,\text{s} \qquad\qquad t = 0.20\,\text{s}$$

Figure 10.5: Recovery of lock. *As the hand begins to move, observations (white dots) succeed initially along all normals. As the accelerating hand leaves the dynamic contour lagging behind, measurements are lost and the search region automatically expands, reflecting increasing uncertainty. The expanded region now contains the edges of the hand and observations recover.*

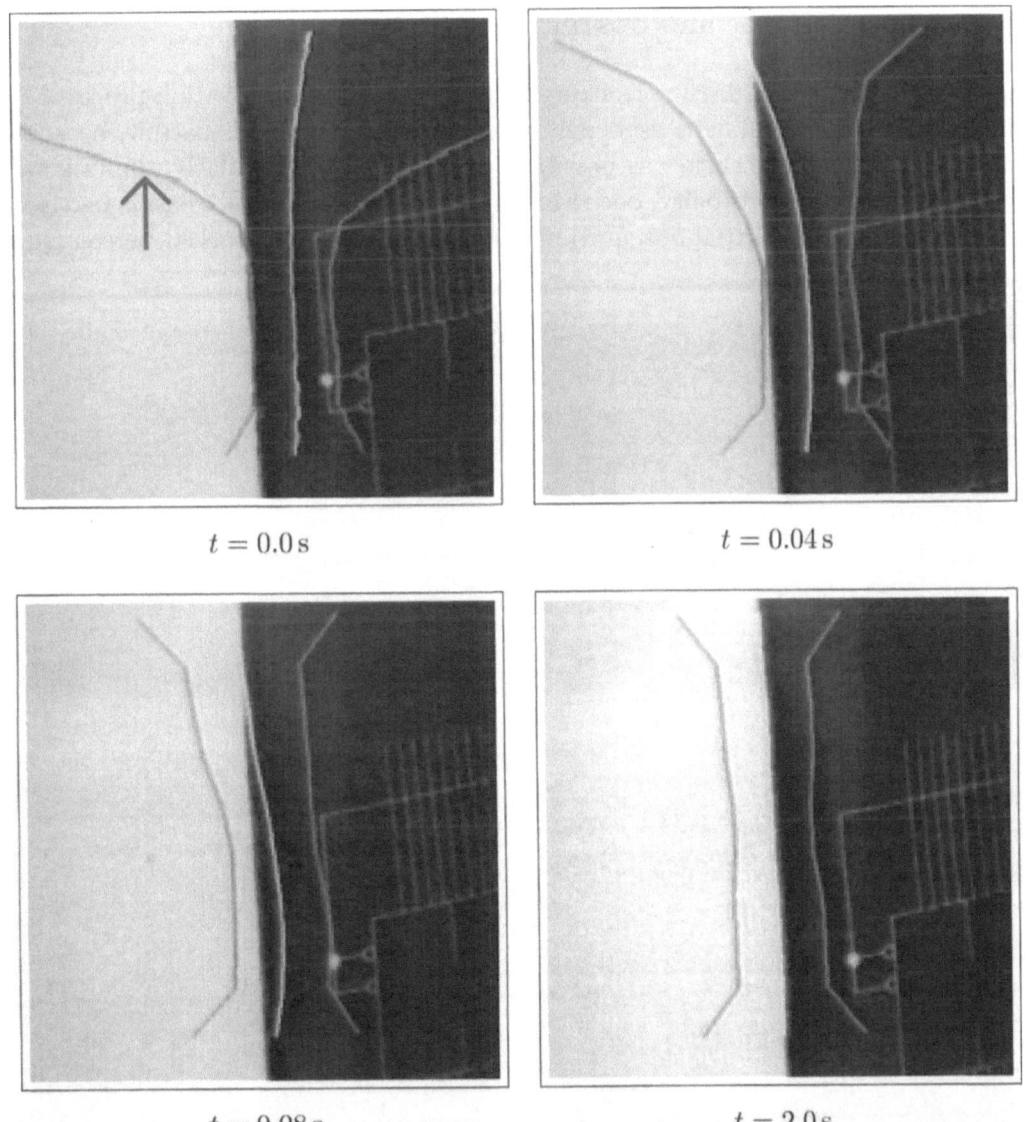

$t = 0.0\,\mathrm{s}$ $t = 0.04\,\mathrm{s}$

$t = 0.08\,\mathrm{s}$ $t = 2.0\,\mathrm{s}$

Figure 10.6: Temporal scale varies inversely with spatial scale *The search region (bounded by grey lines, arrowed in the first frame) is initialised with varying width. Where the search region is wide, temporal scale is short, and the contour deflects more rapidly, as shown. (Figure by courtesy of R. Curwen.)*

Resistance to clutter and obscuration

Background clutter can disrupt tracking by distracting the observation process. The validation gate is helpful here, especially when the search region is narrow, because it excludes as much of the clutter as possible, as in figure 10.7. Similarly, the validation gate helps with loss of visibility, occurring either because the object passes partly out of the image or from partial obscuration when, for instance, a tracked person passes

 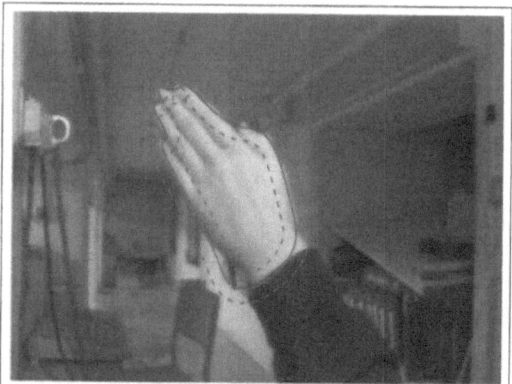

Figure 10.7: Resistance to clutter. *A hand accelerates across a cluttered background. Note that in the second frame the lower left corner of the contour is momentarily distracted by the chair but the disturbance is successfully filtered out over time. (Figure by courtesy of R. Curwen.)*

behind a lamp post. A good example is the tracking of fish viewed underwater, as in figure 10.8. Of course if the obscuring object is known, it can actually be *predicted* that measurements will fail over the corresponding image region. In that case attempts to measure ν_i along normals, in the curve-fitting algorithm, are suspended over the region that is predicted to be obscuring. Where the obscuring object has texture, which could generate false measurements, predicting the obscuration allows those false features to be suppressed.

Agile motion

Agility is a challenge because the continuity of motion is disrupted when sudden changes of speed and direction occur. The validation gate helps here also by signaling the loss of "lock" on the moving object and allowing the tracker to regress towards the "open-loop initialisation" state in which greater positional error is tolerated, as indicated by the widening of the search region. As recovery of lock proceeds, the tracker progressively returns to the steady state associated with continuous motion (figure 10.9). Figure 10.10 illustrates the robustness of agile tracking due the temporal variations of Kalman gain and search-region width.

10.3   Choosing dynamical parameters

This section outlines some design principles for building trackers. How can the parameters of a synthesised AR process of the sort described in section 9.6, and the observation parameters, be chosen to meet performance requirements? The AR parameters to specify are the frequency, damping rate and average displacement

$$f_i \, \mathrm{s}^{-1}, \quad \beta_i \, \mathrm{s}^{-1} \quad \text{and} \quad \overline{p}_i \, \text{pixels},$$

for each shape-subspace $\mathcal{S}_i$ (of dimension N_i) used. For the observation process, the parameter $\overline{p}_f$ must be specified, representing the average displacement error of image measurements along a curve.

AR parameters

In many applications it is reasonable to maintain the critical damping condition $f_i = 0$ so that there is a single, natural time-scale $1/\beta_i$ to be chosen for each space $\mathcal{S}_i$. In other

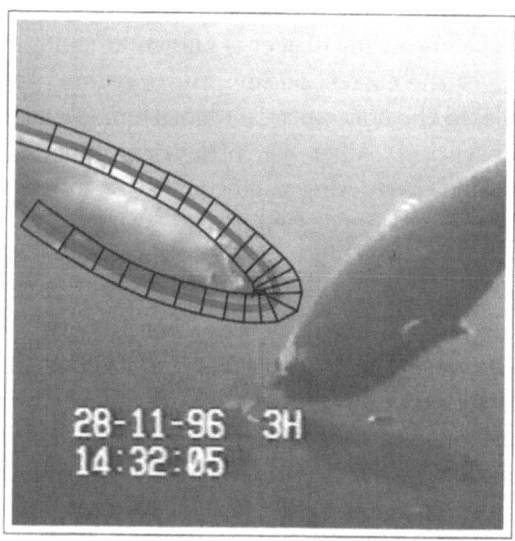

Figure 10.8: Expansion of the search region aids recovery from occlusion. *A tracked fish (top) passes behind another fish (left). The remaining successful observations, together with the dynamical model in shape-space, compensate for lost measurements around the head. Observations around the head resume as it re-emerges (right), one second later.*

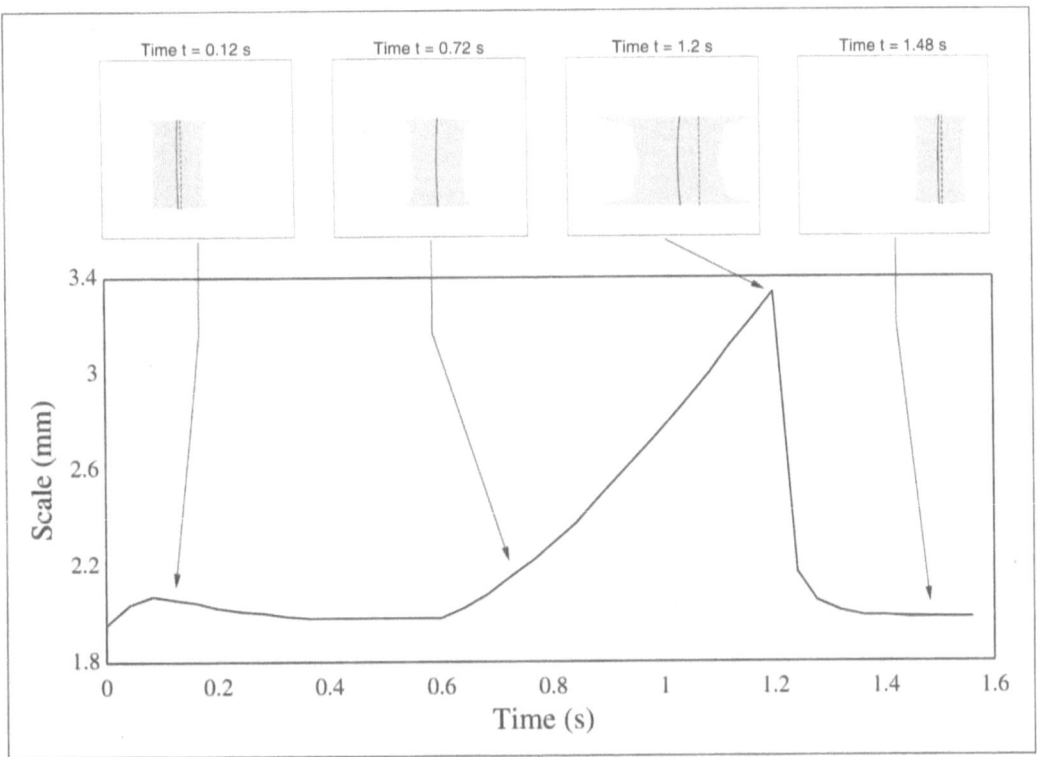

Figure 10.9: Loss and recovery of lock. *When an object falls outside its validation gate, owing to a sudden discontinuity of motion for example, the validation gate widens rapidly. Interestingly, in the case illustrated of a constant-velocity model, it can be shown that gate width grows in proportion to $t^{3/2}$. The expanding gate will therefore catch up with any object fleeing at a constant velocity. Once lock is recovered, the gate collapses rapidly. (Figure courtesy of Rupert Curwen.)*

applications, such as lip motion with its characteristic oscillations during speech, it is more appropriate to set $f_i > 0$ over shape-subspaces associated with lip deformation.

Once dynamical constants f_i and β_i are fixed, it remains to choose the average displacements $\bar{\rho}_i$. That fixes the open-loop search-region width, which is proportional to the average displacement $\bar{\rho}_o$, over the whole shape-space, under open-loop conditions:

$$\bar{\rho}_o = \sqrt{\sum_i \bar{\rho}_i^2}.$$

Snapshot at 10.0 seconds   Snapshot at 17.6 seconds

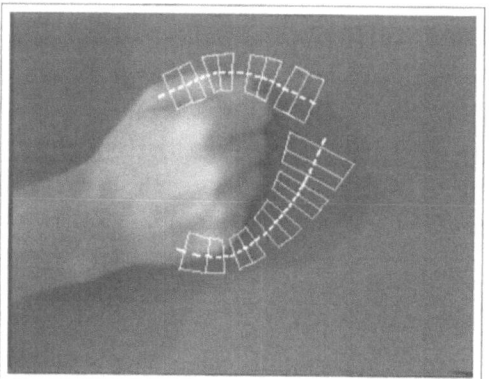 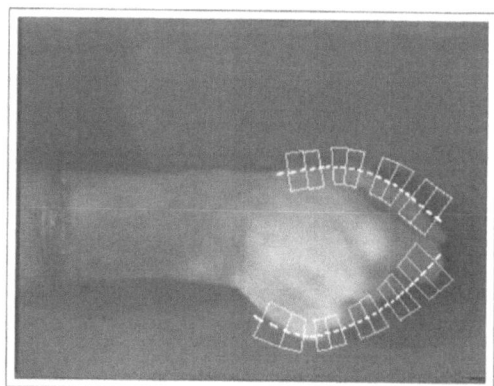

Varying gain and search region

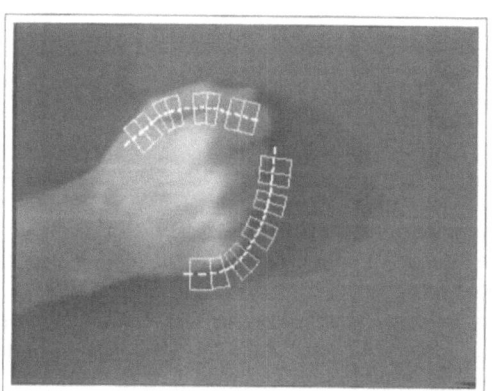 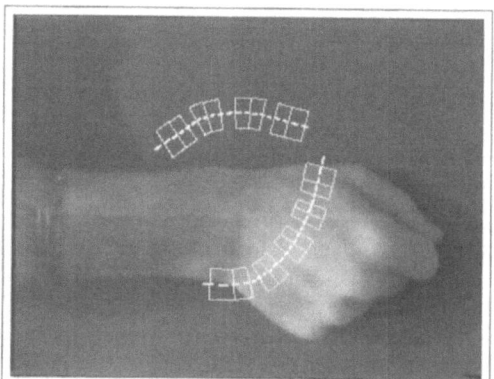

Fixed gain and search region

Figure 10.10: Validation gate strengthens tracking performance for agile motion.
The "chirp" test sequence used here, of 20 seconds duration, involves oscillatory motions of steadily increasing frequency. At around 10 seconds, accelerations are such that the tracked curve lags the hand appreciably. Variation of gain and search-region width allows lost lock to be recovered (left). However, if variation is inhibited by fixing covariance at P_∞ (right), lock is lost irretrievably. The time-course of the translational component of motion is shown in figure 10.11.

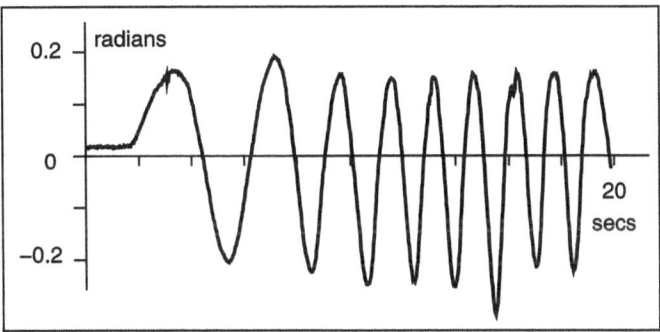

Varying gain and search region

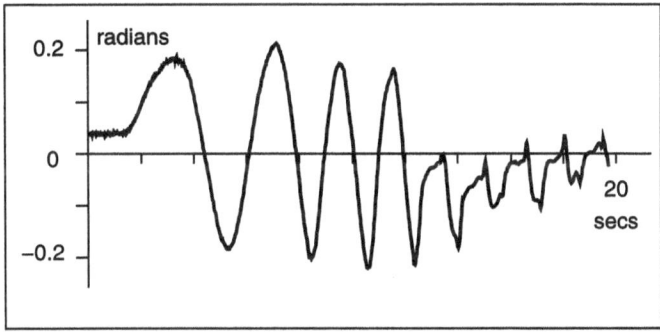

Fixed gain and search region

Figure 10.11: *Time-course of the translational component of motion for the test of figure 10.10. In the fixed gain case, the loss of lock after 12 seconds or so is clearly visible.*

For example, for the hand-mouse it is appropriate to use just two shape-subspaces $\mathcal{S}_s$ and $\mathcal{S}_d$ for translation and deformation respectively. Time constants of the order of a second are appropriate in both subspaces but $\bar{\rho}_s$ might be set an order of magnitude greater than $\bar{\rho}_d$ to reflect relatively large translational excursions across the field of view. The open-loop steady-state distribution will then have a correspondingly large average displacement from the template shape $\bar{\rho}_o \approx \bar{\rho}_s$, dominated by its translational component.

Observation error

Given complete image observations, the search region collapses to the size and shape determined by the prediction covariance $\tilde{P}(t_\infty)$ in the closed-loop steady state. The associated average displacement $\bar{\rho}_c$ relative to the predicted shape, given by

$$\bar{\rho}_c^2 = \text{tr}(\tilde{P}(t_\infty)\mathcal{H}),$$

is determined by the setting of $\bar{\rho}_f$, together with the AR parameters above, and this gives a guide to the size of the search region during steady-state tracking. It is difficult to predict exactly what value $\bar{\rho}_c$ takes but in the case that $\bar{\rho}_f$ is small, including the case $\bar{\rho}_f = 0$,

$$\bar{\rho}_c \approx \sqrt{\text{tr}(B_0 B_0^T \mathcal{H})}, \tag{10.8}$$

approximately independent of the precise value of $\bar{\rho}_f$. For the partitioned harmonic model this gives

$$\bar{\rho}_c \approx \sqrt{\sum_i (b_0^i)^2 N_i}. \tag{10.9}$$

For example, in the case of simple harmonic motion over an unpartitioned shape-space this becomes

$$\bar{\rho}_c \approx b_0 \sqrt{N_X}$$

which can be related to $\bar{\rho}_c$ under the continuous-time assumption (9.27) to give

$$\frac{\bar{\rho}_c}{\bar{\rho}_o} = 2 \left(\beta \left(\beta^2 + 2\pi^2 f^2 \right) \right)^{1/2} \tau^{3/2}. \tag{10.10}$$

This is a useful design rule, giving the factor by which the search region contracts between the open and closed-loop conditions. Note that this case of *exact* measurement is extreme; as measurements become less precise, the ratio between search-region dimensions in the open condition versus the closed one decreases.

Setting $\bar{\rho}_f$ to a small value also ensures accurate position estimates — average positional error is bounded by $\bar{\rho}_f$ in general — but only if observations are successful. If the background is clutter-free, tracking may succeed with a small $\bar{\rho}_f$, and estimated positions will be accurate. If there is significant background clutter, this must be reflected by assuming a larger value of $\bar{\rho}_f$, the average observation error. The effect is to widen the search region, tolerating greater levels of clutter-induced observation error, at the expense of less accurate position estimates.

10.4   Case study

To conclude the chapter, typical, practical settings of dynamical parameters are given for an application — the digital desk. First the case of tracking hand motion over a clean desk is given, and this applies also to a cluttered desk in which the background has been modelled and can be discounted. The aim is to achieve tracking of the agile motions that occur in drawing and pointing gestures. Then the more difficult problem of tracking over unmodelled clutter is addressed.

Three different settings of dynamical parameters, in a shape-space of Euclidean similarities, are illustrated in figures 10.12 and 10.13: "slow," "fast" and "tight."

"Slow" dynamics are:

translation:

$$f = 0, \ \beta = 2\,\mathrm{s}^{-1} \ \text{ and } \ \overline{\rho}_o = 2000\,\text{pixels},$$

which is critically damped, slow given its $\frac{1}{2}$ second time-constant, and very loosely constrained spatially, to allow free rein over the 500 pixel extent of the image.

rotation/scaling:

$$f = 0, \ \beta = 10\,\mathrm{s}^{-1} \ \text{ and } \ \overline{\rho}_o = 50\,\text{pixels},$$

which is tightly constrained, reflecting relatively strong prior knowledge about the shape of the outline.

Observed features are edges detected with a Gaussian operator of width 2 pixels, and with a contrast threshold of 8 grey-levels. Typically several features are detected along each search line; the one with strongest contrast is chosen but with a modest bias for zero innovation. Observation uncertainty is set to

$$\overline{\rho}_f = 5\,\text{pixels},$$

a reasonable value to allow for measurement error, unmodelled shape variations, and the possibility of reporting clutter as an object feature.

"Fast" More agile performance is allowed by increasing the damping rate for translation to

$$\beta = 5\,\mathrm{s}^{-1}.$$

This has the side-effect of increasing the uncertainty in predicted position at each time-step and increasing the width of the search region, increasing the tendency for distraction by spurious features internal to the hand. To compensate, account is taken of contrast polarity, accepting only light to dark transitions on the hand outline. To help further, the contrast threshold is increased to 25 grey-levels, capitalising on the darkness of the background. Now tracking is fast and accurate.

"Tight" A cluttered background (figure 10.13) that is entirely static can be modelled and suppressed, and the problem is essentially no harder than before. Otherwise, clutter seriously distracts the "fast" tracker, so that dynamical parameters need to be tightened up for tracking to work at all. The clutter problem is exacerbated further by the fact that contrast with the desk is no longer so pronounced; the contrast threshold has to be increased to 12 grey-levels, and this increases the effective density of clutter. What is more, polarity of contrast can no longer be relied upon, since some of the desktop is darker than the hand and some lighter. An alternative that works well is to use colour (chapter 5), relying on the distinctive pink hue of skin to discriminate from the background. Sensitivity to clutter needs to be reduced further. This is done by reducing freedom to rotate/scale, by setting

$$\bar{\rho}_o = 25 \, \text{pixels},$$

half its previous value. The search region along each line is artificially restricted to 60% of its normal value by reducing the length factor κ in the validation test of figure 8.6 from the usual $\kappa = 2$ to $\kappa = 1.2$. Now slower hand motions can be tracked satisfactorily over clutter.

To summarise, good performance can be obtained with manually set dynamics, for tracking with a clean background, or a background that is cluttered but static so that it can be discounted. If clutter persists, performance is more limited. In that case, the next chapter shows how to set more acutely tuned dynamical parameters, using examples of motion tracked against a clean background for training. This helps some-what to deal with the cluttered background by strengthening prior knowledge of shape and motion. More radically, the methods of chapter 12 offer powerful, non-Gaussian methods to handle clutter which are very effective but more costly computationally.

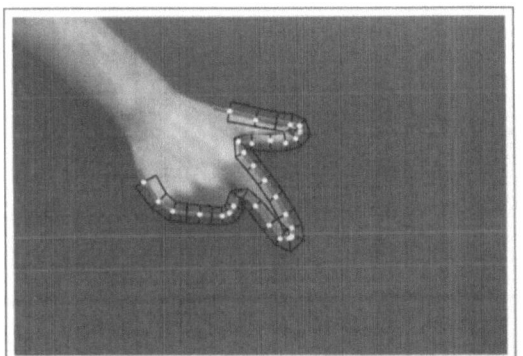 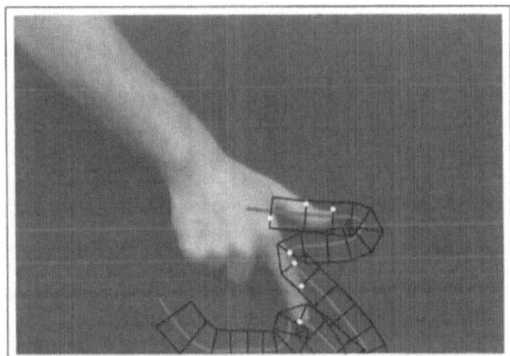

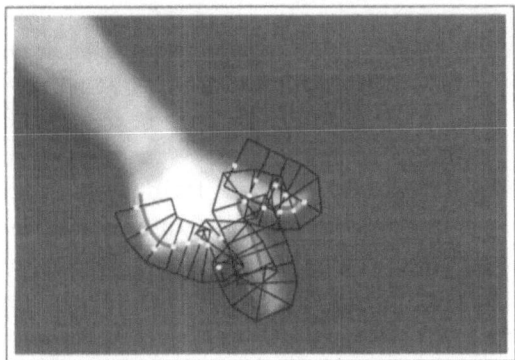

Figure 10.12: Case study: setting up for the digital desk. *"Slow" tuning (top) is stable, works well for gentle motions, but is derailed by faster ones. "Fast" tuning (bottom) is not quite so stable but is capable of following rapid motions — note the motion blur in the still image.*

Bibliographic notes

Established uses of Kalman filtering in machine vision include (Bolles et al., 1987; Broida and Chellappa, 1986; Bolles et al., 1987; Dickmanns and Graefe, 1988b; Matthies et al., 1989; Gennery, 1992). Terzopoulos and Szeliski (1992) recognised the connection between snakes with dynamics and the formalism of the Kalman filter. Low-dimensional curve parameterisations such as the B-spline make it practicable to implement a snake as a Kalman filter (Blake et al., 1993). The idea of using a spatially extended validation gate in a contour tracker was proposed and developed in (Curwen and Blake, 1992; Blake et al., 1993).

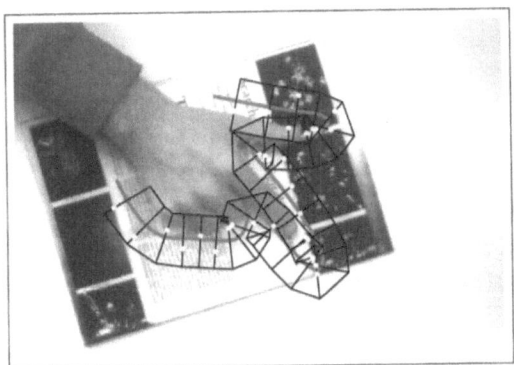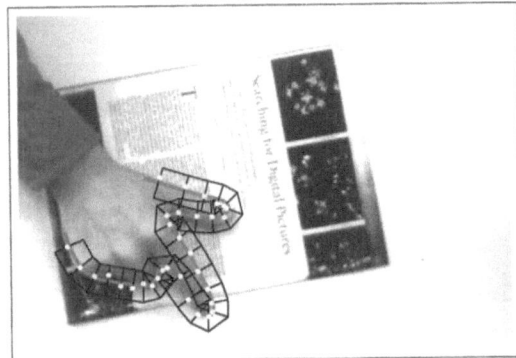

Figure 10.13: Case study: heavy background clutter. *The "fast" tracker of figure 10.12 is too easily distracted by clutter (left) to be used successfully against an unmodelled background of text and pictures. "Tight" parameter settings (right) secure adequate stability, but can only track slower motions.*

In its standard form, the Kalman filter is formulated in terms of Kalman gain (Gelb, 1974), which is derived from the information weighted mean using the "matrix inversion lemma" (Bar-Shalom and Fortmann, 1988). The form of Kalman filter used in this chapter, that avoids singularity problems when measurements are degenerate or fail altogether, was taken from (Harris, 1992b) in which Kalman filtering was applied to the non-linear parameters for position/orientation of a single, polyhedral rigid body. It achieved impressive results in terms of efficiency and agility of tracked motion. Lowe (1992) addressed the related problem of tracking an articulated pair of rigid bodies, in terms of its extended, non-linear parameterisation.

Chapter 11

Learning motion

In the previous chapter, dynamic contour tracking was based on prediction using dynamical models of the kind set out in chapter 9. The parameters of the models were fixed by hand to represent plausible motions such as constant velocity or critically damped oscillation. Experimentation allows these parameters to be refined by hand for improved tracking but this is a difficult and unsystematic business, especially in high-dimensional shape-spaces which may have complex couplings between the dimensions. What is far more attractive is to learn dynamical models on the basis of training sets. Initially, a hand-built model is used in a tracker to follow a training sequence which must be not too hard to track. This can be achieved by allowing only motions which are not too fast, and limiting background clutter or eliminating it using background subtraction (chapter 5). Once a new dynamical model has been learned, it can be used to build a more competent tracker, one that is specifically tuned to the sort of motions it is expected to encounter. That can be used either to track the original training sequence more accurately, or to track a new and more demanding training sequence, involving greater agility of motion. The cycle of learning and tracking is described in figure 11.1. Typically two or three cycles suffice to learn an effective dynamical model.

In mathematical terms, the problem is to estimate the coefficients A_1, A_2, $\overline{\mathbf{X}}$ and B_0 which best model the motion in a training sequence of shapes $\mathbf{X}_1, \ldots, \mathbf{X}_M$, where now $\mathbf{X}_k \equiv \mathbf{X}(t_k)$, gathered at the image sampling frequency. A general algorithm to do this is described below. Note that the learning algorithm as presented treats estimated shape-vectors $\mathbf{X}_k$ in a training sequence as if they were exact observations of the physical process, rather than noisy estimates obtained from a visual tracker. In

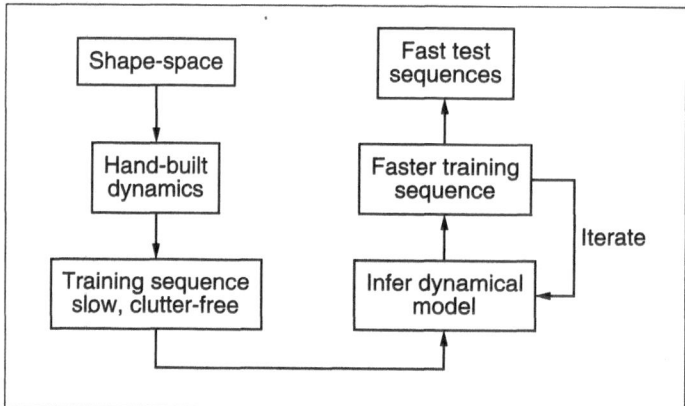

Figure 11.1: Iterative learning of dynamics. *The model acquired in one cycle of learning is installed in a tracker to interpret the training sequence for the next cycle. The process is initialised with a hand-built tracker of the sort described in chapter 9 and 10.*

practice this often works quite well but can give surprising results with highly periodic training motions, and this is discussed later.

11.1   Learning one-dimensional dynamics

First a learning algorithm is described for the simple case of a particle in one dimension — no curve or splines are involved here for the sake of tutorial simplicity, just one number describing the position of the particle along a rail. The problem is to estimate a discrete-time, second order model for particle position x_k so the state-space is defined in terms of $\mathbf{X}_k \equiv x_k$. Then the dynamical coefficients A_1, A_2 and B_0 become scalars denoted a_1, a_2 and b_0. For simplicity, assume that the mean $\bar{x} = 0$ is known, so that the quantity $x_k - a_2 x_{k-2} - a_1 x_{k-1}$ is an independent, zero-mean, scalar, normal variable $b_0 w_k$, for each k, with unknown variance b_0^2. The algorithm is summarised in figure 11.2.

The problem is expressed in terms of a "log-likelihood" function, defined up to an additive constant by

$$L(x_1, \ldots, x_k | a_1, a_2, b_0) \equiv \log p(x_1, \ldots, x_k | a_1, a_2, b_0) + \text{const}$$

where, since the w_k are independent,

$$p(x_1, \ldots, x_k | a_1, a_2, b_0) \propto \prod_k p_{b_0 w_k}(x_k - a_2 x_{k-2} - a_1 x_{k-1})$$

so, using the fact that the $p_{b_0 w_k}(\cdot)$ are standard normal distributions,

$$L(x_1, \ldots, x_k | a_1, a_2, b_0) = -\frac{1}{2 b_0^2} \sum_{k=3}^{M} (x_k - a_2 x_{k-2} - a_1 x_{k-1})^2 - (M-2) \log b_0, \tag{11.1}$$

up to an additive constant. Maximum Likelihood Estimates (MLE) for the coefficients a_1, a_2 and b_0 are obtained by maximising the function L, and this is straightforward because L is quadratic. First, the maximisation over b_0 factors out, leaving estimates $\hat{a}_1$ and $\hat{a}_2$ to be determined by minimising

$$\sum_{k=3}^{M} (x_k - a_2 x_{k-2} - a_1 x_{k-1})^2$$

whose derivatives with respect to a_1 and a_2 are set to 0 to give $\hat{a}_1, \hat{a}_2$ as the solution of the simultaneous equations in step 2 of the algorithm in figure 11.2. Now a_1 and a_2 can be regarded as constants in L, fixed at their estimated values, and L can be maximised with respect to b_0 to give $\hat{b}_0$:

$$\hat{b}_0^{\,2} = \frac{1}{M-2} \sum_{k=3}^{M} (x_k - \hat{a}_2 x_{k-2} - \hat{a}_1 x_{k-1})^2 \tag{11.2}$$

which, it can be shown, can be computed directly in terms of auto-correlation coefficients, as in step 3 of the algorithm.

Exercising the learning algorithm

Talking lips

Learned univariate motion is illustrated by the example used earlier, in chapter 9, of the opening/shutting motion of a pair of talking lips. Once the learning is done, the learned model $(\hat{a}_2, \hat{a}_1, \hat{b}_0)$ can be displayed in spectral form, as was done in chapter 9 for models fit by hand. Figure 11.3 shows that the spectrum of the learned model fits fairly well, having a resonant frequency of around 1 Hz, which is a little higher

---

Dynamical learning problem

Given a training set $\{x_1, \ldots, x_M\}$ of shapes from an image sequence, learn the parameters a_1, a_2, and b_0 for a second-order AR process that describes the dynamics of the moving shape.

Algorithm

1. First, auto-correlation coefficients r_{ij} for $i, j = 0, 1, 2$ are computed:

$$r_{ij} = \sum_{k=3}^{M} x_{k-i} x_{k-j} \ \text{ for } \ i, j = 0, 1, 2$$

2. Estimated parameters $\hat{a}_1$ and $\hat{a}_2$ are obtained by solving the simultaneous equations

$$
\begin{aligned}
r_{02} - \hat{a}_2 r_{22} - \hat{a}_1 r_{12} &= 0 \\
r_{01} - \hat{a}_2 r_{21} - \hat{a}_1 r_{11} &= 0,
\end{aligned}
$$

3. The covariance coefficient b_0 is estimated as

$$\hat{b_0}^2 = \frac{1}{m-2} \left(r_{00} - \hat{a}_2 r_{20} - \hat{a}_1 r_{10} \right)$$

Figure 11.2: **Algorithm for learning one-dimensional dynamics.**

than the peak of $0.8\,\mathrm{Hz}$ in the spectrum of the data. The hand-fitted spectrum in figure 9.11 on page 201 was constrained to have its resonant peak coinciding with that of the data. However, as a trade-off for the shift in its resonant peak, the learned model fits the high frequency spectrum rather better. The learned model should also capture some of the phase coherence of the data, something that was not taken into account in the hand-fitted model. In fact, the learned deterministic parameters are

$$f = 0.95\,\mathrm{Hz}, \quad \beta = 2.9\,\mathrm{s}^{-1},$$

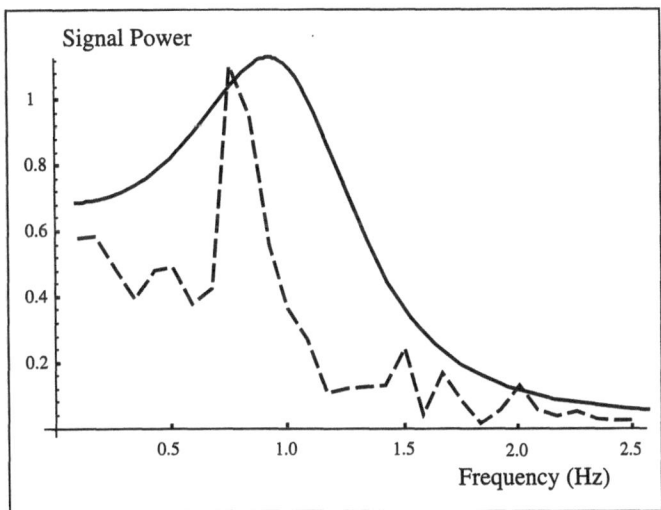

Figure 11.3: Power spectrum of learned lip motion *The power spectrum of the learned model (solid) fits that of the lip-motion training data (dashed) quite well.*

giving the coherence time-constant $1/\beta = 0.35$ s for the process, short enough to indicate that successive cycles in the signal are largely uncorrelated.

As in chapter 9, the learned model can be simulated and compared with the data. The result (figure 11.4) appears, to the naked eye, to be as plausible a replica of the data as the simulation of the hand-fitted model was (figure 9.12 on page 202).

Beating heart

It is instructive to see how the learning algorithm deals with training data that is highly periodic and coherent. Of course, perfectly coherent, periodic data of frequency f_0 can be represented by a second-order ARP with $f = f_0$, $\beta = 0$ and with $b_0 = 0$ — zero driving noise, so that the process is entirely deterministic. The behaviour of such a process is solely determined by initial conditions: once launched, the process follows an entirely predictable trajectory. Such a model, if used as a predictor for tracking, would have the strange effect that the gain of the Kalman filter would fall to 0 in the steady state, because the strength of the predictions would overwhelm the observations. Consequently observations would be ignored.

What is more realistic is *almost* periodic training data such as that from the

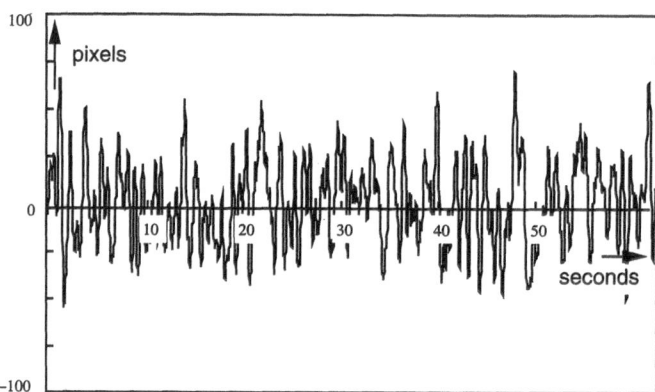

Figure 11.4: Simulation of learned lip motion *A sample path for the learned process is shown here and appears plausibly comparable with the original signal in figure 9.8 on page 198.*

beating heart in figure 9.7 on page 197. It appears to be a regular, periodic process with some additive noise, attributable perhaps to noise in observations. As noted earlier, the learning algorithm neglects to allow for observation noise and it is with highly periodic data that the discrepancy introduced by this erroneous assumption is most evident. Figure 11.5 shows a simulation of a learned dynamical model, exhibiting the right kind of motion over short time-scales (of the order of half a period), but over longer time-scales the phase coherence of the original signal is lost. The deterministic parameters of the learned model are

$$f = 1.12 \, \text{Hz}, \quad \beta = 3.5 \, \text{s}^{-1},$$

representing an oscillation of period $1/f = 0.90$ s which matches closely the average period 0.87 s of the data (based on zero crossings). However, the damping represented by the factor β is strong, representing a decay in amplitude by a factor of 5 over one half period. This corresponds to the loss of coherence apparent in the simulation over times greater than one half period or so. In fact the learned model, despite its imperfection, is still useful for tracking because it works well as a predictor over shorter time-scales, and visual observations carry the signal over longer time-scales. Moreover, the imperfection can be remedied by recourse to a more elaborate learning procedure using "Expectation–maximisation" or EM (see bibliographic notes) which takes explicit account of observation noise.

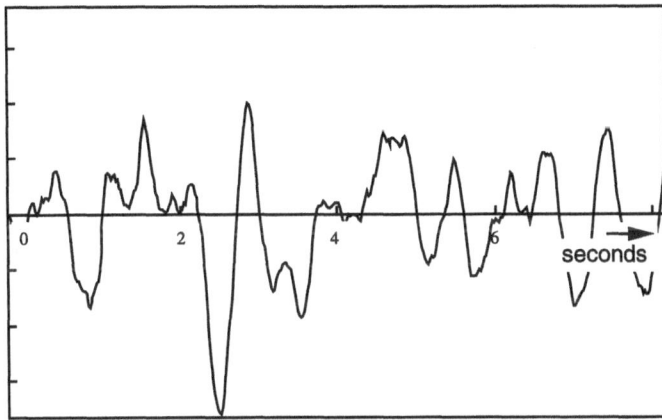

Figure 11.5: Simulating the learned motion of a beating heart. *A dynamical model learned from the training data of figure 9.7 on page 197 is simulated here. Limitations of the learning algorithm (see text) means that although the periodic nature of the training data is represented (each half-cycle is roughly the right size) its phase coherence is lost.*

Learning accuracy and training-sequence duration

Intuitively, a longer training sequence should give more reliable learned dynamical parameters. This is indeed the case, and the nature of the influence of the number of frames M on the accuracy of learned parameters a_1, a_2 and b_0 can be made precise. In fact what is most relevant is to quantify the effect of training-set duration $T = (M-2)\tau$ on the parameters β, f and $\bar{p}$ of the underlying continuous process. In this way, a characterisation of learning performance is obtained that is independent of sampling rate τ. The characterisation is valid under the assumptions that $\beta\tau \ll 1$ and that the process is observed in its steady state. A derivation is given in appendix B.

The principal result is that the proportional error in the parameters β and $\bar{p}$ is of order $1/\sqrt{\beta T}$, and that the proportional error in f relative to β is also of order $1/\sqrt{\beta T}$. As expected, accuracy varies as the square root of training-set size, the usual statistical phenomenon of error averaging. What may be more surprising at first sight is that the error depends not on M directly, but on βT which can be thought of as the number of *independent chunks* in the training signal. This is because $1/\beta$ is a "coherence time," the duration over which parts of the signal are correlated. In the lip-motion example above, for instance, the coherence duration is $1/\beta = 0.35\,\text{s}$ and the total duration of the training set (figure 9.8 on page 198) is 60 seconds. In that case $\beta T = 171$ so that the proportional error in the continuous dynamical parameters should be

$$100\sqrt{\frac{60}{0.35}}\% \approx 7.6\%.$$

For the beating heart the coherence duration is 0.29 s but the training sequence lasts only 8 seconds, giving a bound on error in dynamical parameters of

$$100\sqrt{\frac{8.0}{0.29}}\% \approx 19\%.$$

11.2 Learning AR process dynamics in shape-space

The general multi-variate learning algorithm follows broadly the line of the univariate one, but the separability of the estimation of deterministic and stochastic parameters, although it still holds, is no longer so obvious. Furthermore, it will no longer be assumed that the mean $\overline{\mathbf{X}}$ is known, so that it also must be learned. The log-likelihood function for the multi-variate normal distribution is then, up to a constant:

$$L(\mathbf{X}_1, \dots \mathbf{X}_M | A_1, A_2, C, \overline{\mathbf{X}}) = \tag{11.3}$$

$$-\frac{1}{2}\sum_{k=3}^{M}\left|B_0^{-1}\left(\mathbf{X}'_k - A_2\mathbf{X}'_{k-2} - A_1\mathbf{X}'_{k-1}\right)\right|^2 - (M-2)\log\det B_0$$

where

$$\mathbf{X}'_k = \mathbf{X}_k - \overline{\mathbf{X}}, \tag{11.4}$$

and

$$C = B_0 B_0^T. \tag{11.5}$$

The problem is to estimate A_1, A_2, $\overline{\mathbf{X}}$ and C by maximising the log-likelihood L. This is a non-linear problem because, unlike the simple case considered earlier in which the mean was fixed at $\mathbf{0}$, the mean $\overline{\mathbf{X}}$ now has to be estimated. This means that the likelihood is quartic (rather than quadratic) in the unknowns, owing to the product terms $A_2\overline{\mathbf{X}}$ and $A_1\overline{\mathbf{X}}$ that appear inside the $|\cdot|^2$ term in (11.3). The non-linearity can be removed by using the alternative form for the AR process from chapter 9 (equation (9.6) on page 193) in which

$$\mathbf{D} = (I - A_2 - A_1)\overline{\mathbf{X}}$$

so that the likelihood becomes

$$L(\mathbf{X}_1 \dots \mathbf{X}_M | A_1, A_2, C, \mathbf{D}) = \qquad\qquad (11.6)$$

$$-\frac{1}{2} \sum_{k=3}^{M} \left| B_0^{-1} \left(\mathbf{X}_k - A_2 \mathbf{X}_{k-2} - A_1 \mathbf{X}_{k-1} - \mathbf{D} \right) \right|^2 - (M-2) \log \det B_0$$

which is then quadratic in A_2, A_1 and $\mathbf{D}$.

Minimising the log-likelihood L leads to the learning algorithm of figure 11.6 which estimates the dynamical parameters A_2, A_1, $\mathbf{D}$ and C. It is clearly a generalisation of the univariate algorithm (figure 11.2). A derivation (optional) is given later.

Example: learning the dynamics of writing

As an illustration of multi-variable learning, the written name of figure 9.2 on page 187 is used as training data, to learn a dynamical model for finger-writing. A good test for the plausibility of a learned model is to simulate it randomly, as was done in chapter 9 for synthesised dynamical models. A random simulation of the model for finger-writing is demonstrated in figure 11.7. The resulting scribble has characteristics consistent with the training set in terms of direction and the size, shape and frequency of excursions. Note that the dynamical model was actually learned in the affine space for the finger outline and that the simulation therefore contains (minor) changes of finger shape. The figure shows just the translational components, which account for most of the motion in this case. (The learned model specifies dynamics in the steady state, but not initial conditions; the *speed* at which the hand drifts across the page is left indeterminate by the model. In this demonstration, the handwriting simulation was initialised with zero velocity, and then superimposed on a constant velocity drift roughly matching that in the training set.)

Modular learning: aggregation of training sets

It is often convenient to collect several training sets and to construct a dynamical model which explains the motion in all of the training sets taken jointly. For example, it might be desired to broaden the finger-writing model to cover writing in various directions, not just the single direction in the training set illustrated. Alternatively, when learning lip dynamics (see below) it might be convenient to construct separate training sets for rigid motion of the head, and deformation during speech. In either case it is *incorrect* simply to concatenate the training sets and treat them as one long

---

Dynamical learning problem

Given a training set $\{\mathbf{X}_1, \ldots, \mathbf{X}_M\}$ of shapes from an image sequence, learn the parameters A_1, A_2, B_0 and $\overline{\mathbf{X}}$ for a second-order AR process that describes the dynamics of the moving shape.

Algorithm

1. First, sums R_i, $i = 0, 1, 2$ and auto-correlation coefficients R_{ij} and R'_{ij}, $i, j = 0, 1, 2$ are computed:

$$R_i = \sum_{k=3}^{M} \mathbf{X}_{k-i}, \quad R_{ij} = \sum_{k=3}^{M} \mathbf{X}_{k-i} \mathbf{X}_{k-j}^T, \quad R'_{ij} = R_{ij} - \frac{1}{M-2} R_i R_j^T.$$

2. Estimated parameters $\hat{A}_1$, $\hat{A}_2$ and $\hat{\mathbf{D}}$ are given by

$$\hat{A}_2 = \left(R'_{02} - R'_{01} R'^{-1}_{11} R'_{12} \right) \left(R'_{22} - R'_{21} R'^{-1}_{11} R'_{12} \right)^{-1}$$

$$\hat{A}_1 = \left(R'_{01} - \hat{A}_2 R'_{21} \right) R'^{-1}_{11}$$

$$\hat{\mathbf{D}} = \frac{1}{M-2} \left(R_0 - \hat{A}_2 R_2 - \hat{A}_1 R_1 \right).$$

3. If required for the standard form of the AR process, the mean $\overline{\mathbf{X}}$ is estimated from

$$\hat{\overline{\mathbf{X}}} = (I - \hat{A}_2 - \hat{A}_1)^{-1} \hat{\mathbf{D}}.$$

4. The covariance coefficient B_0 is estimated as a matrix square root $\hat{B}_0 = \sqrt{\hat{C}}$ where

$$\hat{C} = \frac{1}{M-2} \left(R_{00} - \hat{A}_2 R_{20} - \hat{A}_1 R_{10} - \hat{\mathbf{D}} R_0^T \right).$$

Figure 11.6: Algorithm for learning multi-variate dynamics.

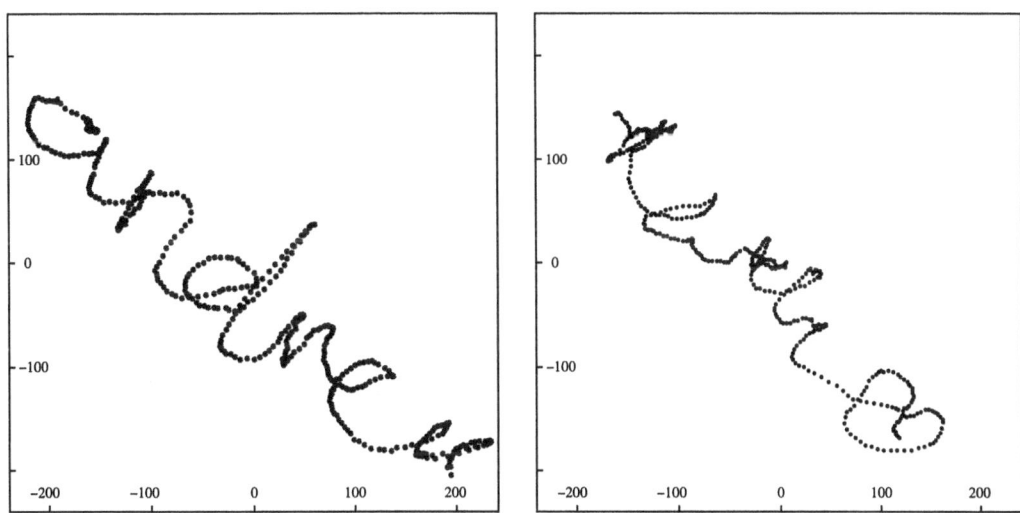

Figure 11.7: Scribbling: simulating a learned model for finger-writing. *A training set (left) consisting of six handwritten letters is used to learn a dynamical model for finger motion. A random simulation from the model (right) exhibits reasonable gross characteristics.*

training set because the abrupt transition at the join of the sets would be treated as genuine data, whereas in fact it is spurious. The solution is to compute autocorrelations individually for the training sets and to combine them in a linear fashion. Details of the combination method are omitted here, but see the bibliographic notes.

Derivation of the learning algorithm of figure 11.6.

Maximising likelihood L (11.6) first with respect to A_1, A_2 and $\mathbf{D}$, it will be shown that separability holds — the resulting estimates $\hat{A}_2$, $\hat{A}_1$ and $\mathbf{D}$ are independent of the value of C. To show this, a lemma is needed.

Lemma: Given a scalar function $f(Y)$ of a matrix Y of the form

$$f = \operatorname{tr}(KZ) \quad \text{with} \quad Z(Y) = YSY^T - YS'^T - S'Y^T,$$

in which K, S and S' are constant matrix coefficients and S and K are non-singular and symmetric, the value $Y = \hat{Y}$ at which $\partial f/\partial Y = 0$ is independent of K.
Proof: $Z(Y)$ can be rewritten

$$Z(Y) = (Y - \hat{Y})S(Y - \hat{Y})^T + \text{const} \quad \text{where} \quad \hat{Y} = S'S^{-1}$$

and then

$$dZ = dY \, S(Y - \hat{Y})^T + (Y - \hat{Y})S \, dY^T$$

and, since K is symmetric,

$$df = 2 \, \mathrm{tr} \left(K \, dY \, S(Y - \hat{Y})^T \right)$$

so that $df = 0$ for all dY iff $Y = \hat{Y}$, independent of the value of K.
□

Now we can proceed with maximising L, which is equivalent to minimising

$$f(A_1, A_2, \mathbf{D}) = \sum_{k=3}^{M} \left| B_0^{-1} \left(\mathbf{X}_k - A_2 \mathbf{X}_{k-2} - A_1 \mathbf{X}_{k-1} - \mathbf{D} \right) \right|^2 \qquad (11.7)$$

with respect to A_1 and A_2. This function can be expressed as

$$f(A_1, A_2, \mathbf{D}) = \mathrm{tr}(ZC^{-1})$$

where

$$Z = \sum_{k=3}^{M} \left(\mathbf{X}_k - A_2 \mathbf{X}_{k-2} - A_1 \mathbf{X}_{k-1} - \mathbf{D} \right) \left(\mathbf{X}_k - A_2 \mathbf{X}_{k-2} - A_1 \mathbf{X}_{k-1} - \mathbf{D} \right)^T$$

and $C = B_0 B_0^T$. Invoking the lemma three times, with Y as each of A_0, A_1 and $\mathbf{D}$ in turn (and with the other two treated as constant), indicates that we can effectively set $B_0 = I$, for the purposes of determining $\hat{A}_1$, $\hat{A}_2$ and $\mathbf{D}$ by finding the minimum of

$$\mathrm{tr}(Z) = \sum_{k=3}^{M} \left| \mathbf{X}_k - A_2 \mathbf{X}_{k-2} - A_1 \mathbf{X}_{k-1} - \mathbf{D} \right|^2 .$$

Setting to 0 the derivatives of $\mathrm{tr}(Z)$ with respect to A_0, A_1 and $\mathbf{D}$ respectively shows that the minimum must satisfy the simultaneous equations

$$\begin{aligned}
R_{02} - \hat{A}_2 R_{22} - \hat{A}_1 R_{12} - \hat{\mathbf{D}} R_2 &= 0 \qquad (11.8) \\
R_{01} - \hat{A}_2 R_{21} - \hat{A}_1 R_{11} - \hat{\mathbf{D}} R_1 &= 0 \\
R_0 - \hat{A}_2 R_2 - \hat{A}_1 R_1 - \hat{\mathbf{D}}(M - 2) &= 0,
\end{aligned}$$

where R_i and R_{ij} are defined as in step 1 of the algorithm. Note that a minimum of f must exist because it is quadratic and bounded below by 0.

Lastly, the simultaneous equations can be simplified by subtracting multiples of the third from the first and second to give

$$\begin{aligned}
R'_{02} - \hat{A}_2 R'_{22} - \hat{A}_1 R'_{12} &= 0 \\
R'_{01} - \hat{A}_2 R'_{21} - \hat{A}_1 R'_{11} &= 0 \\
R_0 - \hat{A}_2 R_2 - \hat{A}_1 R_1 - \hat{\mathbf{D}}(M - 2) &= 0
\end{aligned}$$

where R'_{ij} is defined as in step 1 of the algorithm. Now the first two equations can be solved directly for $\hat{A}_1$ and $\hat{A}_2$ which can then be substituted into the last one to give $\hat{D}$, all as in step 2 of the algorithm.

It remains to estimate B_0 which is obtained as the square root of $C = B_0 B_0^T$. Rewriting (11.6) as

$$L = -\frac{1}{2}\text{tr}(ZC^{-1}) + \frac{1}{2}(m-2)\log\det C^{-1},$$

fixing $A_2 = \hat{A}_2$, $A_1 = \hat{A}_1$ and $\mathbf{D} = \hat{\mathbf{D}}$, and extremising with respect to C^{-1} (using the identity $\partial(\det M)/\partial M \equiv (\det M)M^{-T}$) gives

$$\hat{C} = \frac{1}{M-2} Z(\hat{A}_2, \hat{A}_1, \hat{\mathbf{D}}), \tag{11.9}$$

which simplifies, using the equations of step 2 of the algorithm, to give the formula for $\hat{C}$ in step 4.

11.3   Dynamical modes

One of the advantages of the "state-space" form for the second-order AR process is that the characteristics of the underlying continuous process can readily be identified. The continuous-time interpretation consists of a number of "modes." Some are damped oscillations with the impulse response $\exp{-\beta t}\cos 2\pi f t$ that is characteristic of second-order motion. Others are simple decays, with impulse responses of the form $\exp{-\beta t}$, associated with first-order motion. Each mode exhibits a particular pattern of motion. For example, for a finger tracing out letters (figure 9.2 on page 187) one would expect a slow exponential associated with the gradual, lateral drift, a faster oscillatory mode associated with letter strokes, and various other "minor" modes with relatively rapid decay. Modes characterise the deterministic component of dynamics, indicative of the behaviour of a tracker when the observation process suddenly fails and tracking is left to rely on prediction. That leaves the stochastic component which is relevant under normal tracking conditions, and in particular when observations suddenly resume. The Gaussian envelope following an extended period of prediction is given by the steady-state covariance P_∞, which can be computed from A and B. In fact P_∞ represents the static prior for the dynamical model and as such is similar to the covariance $\overline{P}$ for the training set treated as a static set of shapes, as analysed by PCA in chapter 8.

Modal analysis

Modes of the underlying dynamical process are characterised as follows. First, eigenvalues λ_m and eigenvectors $\mathcal{X}_m^A$ for $m = 1, \ldots, 2N_X$ of the matrix A are computed. Any real, positive eigenvalue with $\lambda_m < 1$ corresponds to a exponentially decaying mode, as for first-order processes, of the form $\exp -\beta_m t$ where

$$\beta_m = \frac{1}{\tau} \log \frac{1}{\lambda_m}. \tag{11.10}$$

Note that if $\lambda_m > 1$ the mode is unstable because $\beta_m < 0$. The mode's pattern of motion in shape-space is given by $\mathbf{X}_m^A$ which is the upper half of the eigenvector

$$\mathcal{X}_m^A = \begin{pmatrix} \mathbf{X}_m^A \\ \mathbf{Y}_m^A \end{pmatrix}. \tag{11.11}$$

(Note that the upper and lower halves are related by $\mathbf{Y}_m^A = \lambda_m \mathbf{X}_m^A$.) Suppose, for an object in Euclidean similarity shape-space, and assuming a sampling interval $\tau = 1/50\,s$, that $\lambda_1 = 0.99$ and $\mathcal{X}_m^A = (1, 1, 0, 0, 0.99, 0.99, 0, 0)^T$. This implies a decaying exponential impulse response with

$$\beta_1 = 50 \log(1/0.99) = 0.50\,\mathrm{s}^{-1}$$

so that the characteristic decay time of the mode is $1/0.50 = 2.0\,\mathrm{s}$. The pattern of displacement for the mode is given by $\mathbf{X}_1^A = (1, 1, 0, 0)^T$ representing translation up and to the right.

Otherwise, eigenvalues can be negative or complex and represent harmonic motion of the form $\exp -\beta_m t \cos 2\pi f_m t$ where

$$\beta_m = \frac{1}{\tau} \log \frac{1}{|\lambda_m|} \tag{11.12}$$

$$f_m = \frac{1}{2\pi\tau} \arg \lambda_m \tag{11.13}$$

(where arg denotes the canonical argument of a complex number). Again, the mode-shape is conveyed by the upper half $\mathbf{X}_m^A$ of the corresponding eigenvector. In general, the elements of $\mathbf{X}_m^A$ are complex, their moduli indicating the amplitude of oscillation in each shape-space component while their complex arguments indicate their relative phases of oscillation. Note that complex eigenvalues must come in conjugate pairs, so each oscillatory mode is expressed in terms of *two* of the eigenvalues out of the $2N_X$ eigenvalues of the matrix A. This means that A can have $2N_X$ exponential modes, or just N_X oscillatory ones, or some combination of the two kinds.

As an example of an oscillatory mode, considering again the object in Euclidean similarity shape-space, $\lambda_2 = 0.95 + 0.24i$ and $\mathbf{X}_2^A = (-1, 1, 0, 0)^T$ imply a damped oscillatory impulse response for which

$$\beta_2 = 50\log\frac{1}{\sqrt{0.95^2 + 0.24^2}} = 1.02\,\mathrm{s}^{-1}$$

$$f_2 = \frac{50}{2\pi}\arctan(\frac{0.24}{0.95}) = 1.97\,\mathrm{Hz}$$

representing an oscillatory period of $1/1.97 = 0.51\,\mathrm{s}$. The decay time-constant is $1/1.02 = 0.98\,\mathrm{s}$, indicating that the oscillation is coherent over approximately two periods; over intervals longer than this, the phase of oscillation would be expected to drift. The corresponding pattern of motion given by $\mathbf{X}^A$ is translational, up and to the left. If instead $\mathbf{X}_2^A = (1, i, 0, 0)$, the horizontal and vertical components are 90° out of phase, indicating circular motion. Note that there must be a conjugate eigenvalue, say $\lambda_3 = \lambda_2^*$, representing the same frequency (up to a change of sign) and damping constant, and with $\mathbf{X}_3^A = (\mathbf{X}_2^A)^*$.

Example: analysis of finger-writing

Earlier, in figure 11.7, the learned dynamical model for finger-writing was displayed by a simulation which depicted a plausible scribble. Modal analysis as above reveals that all the modes are stable ($\beta_m < 0$) and that the two slowest modes have time constants

$$\frac{1}{\beta_m} = 75\,\mathrm{s}, \ 1.15\,\mathrm{s}$$

respectively. The first of these has frequency $f_m = 0$ so that, given the long decay time, this is effectively a constant-velocity mode. It has an (upper) eigenvector

$$\mathbf{X}^A = (1.0, -1.18, -0.10, 0.05, -0.05, -0.05)^T$$

in affine space which represents predominantly translational motion along the upper-left to lower-right diagonal, following the gross flow of finger-writing in the training set. (Note that affine space has been set up as described in chapter 4 so that components have comparable units, and the neglect of the non-translational components for this mode is therefore valid.) The next mode is oscillatory, with frequency 1.01 Hz. This is consistent with the formation of letters, involving one or two strokes each, so that with around 12 strokes in "andrew," written over a duration of 10 seconds, 1 Hz is a very plausible frequency. The decay time of 1.15 s is almost exactly one period of oscillation, suggesting that the stroke sequence is not coherent — successive strokes are sufficiently independent that their phases do not match. The eigenvector for the mode is

$$\mathbf{X}^A = (1, 0.38 - 0.28i, -0.74 + 0.13i, 0.47 - 0.28i, -0.51 + 0.53i, -0.43 + 0.23i)^T$$

which contains translation and also other affine components. The translational part, being complex, represents an elliptical motion at the 1 Hz frequency.

11.4   Performance of trained trackers

Preceding sections have shown, using modal analysis, that trained models capture the oscillatory behaviour of typical motions. This section demonstrates that, as expected, tracking performance is enhanced by replacing default dynamics in the predictor with specific dynamics learned from a training set.

Tuning for lip gestures

First an example is given of tracking the motion of talking lips. Lips may be tracked either in a frontal or in a side-on view. The side-on view, whilst arguably less informative in speech analysis applications, has the advantage that the mouth outline is silhouetted and therefore offers high contrast. Deformations of the lips for two sounds are shown in figure 11.8. Individual dynamical models are learned for each of the sounds "Ooh" and "Pah." For example the "Pah" model is learned from a training sequence in which the "Pah" motion is repeated several times. The resulting selectivity is shown in figure 11.9. It is clear from these results that the tuning effect for individual sounds is strong. Actually to render assistance for speech analysis, it is necessary to learn the repertoire of lip motions that occurs in typical connected speech, and this is addressed next.

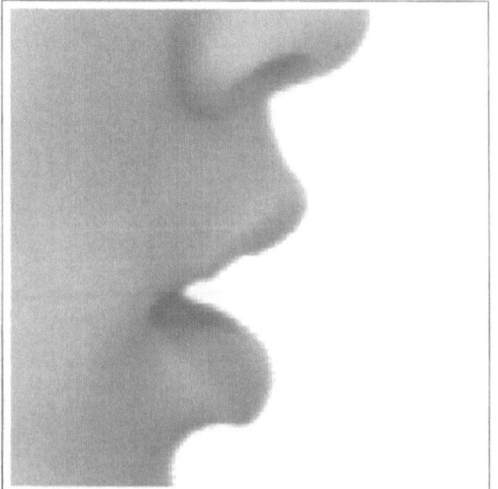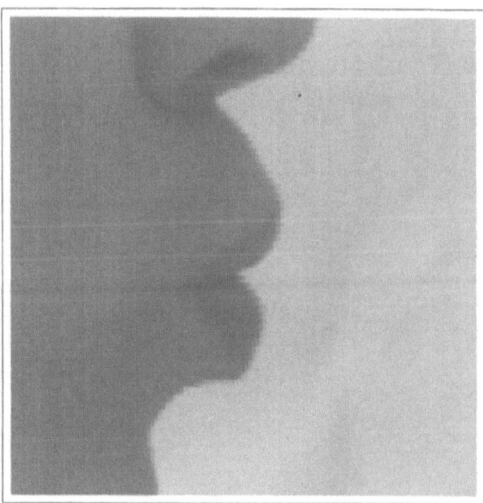

Figure 11.8: Single-syllable training. *Deformations of the mouth are shown corresponding to the sounds a) "Pah" and b) "Ooh."*

Connected speech

Two-stage training is used to learn a dynamical model for connected speech. In the bootstrap stage, the default tracker follows a slow-speech training sequence which is then used, via the learning algorithm, to generate a trained tracker. This "boot-strapped" tracker is capable of following speech of medium speed and is used to follow a medium-speed training sequence, from which dynamics for a full-speed tracker are obtained. The trained tracker is then compared with the default tracker, using a test sequence entirely different from the training sequences. Two deformation components of lip motion are extracted, at 50 Hz, as "lip-reading" signals. The more significant one, in the sense that it accounts for the greater part of the lip motion, corresponds approximately to the degree to which the lips are parted. This component is plotted for the default tracker and the partly and fully trained ones in figure 11.10. It is clear from the figure that the trained filter is considerably more agile. In a demonstration in which the signal was used to animate a head, the untrained filter was clearly able to follow only very slow speech, whereas the trained filter successfully follows speech delivered at normal speed.

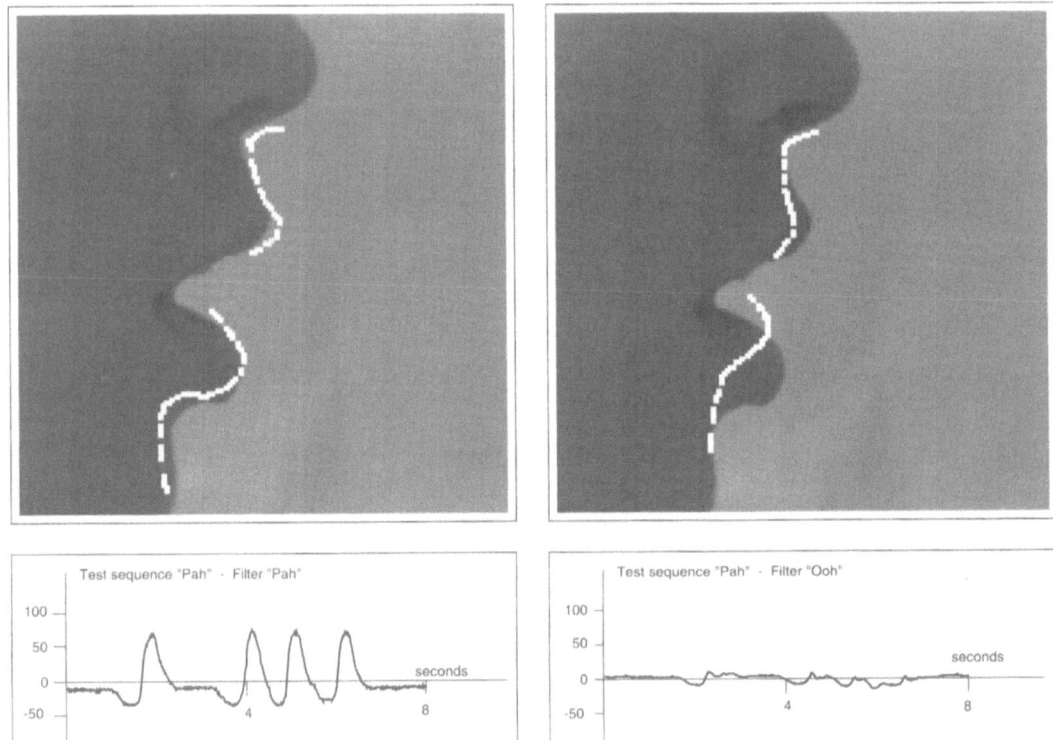

Figure 11.9: Filter selectivity for lip gestures. *The sound "pah" is repeated and tracked by filters trained on the sounds "pah" (left) and "ooh" (right) respectively. Plotted signals refer to the Principal Component of motion in the "Pah" training sequence, which corresponds roughly to the degree to which the mouth is open. Corresponding tracked contours, approximately 4.1 s after the start of the signal, are shown in the snapshots (top left and right). Clearly the filter trained on "Pah" relays the test sequence faithfully, whereas the filter trained on "Ooh" suppresses the test signal almost entirely.*

Bibliographic notes

The dynamical learning algorithm presented here is based on the "Maximum Likelihood" principle (Rao, 1973; Kendall and Stuart, 1979). Maximum likelihood algorithms for estimating dynamics are based on the "Yule-Walker" equations for estimation of the parameters of auto-regressive models (Gelb, 1974; Goodwin and Sin, 1984; Ljung, 1987). A multi-dimensional version of the algorithm which estimates

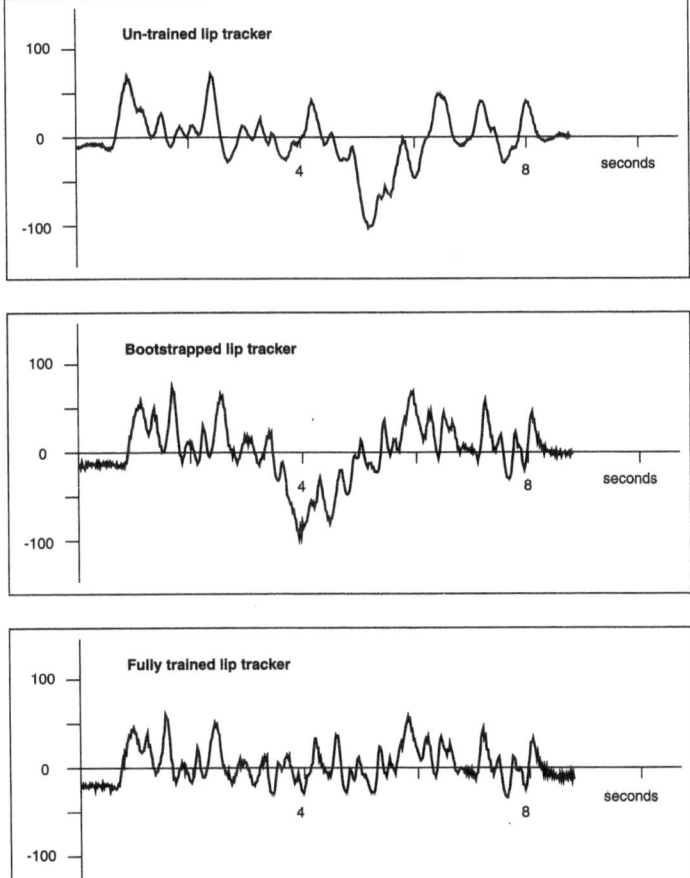

Figure 11.10: Trained lip tracker. *Training a tracker for side-on viewing of speaking lips greatly enhances tracking performance. The graphs show plots from the untrained, default filter, the bootstrapped filter after one training cycle and lastly the filter after a second training cycle. One component of deformation of the lips is shown, corresponding to the degree to which the mouth is open — the space of deformations spanned by the first two templates in figure 4.12 on page 92. Note the considerable loss of detail in the default filter and the overshoots in both default and bootstrapped filters, compared with the fully trained filter. (The sentence spoken here was "In these cases one would like to reduce the dependence of a sensory information processing algorithm on these constraints if possible.")*

not only deterministic parameters A but also the stochastic parameters B is given in (Blake and Isard, 1994; Blake et al., 1995). The particular compact form of the estimator for C in the learning algorithm here is due to Wildenberg (Wildenberg, 1997). A related algorithm for learning deterministic parameters A only is described by Baumberg and Hogg (Baumberg and Hogg, 1995b), who also address the issue of orthogonality constraints on A. The extension of the algorithm to estimate the shape-mean $\overline{\mathbf{X}}$ originated from (Reynard et al., 1996).

An extension of the basic algorithm for *classes* of objects, dealing independently with motion and with variability of mean shape/position over the class, is described in (Reynard et al., 1996). The same algorithm is also used for modular learning — the aggregation of training sets for which a joint dynamical model is to be constructed.

The result that the steady-state covariance P_∞ in a learned dynamic model approximates to the sample covariance $\overline{P}$ for the training set treated as a static set of shapes is described by (Wildenberg, 1997).

The learning algorithm treats the training set as exact whereas in fact it is inferred from noisy observations. Dynamics can be learned directly from the observations using Expectation–maximisation (EM) (Dempster et al., 1977). Learning dynamics by EM is suggested by Ljung (Ljung, 1987) and the detailed algorithm is given in (North and Blake, 1998). It is related to the Baum-Welch algorithm used to learn speech models (Huang et al., 1990; Rabiner and Bing-Hwang, 1993) but with additional complexity because the state-space is continuous rather than discrete.

A number of alternative approaches have been proposed for learning dynamics, with a view to gesture recognition rather than tracking — see for instance (Mardia et al., 1993; Campbell and Bobick, 1995; Bobick and Wilson, 1995).

Chapter 12

Non-Gaussian models and random sampling algorithms

This chapter describes in detail a powerful algorithm for contour tracking that uses random sampling — the CONDENSATION algorithm. It applies to cases where there is substantial clutter in the background. Clutter presents a particular challenge because elements in the background may mimic parts of foreground features. In the most severe case of camouflage, the background may consist of objects similar to the foreground object, for instance when a person is moving past a crowd. The probability density for $\mathcal{X}$ at time t_k is multi-modal and therefore not even approximately Gaussian. The Kalman filter is not suited to this task, being based on pure Gaussian distributions.

The Kalman filter as a recursive linear estimator is a special case, applying only to Gaussian densities, of a more general probability density propagation process. In continuous time the process would be described in terms of diffusion, governed by a "Fokker-Planck" equation, in which the density for $\mathcal{X}(t)$ drifts and spreads under the action of a stochastic model of its dynamics. In the simple Gaussian case, the diffusion is purely linear and the density function evolves as a Gaussian pulse that translates, spreads and is reinforced. It remains Gaussian throughout, as in figure 10.1 on page 214, and its evolution is described analytically and exactly by the Kalman filter. The random component of the dynamical model leads to spreading — increasing uncertainty — while the deterministic component causes the density function to drift bodily. The effect of an external observation $\mathbf{Z}(t)$ is to superimpose a reactive effect on the diffusion in which the density tends to peak in the vicinity of observations. In

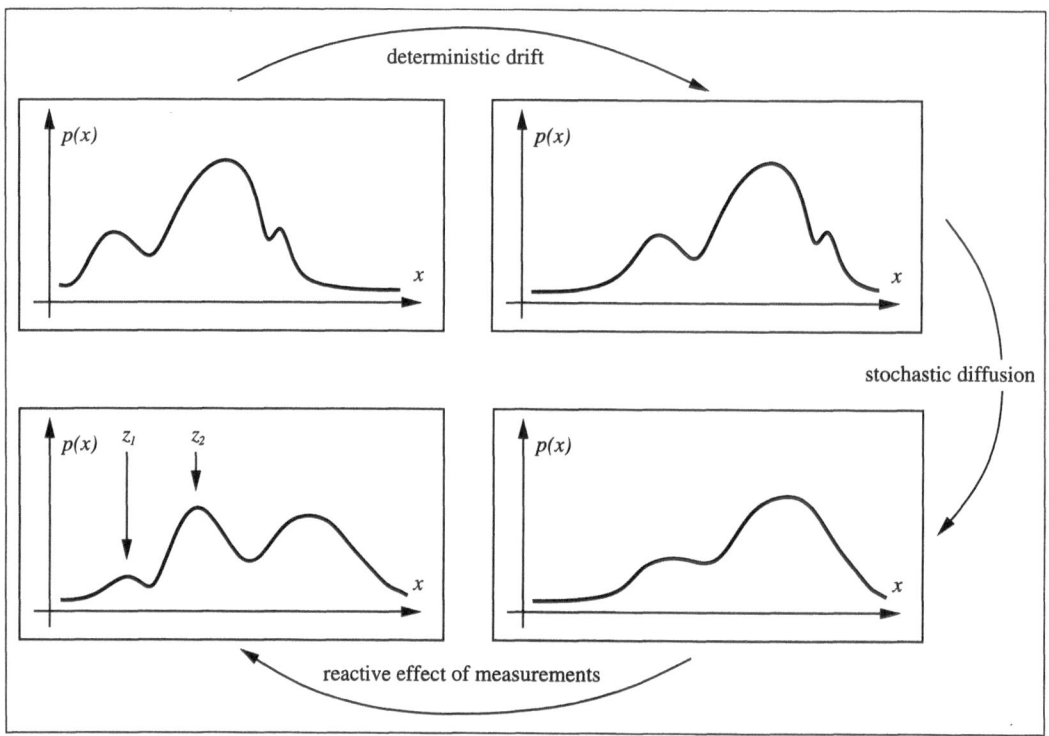

Figure 12.1: Probability density propagation. *In general, the state density describing an object is multi-modal (compare with the Gaussian model used by the Kalman filter in figure 10.1 on page 214). Propagation occurs in three phases: drift due to the deterministic component of object dynamics; diffusion due to the random component; reactive reinforcement due to observations.*

clutter, there are typically several competing observations and these tend to encourage a non-Gaussian state density (figure 12.1).

The CONDENSATION algorithm (CONDitional DENSity propagATION) is designed to address this more general situation. It has the striking property that, generality notwithstanding, it is a considerably simpler algorithm than the Kalman filter. Moreover, despite its use of random sampling which is often thought to be computationally inefficient, the CONDENSATION algorithm runs in near real time. This is because tracking over time maintains relatively tight distributions for shape at successive time-steps, and particularly so given the availability of accurate, learned models of shape and motion.

12.1   Factored sampling

This section describes the factored sampling algorithm, which can be used to search single still images for an object when observations arc non-Gaussian. Then, in the following section, the CONDENSATION algorithm is presented as an extension of factored sampling which handles temporal image sequences.

Chapter 8 introduced the notion of the posterior distribution for an object, and the task of curve-fitting was cast as the problem of maximising $p(\mathbf{X}|\mathbf{Z})$ ((8.19) on page 171), where $\mathbf{Z}$ denoted the "aggregated observation" from least-squares fitting in chapter 6. By restricting the prior and observation densities to be Gaussian, the posterior was also constrained to be Gaussian, described completely by its mean and covariance matrix which were evaluated in closed form using the recursive fitting algorithm on pages 127 and 174.

In clutter, $\mathbf{Z}$ has to incorporate all of the information in an image. It is no longer valid to assume that the features consist of one measurement on each normal of a single curve, and details of the form of $\mathbf{Z}$ to be used will be given in section 12.3. In the general case the prior $p(\mathbf{X})$ and the observation density $p(\mathbf{Z}|\mathbf{X})$ are non-Gaussian, and there is no closed-form algorithm to evaluate the posterior. Factored sampling provides a way of approximating the posterior, using a random number generator to sample from a prior for curve-shape. Random sampling methods were used in chapter 8 as a way of visualising distributions; here they form an integral part of the algorithm.

The factored sampling algorithm generates a random variate $\tilde{\mathbf{X}}$ from a distribution $\tilde{p}(\mathbf{X})$ that approximates the posterior $p(\mathbf{X}|\mathbf{Z})$. First a sample set $\{\mathbf{s}^{(1)}, \ldots, \mathbf{s}^{(N)}\}$ is generated[1] from the prior density $p(\mathbf{X})$ and then an index $n \in \{1, \ldots, N\}$ is chosen with probability $\pi^{(n)}$, where

$$\pi^{(n)} = \frac{p_z(\mathbf{s}^{(n)})}{\sum_{j=1}^{N} p_z(\mathbf{s}^{(j)})}$$

and

$$p_z(\mathbf{s}) = p(\mathbf{Z}|\mathbf{X} = \mathbf{s}),$$

---

[1]This can be done for example using (8.8) on page 164. Note that the presence of clutter causes $p(\mathbf{Z}|\mathbf{X})$ to be non-Gaussian, but the prior $p(\mathbf{X})$ may still happily be Gaussian, and that is what is assumed in later examples.

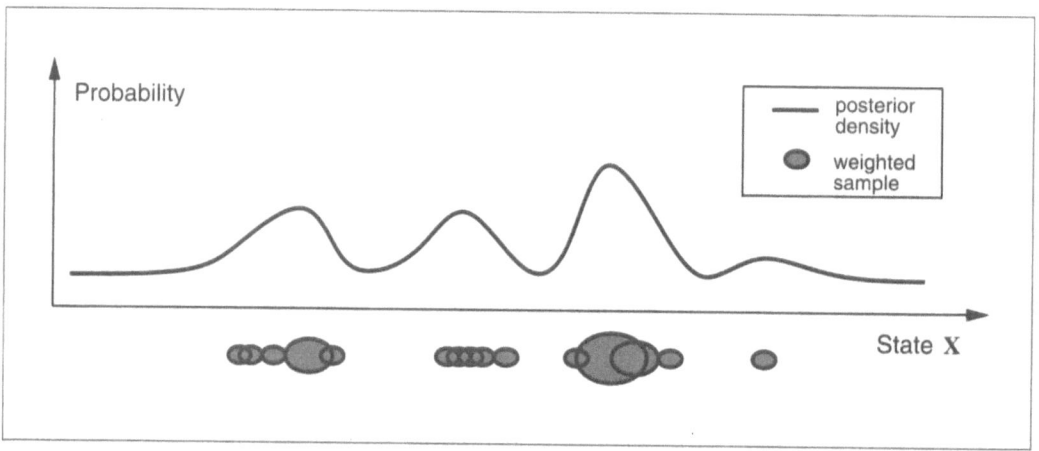

Figure 12.2: Factored sampling. *A set of points* $s^{(n)}$, *the centres of the blobs in the figure, is sampled randomly from a prior density* $p(\mathbf{X})$. *Each sample is assigned a weight* $\pi^{(n)}$ *(depicted by blob area) in proportion to the value of the observation density* $p(\mathbf{Z}|\mathbf{X} = s^{(n)})$. *The weighted point set then serves as a representation of the posterior density* $p(\mathbf{X}|\mathbf{Z})$, *suitable for sampling. The one-dimensional case illustrated here extends naturally to the practical case that the density is defined over several position and shape variables.*

the conditional observation density. The value $\tilde{\mathbf{X}} = s^{(n)}$ chosen in this fashion has a distribution which approximates the posterior $p(\mathbf{X}|\mathbf{Z})$ increasingly accurately as N increases (figure 12.2).

Note that arbitrary posterior expectations $\mathcal{E}[g(\mathbf{X})|\mathbf{Z}]$ can be generated directly from the samples $\{s^{(n)}\}$ by weighting with $p_z(s)$ to give:

$$\mathcal{E}[g(\mathbf{X})|\mathbf{Z}] \approx \frac{\sum_{n=1}^{N} g(s^{(n)}) p_z(s^{(n)})}{\sum_{j=1}^{N} p_z(s^{(j)})}. \tag{12.1}$$

For example, the mean can be estimated using $g(\mathbf{X}) = \mathbf{X}$ (illustrated in figure 12.3) and the second moment using $g(\mathbf{X}) = \mathbf{X}\mathbf{X}^T$. In the case that the density $p(\mathbf{Z}|\mathbf{X})$ is normal, the mean obtained by factored sampling is consistent with an estimate obtained more conventionally, and efficiently, from linear least-squares estimation. An example of the application of the factored sampling algorithm to locate objects is given in figure 12.4. A prior distribution is used which is quite diffuse, covering most of the image area, and allowing a wide range of variation of orientation and shape. Given a suitable observation density, the posterior distribution shown is strongly peaked at

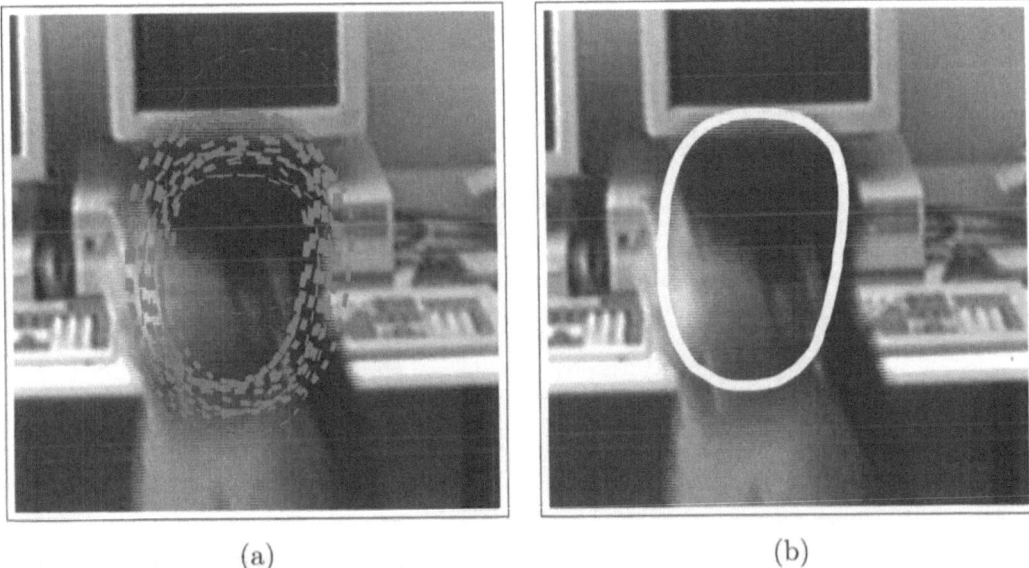

(a)   (b)

Figure 12.3: Sample-set representation of shape distributions. *The sample-set representation of probability distributions, illustrated in one dimension in figure 12.2, is illustrated here (a) as it applies to the distribution of a multi-dimensional curve parameter* **X**. *Each sample* $\mathbf{s}^{(n)}$ *is shown as a curve (of varying position and shape) with a thickness proportional to the weight* $\pi^{(n)}$. *The weighted mean of the sample set (b) serves as an estimator of the distribution mean.*

the locations of several genuine objects and also a few fairly convincing frauds.

12.2 The CONDENSATION algorithm

The CONDENSATION algorithm is based on factored sampling but extended to apply iteratively to images in a sequence, taken at successive times t_k. In chapter 10 the Kalman filter was used to incorporate prediction, using a dynamical model, into the curve-fitting process. CONDENSATION is the analogous extension to factored sampling, and in the examples described later the dynamical model used is exactly that developed in chapters 9 and 11. In fact much more general classes of dynamical models can be used within the algorithm and these are discussed in the bibliographic notes.

The state of the modelled object at time t_k, denoted $\mathcal{X}(t_k)$ earlier, will now be

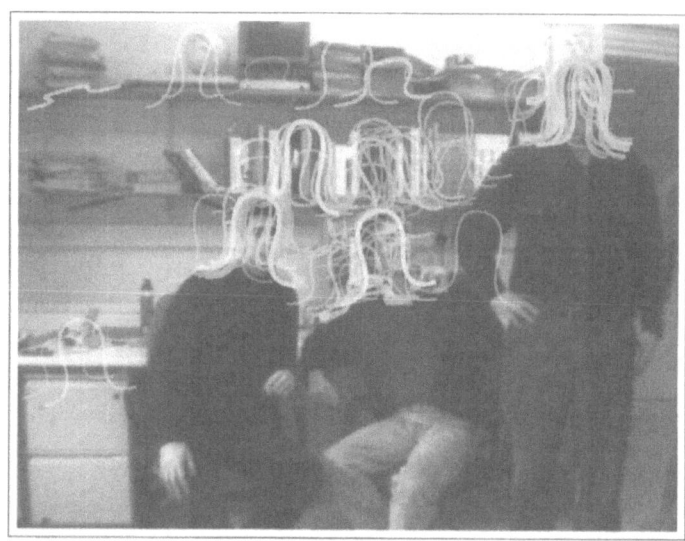

Figure 12.4: Sample-set representation of posterior shape distribution *for a curve with parameters* **X**, *modelling a head outline. Each sample* $\mathbf{s}^{(n)}$ *is shown as a curve (of varying position and shape) with a thickness and intensity proportional to the weight* $\pi^{(n)}$. *The prior is uniform over translation, and a constrained Gaussian in the remainder of its affine shape-space. (Figure taken from (MacCormick and Blake, 1998).)*

denoted as $\mathcal{X}_k$, for compactness. Its "history" is $\underline{\mathcal{X}}_k = \{\mathcal{X}_1, \ldots, \mathcal{X}_k\}$. Similarly the image observation at time t_k, previously denoted $\mathbf{Z}(t_k)$, will be denoted $\mathbf{Z}_k$ with history $\underline{\mathbf{Z}}_k = \{\mathbf{Z}_1, \ldots, \mathbf{Z}_k\}$.

The process at each time-step is a self-contained iteration of factored sampling, so the output of an iteration will be a weighted, time-stamped sample set, denoted $\{\mathbf{s}_k^{(n)}, \ n = 1, \ldots, N\}$ with weights $\pi_k^{(n)}$, representing approximately the conditional state density $p(\mathcal{X}_k | \underline{\mathbf{Z}}_k)$ at time t_k. How is this sample set obtained? The process must begin with a prior density and the effective prior for time-step t_k should be $p(\mathcal{X}_k | \underline{\mathbf{Z}}_{k-1})$. This prior is multi-modal in general and no functional representation of it is available. It is derived from the sample-set representation $\{(\mathbf{s}_{k-1}^{(n)}, \pi_{k-1}^{(n)}), \ n = 1, \ldots, N\}$ of $p(\mathcal{X}_{k-1} | \underline{\mathbf{Z}}_{k-1})$, the output from the previous time-step, to which prediction must then be applied.

The iterative process as applied to sample sets, depicted in figure 12.5, mirrors the continuous diffusion process in figure 12.1. At the top of the diagram, the output

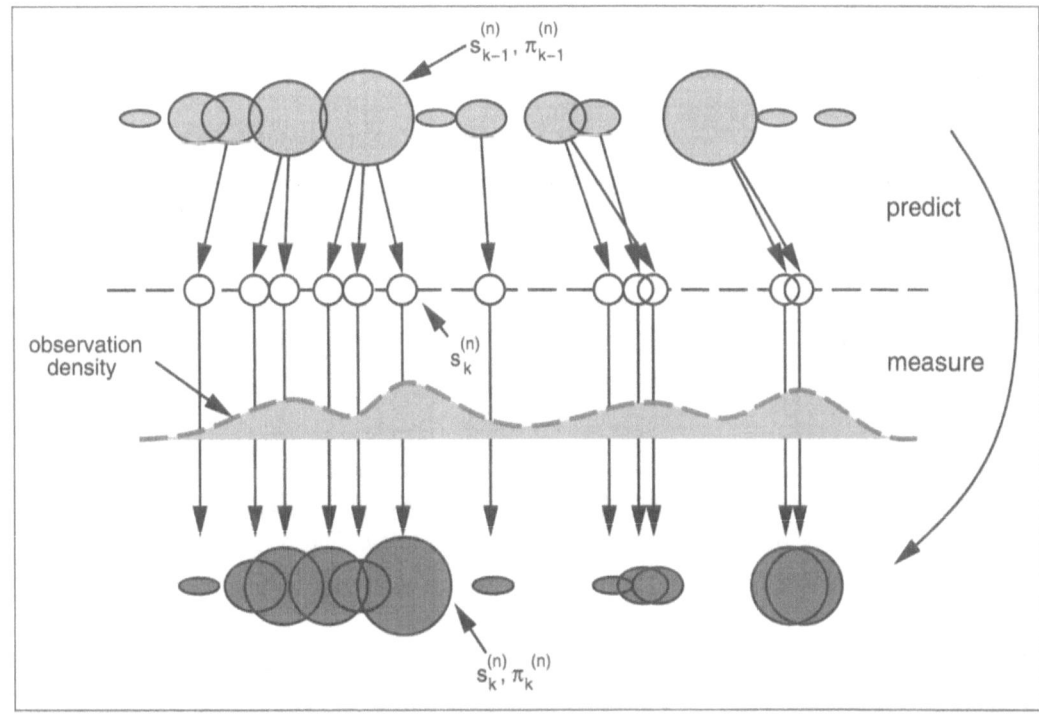

Figure 12.5: One time-step in the CONDENSATION **algorithm.** *The drift and diffusion steps of the probabilistic propagation process of figure 12.1 are combined into a single prediction stage.*

from time-step t_{k-1} is the weighted sample set $\{(\mathbf{s}_{k-1}^{(n)}, \pi_{k-1}^{(n)}),\ n = 1, \ldots, N\}$. The aim is to maintain, at successive time-steps, sample sets of fixed size N, so that the algorithm can be guaranteed to run within a given computational resource. The first operation therefore is to sample (with replacement) N times from the set $\{\mathbf{s}_{k-1}^{(n)}\}$, choosing a given element with probability $\pi_{k-1}^{(n)}$. Some elements, especially those with high weights, may be chosen several times, leading to identical copies of elements in the new set. Others with relatively low weights may not be chosen at all.

Each element chosen from the new set is now subjected to a predictive step. This corresponds to sampling from the distribution $p(\mathcal{X}_k|\mathcal{X}_{k-1})$, and for the second-order AR models described in chapter 9 the sampling equation is given by (9.17) on page 204. At this stage, the sample set $\{\mathbf{s}_k^{(n)}\}$ for the new time-step has been generated but, as yet, without its weights; it is approximately a fair random sample from the effective prior density $p(\mathcal{X}_k|\underline{\mathbf{Z}}_{k-1})$ for time-step t_k. Finally, the observation step from factored sampling is applied, generating weights from the observation density $p(\mathbf{Z}_k|\mathcal{X}_k)$ to obtain the sample-set representation $\{(\mathbf{s}_k^{(n)}, \pi_k^{(n)})\}$ of state density for time t_k.

Figure 12.6 gives a synopsis of the algorithm. Note the use of *cumulative* weights $c_{k-1}^{(j)}$ (constructed in step 3) to achieve efficient sampling in step 1. After any time-step, it is possible to "report" on the current state, for example by evaluating some moment of the state density:

$$\mathcal{E}[g(\mathcal{X}_k)] = \sum_{n=1}^{N} \pi_k^{(n)} g\left(\mathbf{s}_k^{(n)}\right).$$

In later examples the mean position is displayed using $g(\mathcal{X}) = \mathcal{X}$.

One of the striking properties of the CONDENSATION algorithm is its simplicity, compared with the Kalman filter, despite its generality. Largely this is because it is not necessary in the CONDENSATION framework to propagate covariance explicitly.

12.3   An observation model

The observation process defined by $p(\mathbf{Z}_k|\mathcal{X}_k)$ is assumed here to be stationary in time (though the CONDENSATION algorithm does not necessarily demand this) so a static function $p(\mathbf{Z}|\mathcal{X})$ needs to be specified. It is also assumed here that observations depend only on position and shape, not on velocities, so that

$$p(\mathbf{Z}|\mathcal{X}) = p(\mathbf{Z}|\mathbf{X}).$$

One-dimensional observations in clutter

The observation density $p(\mathbf{Z}|\mathbf{X})$ for curves in clutter is quite different to the Gaussian approximation used in chapter 8, so for clarity a simplified one-dimensional form is described first. In one dimension, observations reduce to a set of scalar positions $\{\mathbf{z} = (z_1, z_2, \ldots, z_M)\}$ and the observation density has the form $p(\mathbf{z}|x)$ where x is one-dimensional position. The multiplicity of measurements reflects the presence of clutter so either one of the events

$$\phi_m = \{\text{true measurement is } z_m\}, \ m = 1, \ldots, M$$

Iterate

From the "old" sample set $\{\mathbf{s}_{k-1}^{(n)}, \pi_{k-1}^{(n)}, c_{k-1}^{(n)}, n-1, \ldots, N\}$ at time-step t_{k-1}, construct a "new" sample set $\{\mathbf{s}_k^{(n)}, \pi_k^{(n)}, c_k^{(n)}, n = 1, \ldots, N\}$ for time t_k.

Construct the nth of N new samples as follows:

1. **Select** a sample $\mathbf{s}'^{(n)}_k$ as follows:

   (a) generate a random number $r \in [0,1]$, uniformly distributed.

   (b) find, by binary subdivision, the smallest j for which $c_{k-1}^{(j)} \geq r$

   (c) set $\mathbf{s}'^{(n)}_k = \mathbf{s}_{k-1}^{(j)}$

2. **Predict** by sampling from

$$p(\mathcal{X}_k | \mathcal{X}_{k-1} = \mathbf{s}'^{(n)}_k)$$

   to choose each $\mathbf{s}_k^{(n)}$. For instance, in the case that the dynamics are governed by a linear AR process, the new sample value may be generated as: $\mathbf{s}_k^{(n)} = A\mathbf{s}'^{(n)}_k + (I-A)\overline{\mathcal{X}} + B\mathbf{w}_k^{(n)}$ where $\mathbf{w}_k^{(n)}$ is a vector of standard normal random variates, and BB^T is the process noise covariance.

3. **Measure** and weight the new position in terms of the measured features $\mathbf{Z}_k$:

$$\pi_k^{(n)} = p(\mathbf{Z}_k | \mathcal{X}_k = \mathbf{s}_k^{(n)})$$

   then normalise so that $\sum_n \pi_k^{(n)} = 1$ and store together with cumulative probability as $(\mathbf{s}_k^{(n)}, \pi_k^{(n)}, c_k^{(n)})$ where

$$
\begin{aligned}
c_k^{(0)} &= 0, \\
c_k^{(n)} &= c_k^{(n-1)} + \pi_k^{(n)} \quad \text{for} \quad n = 1, \ldots, N.
\end{aligned}
$$

Figure 12.6: The CONDENSATION algorithm.

occurs, or else the target object is not visible with probability $q = 1 - \sum_m P(\phi_m)$. Now the observation density can be expressed as

$$p(\mathbf{z}|x) = q\,p(\mathbf{z}|\text{clutter}) + \sum_{m=1}^{M} p(\mathbf{z}|x, \phi_m)P(\phi_m).$$

A reasonable functional form for this can be obtained by making some specific assumptions: that $P(\phi_m) = p$ for all m features[2], that the clutter is a Poisson process along the line with spatial density λ and that any true target measurement is unbiased and normally distributed with standard deviation σ. This leads to

$$p(\mathbf{z}|x) \propto 1 + \frac{1}{\sqrt{2\pi}\sigma\alpha} \sum_m \exp -\frac{\nu_m^2}{2\sigma^2} \tag{12.2}$$

where $\alpha = q\lambda$ and $\nu_m = z_m - x$, and is illustrated in figure 12.7. Peaks in the density function correspond to measured features and the state density will tend to be reinforced in the CONDENSATION algorithm at such points. The background level reflects the possibility that the true target has not been detected at all, and allows a good hypothesis to survive a transitory failure of observations due, for example, to occlusion of the tracked object. The parameters σ (units of distance) and α (units of inverse distance) must be chosen, though in principle they could be estimated from data by observing measurement error σ and both the density of clutter λ and probability of non-detection q.

Considerable economy can be applied, in practice, in the evaluation of the observation density. Given a hypothesised position x in the "observation" step (figure 12.6) it is not necessary to attend to all features $z_1, \ldots, z_M$. Any ν_m for which

$$\frac{1}{\sqrt{2\pi}\sigma\alpha} \exp -\frac{\nu_m^2}{2\sigma^2} \ll 1$$

can be neglected and this sets a search window around the position x outside which measurements can be ignored. For practical values of the constants the search window will have a width of a few σ.

Note that the density $p(\mathbf{z}|x)$ represents the information about x given a fixed number M of measurements. Potentially, the *event* ψ_M that there are M measurements,

---

[2]There could be some benefit in allowing the $P(\phi_m)$ to vary with m to reflect varying degrees of feature affinity, based on contrast, colour or orientation.

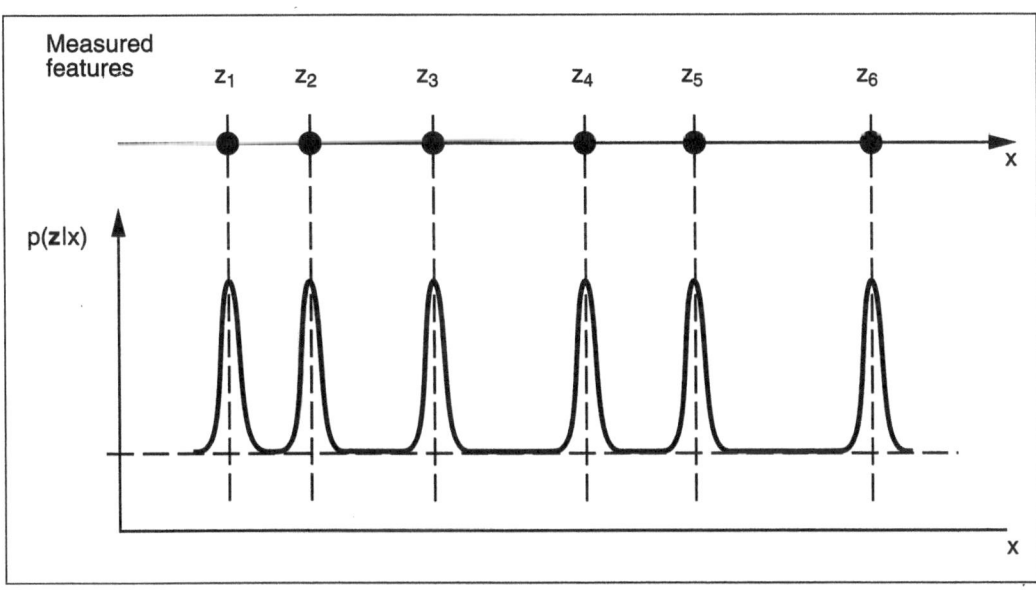

Figure 12.7: One-dimensional observation model. *A probabilistic observation model allowing for clutter and the possibility of missing the target altogether is specified here as a conditional density $p(\mathbf{z}|x)$.*

regardless of the actual *values* of those measurements, also constitutes information about x. However, we can reasonably assume that

$$P(\psi_M|x) = P(\psi_M),$$

for instance because x is assumed to lie always within the image window. In that case, by Bayes' theorem,

$$p(x|\psi_M) = p(x)$$

— the event ψ_M provides no additional information about the position x. (If x is allowed also to fall outside the image window then the event ψ_M *is* informative: a value of M well above the mean value for the background clutter enhances the probability that x lies within the window.)

Two-dimensional observations

In a two-dimensional image, the set of observations $\mathbf{Z}$ is, in principle, the entire set of features visible in the image. However, an important aspect of achieving real-time

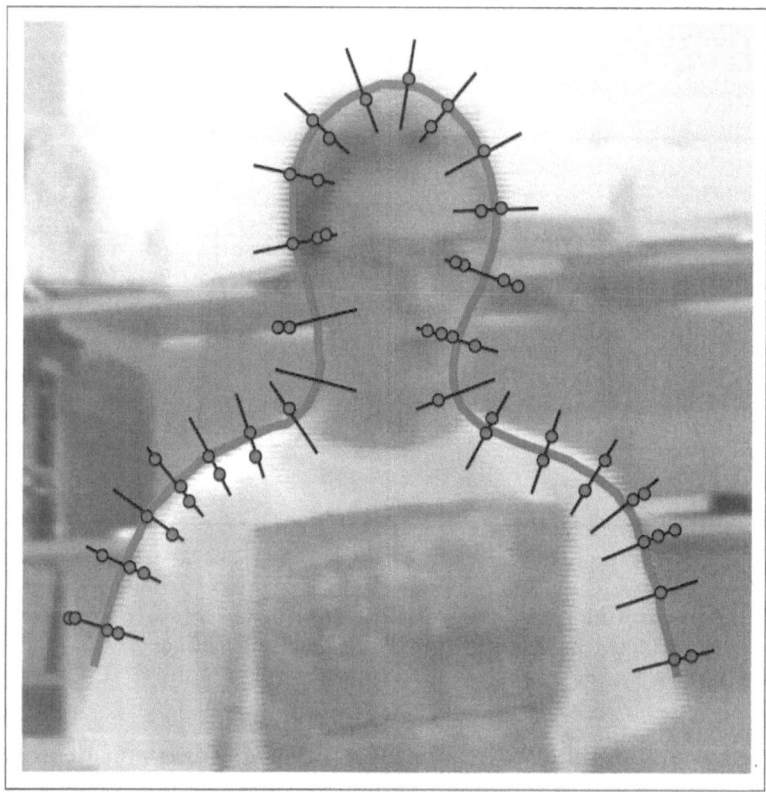

Figure 12.8: Observation process. *The thick line is a hypothesised shape, represented as a parametric spline curve. The spines are curve normals along which high-contrast features (dots) are sought.*

performance has been the restriction of measurement to a sparse set of lines normal to the tracked curve. These two apparently conflicting ideas are hard to resolve and some discussion is given elsewhere (see bibliographic notes). One good choice is simply to construct the two-dimensional observation density as the product of one-dimensional densities (12.2), evaluated independently along M curve normals as in figure 12.8. Note that *locally* $p(\mathbf{Z}|\mathbf{X})$ is approximately Gaussian, especially if constants are chosen so that σ is small and only one feature typically falls within each search region. As $\mathbf{X}$ varies across the entire image, however, the multi-modality of the observation density is apparent.

12.4   Applications of the CONDENSATION algorithm

Four examples are shown here of the practical efficacy of the CONDENSATION algorithm. MPEG versions of some results are available on the web page for the book.

Tracking a multi-modal distribution

The ability of the CONDENSATION algorithm to represent multi-modal distributions is demonstrated in a sequence of a cluttered room containing three people each facing the camera (figure 12.9). One of the people moves from right to left, in front of

Figure 12.9: Tracking three people in a cluttered room. *The first frame of a sequence in which one figure moves from right to left in front of two stationary figures.*

the other two. The shape-space for tracking is built from a hand-drawn template of head and shoulders (figure 12.8) which is then allowed to deform via planar affine transformations. A Kalman filter contour-tracker with default motion parameters is able to track a single moving person just well enough to obtain a sequence of outline curves that is usable as training data. Given the high level of clutter, adequate performance with the Kalman filter is obtained here by means of statistical background

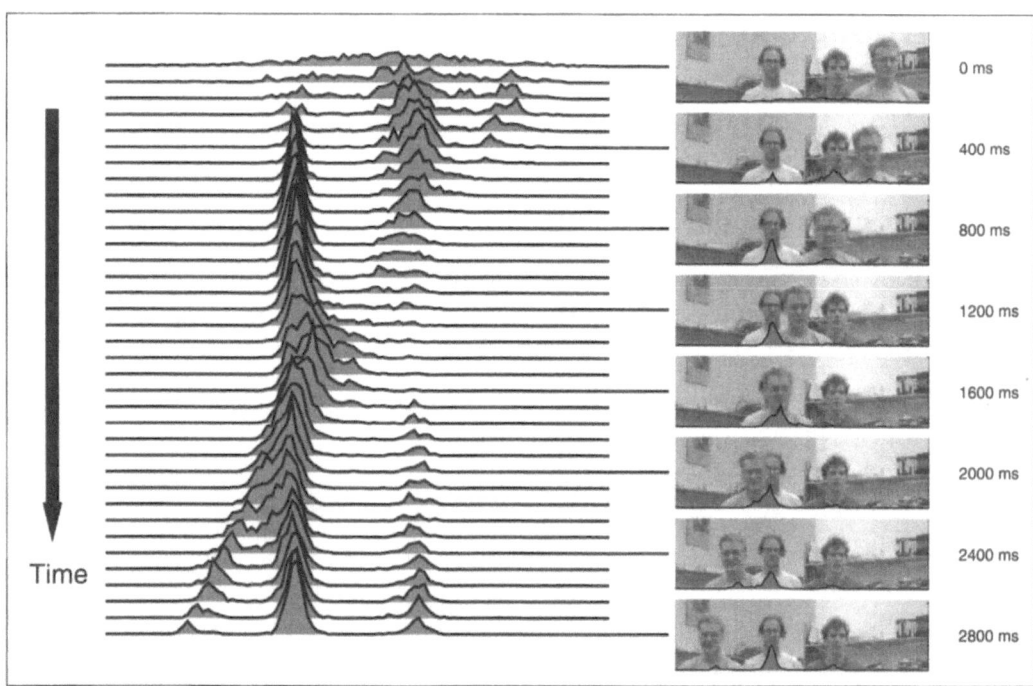

Figure 12.10: Tracking with a multi-modal state density. *An approximate depiction of the state density is shown, computed by smoothing the distribution of point masses $s_k^{(1)}, s_k^{(2)}, \ldots$ in the* CONDENSATION *algorithm. The density is, of course, multi-dimensional; its projection onto the horizontal translation axis is shown here. The initial distribution is roughly Gaussian but this rapidly evolves to acquire peaks corresponding to each of the three people in the scene. The rightmost peak drifts leftwards, following the moving person, coalescing with and separating from the other two peaks as it moves. Having specified a tracker for one* person *we effectively have, for free, a multi-person tracker, owing to the innate ability of the* CONDENSATION *algorithm to maintain multiple hypotheses.*

subtraction. It transpires, for this particular training set, that the learned motions comprise primarily horizontal translation, with vertical translation and horizontal and vertical shear present to a lesser degree.

The learned shape and motion model can be installed as $p(\mathcal{X}_k | \mathcal{X}_{k-1})$ in the CONDENSATION algorithm which is now run on a test sequence but *without* the benefit of background modelling, so that the background clutter is now visible to the tracker. Figure 12.10 shows how the state density evolves as tracking progresses. Initialisation is performed simply by iterating the stochastic model, in the absence of measurements, to its steady state and it can be seen that this corresponds, at time t_0, to a roughly Gaussian distribution, as expected. The distribution rapidly collapses down to three peaks which are then maintained appropriately even during temporary occlusion. Although the tracker was designed to track just one person, the CONDENSATION algorithm takes account of the motion of all three; the ability to represent multi-modal distributions effectively provides multiple-hypothesis capability. Tracking is based on frame rate (40 ms) sampling in this demonstration and distributions are plotted in the figure for alternate frames. A distribution of $N = 1000$ samples per time-step is used.

Tracking rapid motions through clutter

The ability to track more agile motion, still against clutter, is demonstrated by a sequence of a girl dancing vigorously to a Scottish reel. The shape-space for tracking is planar affine, based on a hand-drawn template curve for the head outline. The training sequence consists of dancing against a largely uncluttered background, tracked by a Kalman filter contour-tracker with default dynamics to record 140 fields (2.8 seconds) of tracked head positions, the most that can be tracked before losing lock. Those 140 fields are sufficient to learn a bootstrap motion model which then allows the Kalman filter to track the training data for 800 fields (16 seconds) before loss of lock. The motion model obtained from these 800 fields can now be applied to test data that includes clutter.

Figure 12.11 shows some stills from the test sequence, with a trail of preceding head positions to indicate motion. The motion is primarily translation, with some horizontal shear apparent as the dancer turns her head. Representing the state density with $N = 100$ samples at each time-step proves just sufficient for successful tracking. As in the previous example, a prior density can be computed as the steady state of the motion model and, in this case, that yields a prior for position that spreads across most of the image area, as might be expected given the range of the dance. Such a broad distribution cannot effectively be represented by just $N = 100$ samples. One alternative is to increase N in the early stages of tracking, and this is demonstrated later. Alternatively, the prior can be based on a narrower distribution whose centre is positioned by hand over the object at time t_0, and that is what has been done here. (Observation parameters were $\alpha = 0.005$, $\sigma = 7$ with 18 normals.)

Figure 12.12 shows the motion of the centroid of the estimates head position as tracked both by the CONDENSATION algorithm and by a Kalman filter using the

field 91 (1820 ms)

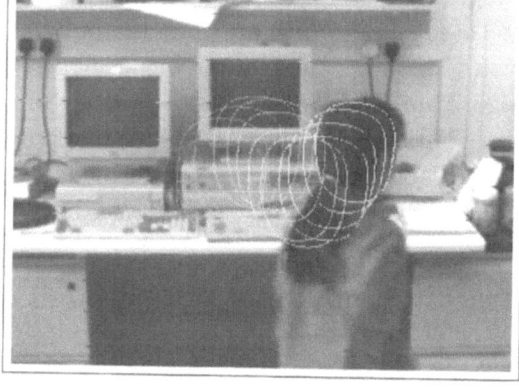

field 121 (2420 ms)

field 221 (4420 ms)

field 265 (5300 ms)

Figure 12.11: Tracking agile motion in clutter. *The test sequence consists of 500 fields (10 seconds) of agile dance against a cluttered background. The dancer's head is tracked through the sequence. Several representative fields are shown here, each with a trail of successive mean tracked head positions at intervals of 40 ms. The CONDENSATION algorithm used $N = 100$ samples per time-step to obtain these results.*

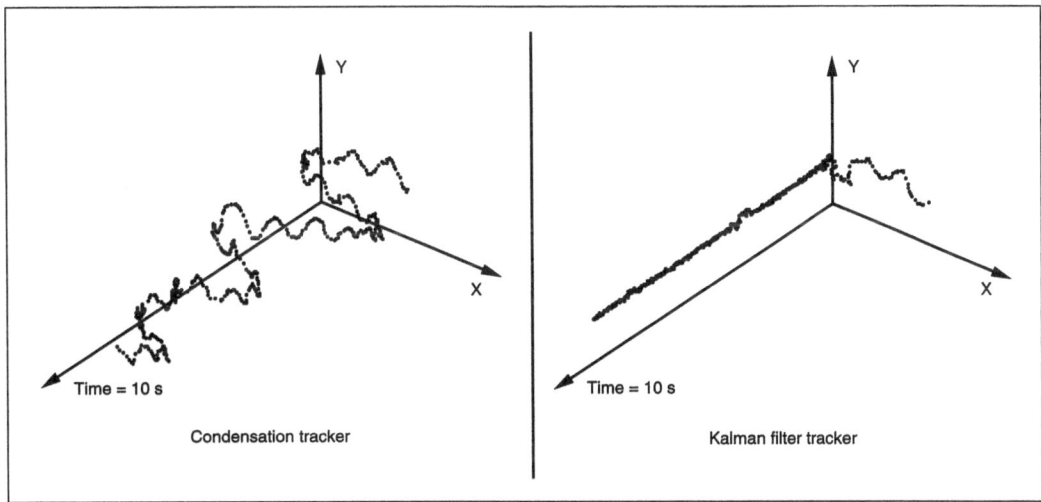

Figure 12.12: The Condensation tracker succeeds where a Kalman filter fails.
The estimated centroid for the sequence shown in figure 12.11 is plotted against time for the entire 500 field sequence, as tracked first by the CONDENSATION tracker, then by a comparable Kalman filter tracker. The CONDENSATION algorithm correctly estimates the head position throughout the sequence. The Kalman filter tracks briefly, but is soon distracted by clutter.

same motion model. The CONDENSATION tracker correctly estimated head position throughout the sequence, but after about 40 fields (0.80 s), the Kalman filter is distracted by clutter, never to recover.

Given that there is only one moving person now, unlike the previous example in which there were three, it might seem that a uni-modal representation of the state density would suffice. This is emphatically not the case. The facility to represent multiple modes is crucial to robustness as figure 12.13 illustrates. The figure shows how the distribution becomes misaligned (at 900 ms), reacting to the distracting form of the computer screen. After 20 ms the distribution splits into two distinct peaks, one corresponding to clutter (the screen), one to the dancer's head. At this point the clutter peak actually has the higher posterior probability — a uni-modal tracker, for instance a Kalman filter, would almost certainly discard the lower peak, rendering it unable to recover. The CONDENSATION algorithm however, capable as it is of carrying several hypotheses simultaneously, does recover rapidly as the clutter peak decays for lack of confirmatory observation, leaving just one peak corresponding to the dancer after 60 ms.

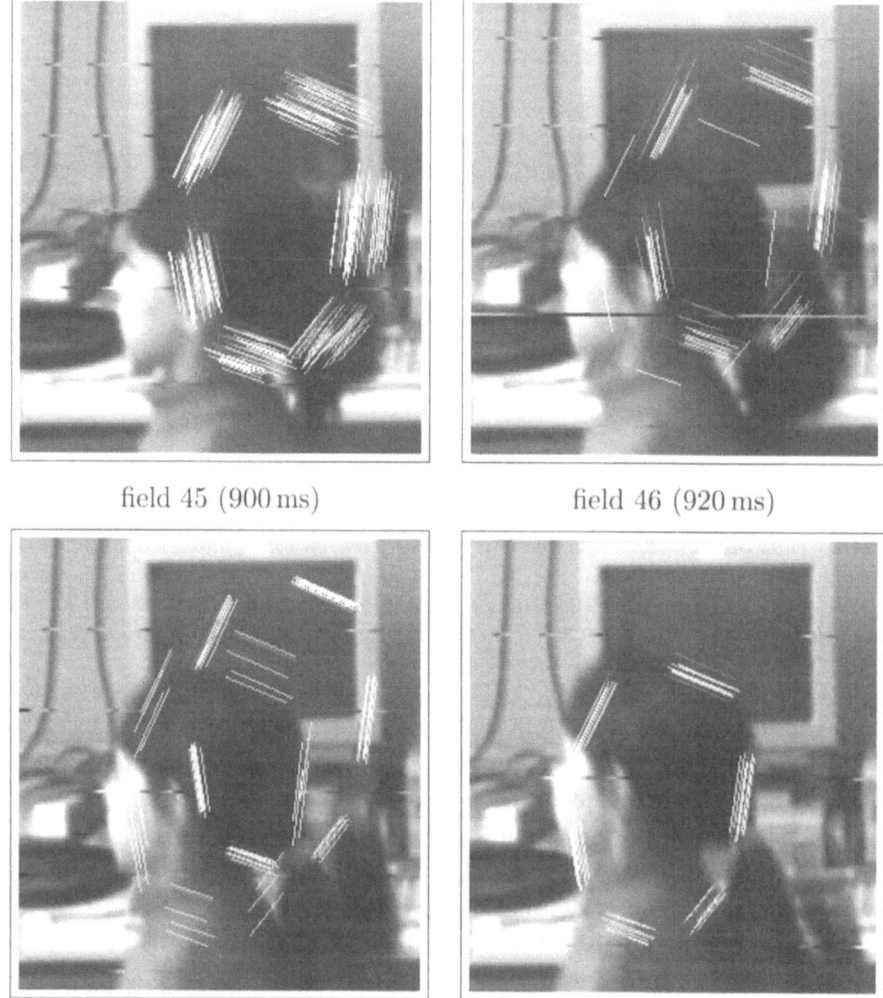

field 45 (900 ms)   field 46 (920 ms)

field 47 (940 ms)   field 48 (960 ms)

Figure 12.13: Recovering from tracking failure. *Detail from 4 consecutive fields of the sequence illustrated in figure 12.11. Each sample from the distribution is plotted on the image, with intensity scaled to indicate its posterior probability. (Most of the $N = 100$ samples have too low a probability to be visible in this display.) At field 45 the distribution is misaligned, and has begun to diverge. At fields 46 and 47 it has split into two distinct peaks, the larger attracted to background clutter, but converges back onto the dancer at field 48.*

Tracking an articulated object

The preceding sequences show motion taking place in affine shape-spaces of just 6 dimensions. High dimensionality is one of the factors, in addition to agility and clutter, that makes tracking hard. In order to demonstrate tracking performance in higher dimensions, we use a test sequence of a hand translating, rotating, and flexing its fingers independently, over a highly cluttered desk scene (figure 12.14). Figure 12.15

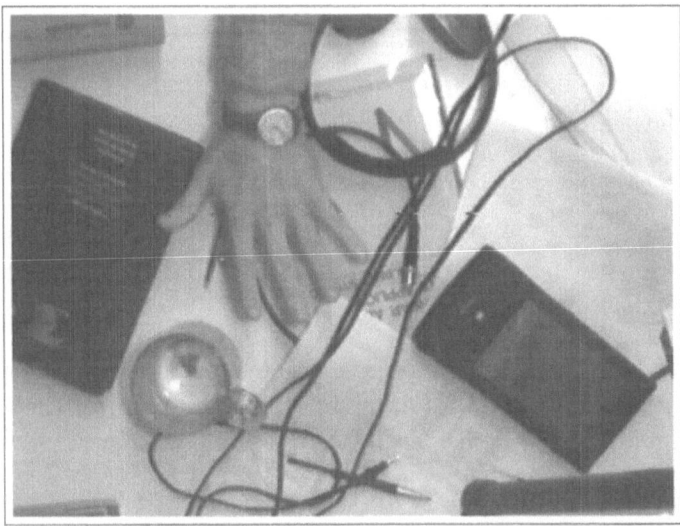

Figure 12.14: A hand moving over a cluttered desk. *Field 0 of a 500 field (10 second) sequence in which the hand translates, rotates, and the fingers and thumb flex independently.*

shows just how severe the clutter problem is — the hand is immersed in a dense field of edges.

A model of shape and motion has been learned from training sequences of hand motion against a plain background, tracked by Kalman filter (using signed edges to help to disambiguate finger boundaries). The procedure comprised several stages, a creative assembly of methods from the available toolkit for generating a shape-space and learning dynamics.

1. **Shape-space** is constructed from 6 templates drawn around the hand with the palm in a fixed orientation and with the fingers and thumb in various configurations. The 6 templates combine linearly to form a 5-dimensional space

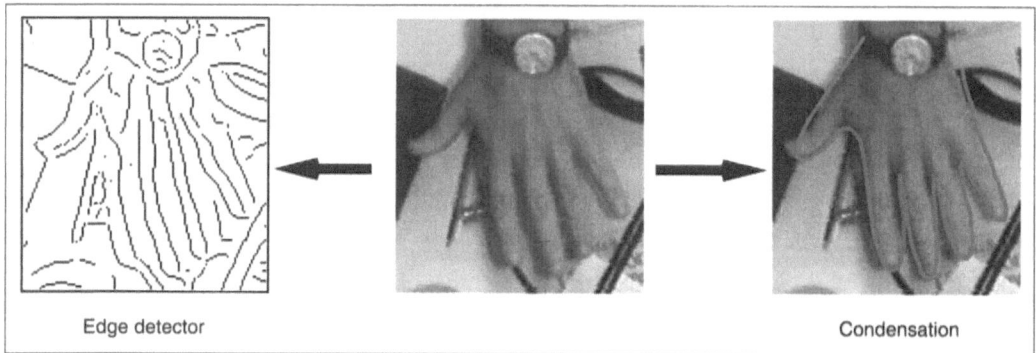

Figure 12.15: Severe clutter. *Detail of one field from the test sequence shows the high level of potential ambiguity. Output from a directional Gaussian edge detector shows that there are many clutter edges present as potential distractors.*

of deformations which are then added to the space of translations to form a 7-dimensional shape-space.

2. **Default dynamics** in the shape-space above are adequate to track a clutter-free training sequence of 600 frames in which the palm of the hand maintains an approximately fixed attitude.

3. **Principal components analysis:** the sequence of 600 hand outlines is replicated with each hand contour rotated through 90 degrees, and the sequences concatenated to give a sequence of 1200 deformations. Projecting out the translational component of motion, the application of Principal Component Analysis (PCA) to the sequence of residual deformations of the 1200 contours establishes a 10-dimensional space that accounted almost entirely for deformation. This is then combined with the translational space to form a 12-dimensional shape-space that accounts both for the flexing of fingers and thumb and also for rotations of the palm.

4. **Bootstrapping:** a Kalman filter with default dynamics in the 12-dimensional shape-space is sufficient to track a training sequence of 800 fields of the hand translating, rotating, and flexing fingers and thumb slowly. This is used to learn a model of motion.

5. **Re-learning:** that motion model is installed in a Kalman filter and used to track

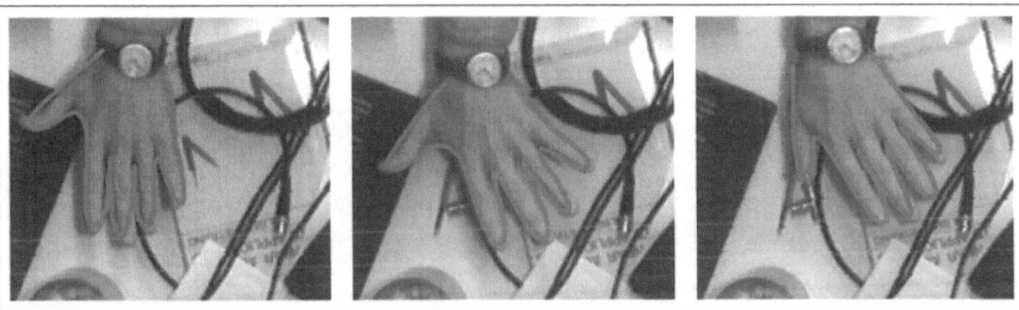

Figure 12.16: Tracking a flexing hand across a cluttered desk. *Representative stills from a 500 field (10 second) sequence show a hand moving over a highly cluttered desk scene. The fingers and thumb flex independently, and the hand translates and rotates. Here the* CONDENSATION *algorithm uses* $N = 1500$ *samples per time-step initially, dropping over 4 fields to* $N = 500$ *for the tracking of the remainder of the sequence. The mean configuration of the contour is displayed.*

another, faster training sequence of 800 fields. This allows a model for more agile motion to be learned, which is then used in a high-performance CONDENSATION tracker.

Figure 12.16 shows detail of a series of images from a tracked, 500 field test sequence. The initial state density is simply the steady state of the motion model, obtained by allowing the filter to iterate in the absence of observations. Tracker initialisation is facilitated by using more samples per time-step ($N = 1500$) at time t_0, falling to 500 over the first 4 fields. The rest of the sequence is tracked using $N = 500$. As with the previous example of the dancer, clutter may distract the tracker but the ability to represent multi-modal state density means that tracking can recover.

Tracking a camouflaged object

Finally, the ability of the algorithm to track rapid motion against background distraction is demonstrated in an extreme case: that background objects actually mimic the tracked object. A 12 second (600 field) sequence shows a bush blowing in the wind, the task being to track one particular leaf. A template is drawn by hand around a still of one chosen leaf and allowed to undergo affine deformations during tracking. Given that a clutter-free training sequence cannot be obtained in this case, the motion model is again learned by means of a bootstrap procedure. A tracker with hand-specified

1.46 s

2.66 s

5.54 s

7.30 s

Figure 12.17: Tracking with camouflage. *The aim is to track a single camouflaged moving leaf in this 12 second sequence of a bush blowing in the wind. Despite the heavy clutter of distractors which actually mimic the foreground object, and occasional violent gusts of wind, the chosen foreground leaf is successfully tracked throughout the sequence. Representative stills depict mean contour configurations, with preceding tracked leaf positions plotted at 40 ms intervals to indicate motion.*

dynamics proves capable of tracking the first 150 fields of a training sequence before losing the leaf, and those tracked positions allow a first approximation to the model to be learned. Installing that in a CONDENSATION tracker, the entire sequence can be tracked, though with occasional misalignments. Finally, a second learned model is capable of tracking accurately the entire 12 second training sequence. Despite occasional violent gusts of wind and temporary obscuration by another leaf, the CON-DENSATION algorithm successfully follows the object, as figure 12.17 shows. In fact, tracking is accurate enough using $N = 1200$ samples to separate the foreground leaf from the background reliably, an effect which can otherwise only be achieved using "blue-screening." Having obtained the model iteratively as above, further sequences can be tracked without further training. With $N = 1200$ samples per time-step the tracker runs at 6.5 Hz on a SGI Indy SC4400 200 MHz workstation. Reducing this to $N = 100$ increases processing speed to video field rate (50 Hz), at the cost of occasional misalignments in the mean configuration of the contour. (Observation parameters are $\alpha = 0.022$, $\sigma = 3$ with 21 normals.)

Bibliographic notes

There has been much written about non-linear filtering algorithms, for handling non-Gaussian probability densities. The Extended Kalman filter (Gelb, 1974; Bar-Shalom and Fortmann, 1988; Jacobs, 1993), is a linearised approximation designed to deal with non-linearities in dynamics and/or sensors. It effectively approximates the non-Gaussian state density as a Gaussian. Bayesian multiple-hypothesis filters and approximate variants include the PDAF and JPDAF (Bar-Shalom and Fortmann, 1988) and can be applied to motion correspondence (Cox, 1993) and visual tracking of discrete features (Rao et al., 1993). The RANSAC algorithm is an alternative mechanism for dealing with ambiguous association (Fischler and Bolles, 1981), in which hypotheses are generated bottom-up, from subsets of image features. Bucy's numerical integration of Bayes' rule for one-dimensional state (Bucy, 1969) is very general but feasible only in one or two dimensions, inapplicable to the state-spaces used in contour tracking whose dimensionality is typically 8–30. Additive Gaussian mixtures as used for representation of non-Gaussian densities in pattern recognition theory (Duda and Hart, 1973; Bishop, 1995) can be used in temporal filtering by dynamically re-weighting the mixture (Sorenson and Alspach, 1971). For approximate methods using additive Gaussian mixtures see also (Anderson and Moore, 1979).

The Fokker-Planck equation (Astrom, 1970) governs the evolution of the probability distribution for point particles following a random walk. The equation describes a diffusing probability density and its coefficients depend on the deterministic and stochastic components of the random walk, as described by the A and B coefficients in the case of an AR process.

One of the best known uses of iterative sampling in image processing is in the statistical restoration algorithms of Geman and Geman (Geman and Geman, 1984) in which a "Gibbs sampler" is used to sample fairly from the posterior distribution for the restored image. Their idea has been generalised and named "Markov Chain Monte-Carlo" (MCMC) (Gelfand and Smith, 1990). Random sampling as a means of image construction has also been applied to curves rather than image intensities (Ripley and Sutherland, 1990; Grenander et al., 1991), to sweep out a posterior distribution for an object outline, and also extended using simulated annealing (Storvik, 1994) to converge to a particular curve estimate. The term "factored sampling" is due to Grenander *et al.* (1991).

Importance sampling is a general technique (Ripley, 1987) for Monte-Carlo methods to bias generation of variates which would otherwise be generated from the stated prior. The purpose of the bias is to increase efficiency by concentrating on areas in which the observation density is likely to be non-negligible. The method includes a correction factor for the bias so that generated samples continue to be drawn fairly from the posterior density for the problem.

Fuller details of the CONDENSATION algorithm are given in (Isard and Blake, 1998a), including a proof of the (asymptotic) correctness of the algorithm. A similar extension of sampling methods to work over time has been reported also by (Gordon et al., 1993) who refer to the "bootstrap" algorithm and derive it via MCMC, rather than factored sampling. Another account (Kitagawa, 1996) elegantly extends the idea to a "smoothing" algorithm which is applicable to off-line applications. It is a two-pass method in which the estimate at each time-step is derived not only from preceding observations but also takes into account all following observations. This is analogous to the smoothing algorithm for Gaussians (Gelb, 1974) which is a two-pass extension of the Kalman filter.

Reasoning about clutter and false alarms of the sort given in section 12.3 is commonly used in target tracking (Bar-Shalom and Fortmann, 1988). Some discussion of the problem of extending the observation model from a single line to a curve of normals is given in (Isard and Blake, 1998a). An extension to account for object opacity is given in (MacCormick and Blake, 1998).

Some progress has been made recently in extending the scope of the motion models used in the CONDENSATION algorithm to include mixed continuous/discrete states (Isard and Blake, 1998b). This is related to Hidden Markov Modelling which is the dominant paradigm for speech recognition algorithms, and a comprehensive introduction is given in (Rabiner and Bing-Hwang, 1993).

Appendix A

Mathematical background

A.1 Vectors and matrices

A brief summary of vector and matrix conventions and operations is given here. An excellent handbook for vector and matrix computation is (Barnett, 1990) and readers should refer to it for details. An alternative that may also be found helpful is (Golub and van Loan, 1989). The vector and matrix operations described should be available in appropriate programming languages such as **matlab**.

Vectors Throughout the book, vectors are denoted in bold, for example

$$\mathbf{r} = \begin{pmatrix} x \\ y \end{pmatrix} \quad \text{or} \quad \mathbf{a} = \begin{pmatrix} a_1 \\ a_2 \\ a_3 \end{pmatrix}.$$

Scalar product and vector product take their usual meanings and are denoted

$$\mathbf{a} \cdot \mathbf{b} \quad \text{and} \quad \mathbf{a} \times \mathbf{b}$$

respectively. In three dimensions, for instance,

$$\mathbf{a} \cdot \mathbf{b} = a_1 b_1 + a_2 b_2 + a_3 b_3.$$

and

$$\mathbf{a} \times \mathbf{b} = (a_2 b_3 - a_3 b_2, \ a_3 b_1 - a_1 b_3, \ a_3 b_1 - a_1 b_3)^T.$$

The magnitude of a vector may be measured via its **Euclidean norm**:

$$|\mathbf{r}| = \sqrt{\mathbf{r} \cdot \mathbf{r}}$$

and a vector $\mathbf{r}$ for which $|\mathbf{r}| = 1$ is said to be normalised, or a "unit" vector.

Matrices   Matrices are generally non-bold capitals, for example A, with components denoted A_{ij}. The transpose A^T is defined by

$$A^T_{ij} = A_{ji}.$$

The **rank** of a matrix A is the number of linearly independent vectors that comprise its columns.

A matrix operation that is frequently useful is the **Kronecker product**

$$A \otimes B = \begin{pmatrix} A_{11}B & A_{12}B & \dots \\ A_{21}B & A_{22}B & \dots \\ \dots & \dots & \dots \end{pmatrix} \tag{A.1}$$

which combines two arrays of dimension $M_1 \times N_1$ and $M_2 \times N_2$ to make a larger one of dimension $M_1 M_2 \times N_1 N_2$.

Linear equations   The linear simultaneous equations

$$A\mathbf{x} = \mathbf{b}$$

have a unique solution when A is square and is non-singular — that is, $\det A \neq 0$, where $\det A$ is the **determinant** of A. Then a solution can be found using the standard inverse

$$\mathbf{x} = A^{-1}\mathbf{b}.$$

If there is no solution, as may happen when A is not square, there may nonetheless be a unique, optimal, approximate solution which is expressed using a **pseudo-inverse** A^+:

$$\mathbf{x} = A^+\mathbf{b} \quad \text{where} \quad A^+ = (A^T A)^{-1} A^T.$$

(More general definitions of pseudo-inverse can be made, but are not used in this book.)

Rotation matrices Matrices for rotation about x, y and z axes are respectively denoted R_x, R_y and R_z where, for example,

$$R_z(\theta) = \begin{pmatrix} \cos\theta & -\sin\theta & 0 \\ \sin\theta & \cos\theta & 0 \\ 0 & 0 & 1 \end{pmatrix} \tag{A.2}$$

so that a point in three dimensions given by a vector $\mathbf{r} = (x, y, z)^T$ is transformed to a rotated point

$$\mathbf{r}' = R_z(\theta)\mathbf{r}.$$

In two dimensions, a rotation is a 2×2 matrix

$$R(\theta) = \begin{pmatrix} \cos\theta & -\sin\theta \\ \sin\theta & \cos\theta \end{pmatrix}.$$

Rotation matrices have the property that they are **orthogonal**, satisfying $R^T R = I$. Generally, an orthogonal matrix U satisfies $U^T U = I$ and, in three dimensions, can be interpreted as a rotation (when $\det U = 1$) or a combination of rotation and reflection ($\det U = -1$).

Eigenvalues and eigenvectors The eigenvalues λ_n and eigenvectors $\mathbf{u}_n$ of a square $N \times N$ matrix A are defined as satisfying

$$A\mathbf{u}_n = \lambda_n \mathbf{u}_n, \quad n = 1, \dots, N.$$

Eigenvalues and eigenvectors may have complex values, unless A is **symmetric** ($A^T = A$) in which case they are guaranteed to be real.

The **trace** of the matrix is defined to be

$$\mathrm{tr}(A) = \sum_{n=1}^{N} A_{nn}$$

and has the property that

$$\mathrm{tr}(A) = \sum_{n=1}^{N} \lambda_n.$$

Diagonalisation Eigenvalues and eigenvectors can be used to decompose a square matrix A as

$$A = UDU^{-1}$$

where D is the "diagonal" matrix

$$D = \text{diag}(\lambda_1, \dots, \lambda_N)$$

with the eigenvalues along the diagonal and zeros elsewhere. The matrix U consists of columns which are normalised eigenvectors of A. One important application of the diagonal form is in computing powers of A:

$$A^p = UD^pU^{-1} \quad \text{where} \quad D^p = \text{diag}(\lambda_1^p, \dots, \lambda_N^p).$$

Setting $p = \frac{1}{2}$ allows a **square root** of A to be computed.

Singular value decomposition (SVD) An alternative form of decomposition of a matrix A is the SVD, which applies not only to square matrices but also to rectangular ones of size $M \times N$. It has the form

$$A = UDV$$

where U is an $M \times M$ matrix, D is a diagonal $M \times N$ matrix and V is an $N \times N$ matrix. Both U and V are orthogonal matrices. The diagonal values of D are $D_{nn} = \sigma_n$ where $\lambda_n = \sigma_n^2$ are eigenvalues of the symmetric matrix $A^T A$ and are hence guaranteed to be positive.

A measure of the "size" of A is its **spectral radius** $\|A\|_2 = \sqrt{\lambda_1}$ where λ_1 is the largest eigenvalue of $A^T A$. The condition for an iterative process involving A, in which arbitrarily large powers A^n of A are applied to vectors, to be stable is that $\|A\|_2 < 1$.

A.2 B-spline basis functions

Useful introductory reference books on splines are (Faux and Pratt, 1979; Foley et al., 1990). An excellent, comprehensive reference is (Bartels et al., 1987).

In chapter 3, spline functions are written as a linear combination of a number of spline "basis functions." Basis functions are constructed using the following general rule which can be used to define any arbitrary set of polynomial splines. Let $B_{n,d}$ be the nth basis function for a spline of order d. Then for a spline with single knots of unit spacing, the following recursive rule applies:

Ground instance

$$B_{n,1}(s) = \begin{cases} 1 & \text{if } n < s < n+1 \\ 0 & \text{otherwise} \end{cases}$$

Inductive step

$$B_{n,d}(s) = \frac{(s-n)B_{n,d-1}(s) + (n+d-s)B_{n+1,d-1}(s)}{d-1}$$

and some examples are shown in figure A.1. These functions satisfy the following conditions:

Support $B_{n,d}(s) = 0$ for $s \notin [n, n+d)$

Positivity $B_{n,d}(s) \geq 0$ for all s

Normalisation $\sum_{-\infty}^{\infty} B_{n,d}(s) = 1$ for all s

Translational invariance $B_{n+1,d}(s) = B_{n,d}(s-1)$ for all s

and further, there is a smoothness constraint for $d > 1$, namely that $B_{n,d}$ has continuous $(d-2)$th derivative for all s and all $d > 1$.

Non-uniform B-spline functions

The spline basis functions generated above, for which the knots are uniformly spaced at unit intervals, can be generalised to produce spline functions with arbitrary knot spacing. Consider a spline with N_K knots at positions $k_0 \leq k_1 \leq \ldots \leq k_{N_K-1}$, then the recursive rule becomes

Ground instance

$$B_{n,1}(s) = \begin{cases} 1 & \text{if } k_n \leq s < k_{n+1} \\ 0 & \text{otherwise} \end{cases}$$

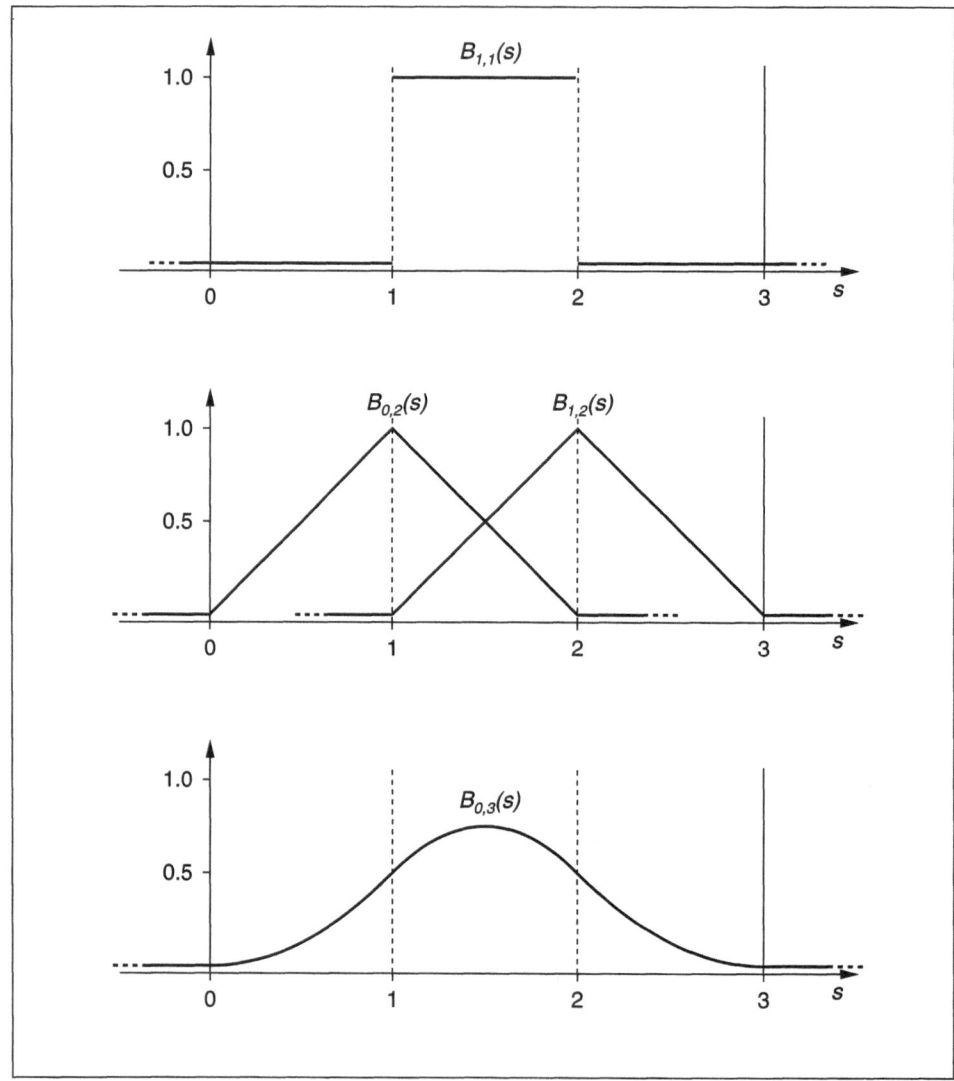

Figure A.1: A spline basis function $B_{n,d}$ of order d is built up recursively from basis functions of lower order.

Inductive step

$$B_{n,d}(s) = \frac{(s - k_n)B_{n,d-1}(s)}{k_{n+d-1} - k_n} + \frac{(k_{n+d} - s)B_{n+1,d-1}(s)}{k_{n+d} - k_{n+1}}$$

which reduces to the uniform case when $k_n = n$. The rule can be used to generate spline functions with knots of multiplicity m by setting m consecutive k_n to be equal. (Terms in the inductive step are zero when the denominator is zero. The validity of this can be shown by taking a limit as the knots approach one another — it is easy to see by induction that the basis function in the numerator is identically zero whenever the denominator is zero). The conditions of positivity and normalisation still hold in general, and now the support of the basis function $B_{n,d}$ is $[k_n, k_{n+d})$. The basis functions are clearly no longer necessarily translated copies of each other, however, and the introduction of a multiple knot reduces the smoothness of a basis function; the function is C^{d-1-m} at a knot of multiplicity m. This weakening of the smoothness property is the motivation for using multiple knots; it permits B-spline functions, and therefore curves, with sharp corners and discontinuities.

An implementation of B-spline functions

The recursive rule for generating $B_{n,d}$ can be converted into an algorithm by expressing each basis function as a sequence of polynomials $p_n(s)$ defined over the intervals $[k_n, k_{n+1})$. Since the support[1] of $B_{n,d}$ is $[k_n, k_{n+d})$, any spline basis function of order d can be represented using just d polynomials $B_{n,d}^\sigma$, one for each of the d spans S_σ in the support of $B_{n,d}$. Now the inductive step of the rule can be applied, over each interval in turn, to obtain each of the $B_{n,d}^\sigma$.

Where a B-spline contains multiple knots, some of the inter-knot intervals have zero length, so it is convenient to introduce the concept of "spans". These correspond to the non-empty inter-knot intervals above, and the span ends are called "breakpoints". A B-spline function, therefore, is a piecewise polynomial curve made up of a series of L spans $S_0 \ldots S_{L-1}$ connected at breakpoints $s_0 < s_1 < \ldots < s_L$. We adopt the convention that all spans are unit length, ($s_i = i$), so the basis functions making up a spline are uniquely determined by the knot multiplicities $m_0 \ldots m_L$ at the breakpoints. A periodic B-spline function is constructed by considering the basis functions to be

---

[1]The support of a function is the interval over which it is non-zero.

periodic over the interval $[0, L]$. A periodic B-spline function must have $m_0 = m_L$ and it has "multiple knot count"

$$M = \sum_1^L (m_i - 1)$$

while for an aperiodic spline $m_0 = m_L = d$ to control the boundary conditions of the spline, and

$$M = \sum_0^L (m_i - 1).$$

An L span B-spline is a linear combination of N_B basis functions, where

$$N_B = L + M = N_K - m_0$$

for a periodic spline, and

$$N_B = L + M + 1 - d = N_K - d$$

for an aperiodic spline. Thus, for example, a simple L span aperiodic quadratic B-spline is a linear combination of $N_B = L+2$ basis functions (see figure 3.6 on page 48). The relationship between spans, knots and basis functions is illustrated for two cases in figures A.2 and A.3.

Building a spline function from basis functions

B-spline functions can be evaluated efficiently using "span matrices." Over the span S_σ, any spline function is a linear combination of the basis functions $B_{b_\sigma,d} \ldots B_{b_\sigma+d-1,d}$ where

$$b_\sigma = \left(\sum_{i=0}^{\sigma} m_i \right) - d$$

so

$$x(s)^{[\sigma,\sigma+1)} = x(s)^\sigma = \sum_{b_\sigma}^{b_\sigma+d-1} x_i B_{i,d}(s)$$

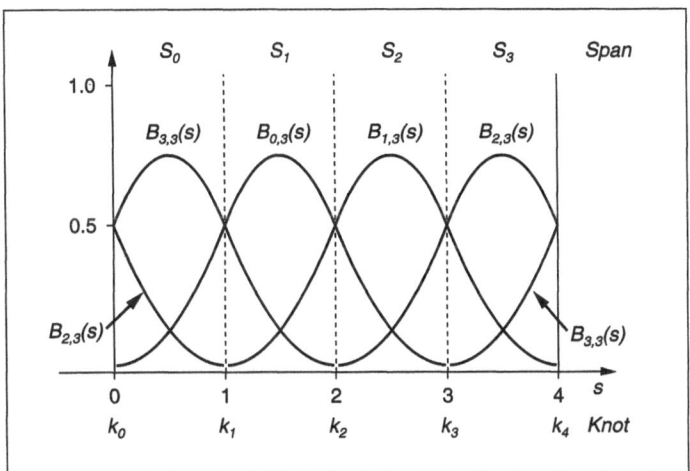

Figure A.2: A simple periodic B-spline with no multiple knots has $L = 4$ spans, $N_K = 5$ knots, and is a combination of $N_B = 4$ (periodic) basis functions.

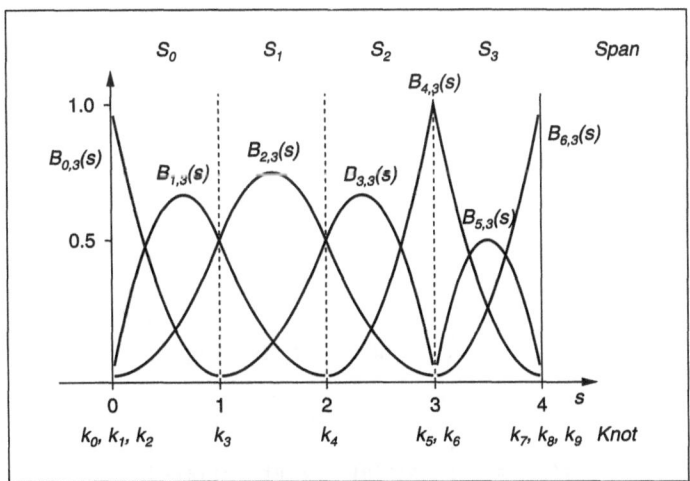

Figure A.3: An aperiodic spline must have knots of multiplicity d at its endpoints.
Here there is also a double knot between the third and fourth spans, leading to a discontinuity in the first derivative of the function. Here $L = 4$, $N_K = 10$, $N_{\dot{B}} = 7$.

(with obvious variations for periodic splines). For each span, therefore, we can compute a $d \times d$ span matrix B_σ^S such that

$$x(s+\sigma)^\sigma = (1 \ s \ \ldots \ s^{d-1})B_\sigma^S \begin{pmatrix} x_{b_\sigma} \\ x_{b_\sigma+1} \\ \vdots \\ x_{b_\sigma+d-1} \end{pmatrix}$$

where the ith column of the span matrix corresponds to the polynomial coefficients of the basis function $B_{b_\sigma+i-1,d}$ over the interval of that span (in practice it is convenient to define each span matrix over the interval $[0,1)$). The span matrices for the spline in figure A.3 are as follows:

$$B_0^S = \begin{pmatrix} 1.00 & 0.00 & 0.00 \\ -2.00 & 2.00 & 0.00 \\ 1.00 & -1.50 & 0.50 \end{pmatrix}$$

$$B_1^S = \begin{pmatrix} 0.50 & 0.50 & 0.00 \\ -1.00 & 1.00 & 0.00 \\ 0.50 & -1.00 & 0.50 \end{pmatrix}$$

$$B_2^S = \begin{pmatrix} 0.50 & 0.50 & 0.00 \\ -1.00 & 1.00 & 0.00 \\ 0.50 & -1.50 & 1.00 \end{pmatrix}$$

$$B_3^S = \begin{pmatrix} 1.00 & 0.00 & 0.00 \\ -2.00 & 2.00 & 0.00 \\ 1.00 & -2.00 & 1.00 \end{pmatrix}$$

and the algorithm used to calculate them is given in figure A.4. Once span matrices have been computed off-line, the spline can be evaluated efficiently at any values of s

To calculate span matrices for a non-periodic B-spline of order d with knot multiplicities m_i, $0 \leq i \leq L$.

1. **Calculate** the knot values k_i:

   (a) **Initialise:** $p = 0$, $q = 0$

   (b) **For** $i = 0 \ldots L$

      i. **For** $j = 1 \ldots m_i$
         $$k_p = q, \ p = p + 1$$

      ii. $q = q + 1$

2. **For** each span $\sigma = 0 \ldots L - 1$:

   (a) **Find** the index b_σ of the first basis function whose support includes the span.

   $$b_\sigma = \left(\sum_0^\sigma m_i \right) - d$$

   (b) **For** $i = 1 \ldots d$ recursively calculate the basis polynomial $B^\sigma_{b_\sigma + i - 1, d}$ for span σ using the following rule

      i. **Ground instance**

      $$B^\sigma_{n,1}(s) = \begin{cases} 1 & \text{if } k_n \leq \sigma < k_{n+1} \\ 0 & \text{otherwise} \end{cases}$$

      ii. **Recursive rule**

      $$B^\sigma_{n,d}(s) = \frac{(s + \sigma - k_n) B^\sigma_{n,d-1}(s)}{k_{n+d-1} - k_n} + \frac{(k_{n+d} - s - \sigma) B^\sigma_{n+1,d-1}(s)}{k_{n+d} - k_{n+1}}$$

      where terms are zero when the denominator is zero.

      iii. **Store** the coefficients of $B^\sigma_{b_\sigma + i - 1, d}$ as the ith column of the $d \times d$ span matrix B^S_σ, where the top row corresponds to the constant polynomial coefficient.

Figure A.4: Algorithm to calculate span matrices for aperiodic B-splines. *Obvious modifications must be made for the periodic case.*

and $x_0 \ldots x_{N_B-1}$. For notational purposes it is convenient also to define the $d \times N_B$ "placement matrices" G_σ:

$$(G_\sigma)_{ij} = \begin{cases} 1 & \text{if } i - b_\sigma = j \\ 0 & \text{otherwise} \end{cases} \tag{A.3}$$

so that

$$x(s + \sigma) = (1 \ s \ \ldots \ s^{d-1})B_\sigma^S G_\sigma \mathbf{Q}$$

where $0 \leq s < 1$. The derivative of the function can be calculated as

$$x'(s + \sigma) = (0 \ 1 \ \ldots \ (d-1)s^{d-2})B_\sigma^S G_\sigma \mathbf{Q}$$

and so when considering a spline curve,

$$\begin{pmatrix} x \\ y \end{pmatrix} = (1 \ s \ \ldots \ s^{d-1} \ 1 \ s \ \ldots \ s^{d-1}) \begin{pmatrix} B_\sigma^S G_\sigma & 0 \\ 0 & B_\sigma^S G_\sigma \end{pmatrix} \begin{pmatrix} \mathbf{Q}^x \\ \mathbf{Q}^y \end{pmatrix},$$

the tangent to the curve is given by

$$\begin{pmatrix} x' \\ y' \end{pmatrix} = (0 \ 1 \ \ldots \ (d-1)s^{d-2} \ 0 \ 1 \ \ldots \ (d-1)s^{d-2}) \begin{pmatrix} B_\sigma^S G_\sigma & 0 \\ 0 & B_\sigma^S G_\sigma \end{pmatrix} \begin{pmatrix} \mathbf{Q}^x \\ \mathbf{Q}^y \end{pmatrix},$$

and the normal is given by

$$\begin{pmatrix} n_x \\ n_y \end{pmatrix} = \begin{pmatrix} -y' \\ x' \end{pmatrix}.$$

Calculating the spline metric matrix

Using the span matrices it is straightforward to compute the spline metric matrix $\mathcal{B}$, where

$$\mathcal{B} = \frac{1}{L} \int_0^L \mathbf{B}(s)\mathbf{B}(s)^T \, ds \tag{A.4}$$

as described on page 50.

$$\mathcal{B} = \frac{1}{L}\sum_{\sigma=0}^{L-1}\left(\int_0^1 \mathbf{B}(s+\sigma)\mathbf{B}(s+\sigma)^T\,ds\right)$$

$$= \frac{1}{L}\sum_{\sigma=0}^{L-1}G_\sigma^T(B_\sigma^S)^T\mathcal{P}B_\sigma^S G_\sigma$$

where

$$\mathcal{P} = \int_0^1 \begin{pmatrix} 1 \\ \vdots \\ s^{d-1} \end{pmatrix}\begin{pmatrix} 1 & \cdots & s^{d-1} \end{pmatrix}\,ds,$$

the "Hilbert" matrix (Barnett, 1990) whose coefficients are

$$\mathcal{P}_{ij} = \frac{1}{i+j-1}.$$

Similarly, the matrix $\mathcal{B}'$ used on page 65 to define the area coefficients $\mathcal{A}$, was defined as

$$\mathcal{B}' = \frac{1}{L}\int_0^L \mathbf{B}(s)\mathbf{B}'^T(s)\,ds$$

and may be calculated as follows:

$$\mathcal{B} = \frac{1}{L}\sum_{\sigma=0}^{L-1}G_\sigma^T(B_\sigma^S)^T\mathcal{P}'B_\sigma^S G_\sigma$$

where

$$\mathcal{P}' = \int_0^1 \begin{pmatrix} 1 \\ \vdots \\ s^{d-1} \end{pmatrix}\begin{pmatrix} 0 & \cdots & (d-1)s^{d-2} \end{pmatrix}\,ds, \quad \text{so}$$

$$\mathcal{P}'_{ij} = \begin{cases} 0 & \text{if } i = j = 1 \\[2mm] \frac{j-1}{i+j-2} & \text{otherwise} \end{cases}.$$

A.3   Probability

An excellent introductory text on probability is (Papoulis, 1990). It is impossible to cover the necessary ground here, but since much of the argument in the book is probabilistic, a few basic concepts are reviewed here.

Probability distributions   A continuous random variable x taking real values $x \in \mathbb{R}$ has a probability distribution defined by its **density function** $p(x) \geq 0$. Its interpretation is that, for an interval $\mathcal{I} = [a, b]$:

$$P(x \in \mathcal{I}) = \int_a^b p(x) \, dx.$$

This definition extends to a multi-dimensional random variable $\mathbf{X} \in \mathbb{R}^{N_X}$ so that, for a subset $\mathcal{I} \in \mathbb{R}^{N_X}$:

$$P(\mathbf{X} \in \mathcal{I}) = \int_{\mathcal{I}} p(\mathbf{X}) \, d\mathbf{X}.$$

Since $\mathbf{X}$ has to take some value, p must satisfy the normalisation property that

$$\int_{\mathbb{R}^{N_X}} p(\mathbf{X}) \, d\mathbf{X} = 1.$$

A **conditional** distribution for X specifies the probable values of $\mathbf{X}$ *given* that the value of some related variable $\mathbf{Y}$ is known and is defined by the density $p(X|Y)$. This is interpreted, as before, via integration:

$$P(\mathbf{X} \in \mathcal{I}|\mathbf{Y}) = \int_{\mathcal{I}} p(\mathbf{X}|\mathbf{Y}) \, d\mathbf{X}.$$

The associated normalisation property is

$$\int_{\mathbb{R}^{N_X}} p(\mathbf{X}|\mathbf{Y}) \, d\mathbf{X} = 1.$$

Mean and variance   The expectation or mean of the random variable $\mathbf{X}$, denoted $\mathcal{E}[\mathbf{X}]$, is

$$\mathcal{E}[\mathbf{X}] = \int_{\mathbb{R}^{N_X}} p(\mathbf{X}) \, d\mathbf{X}$$

which is a linear operation so that

$$\mathcal{E}[A\mathbf{X} + \mathbf{b}] = A\mathcal{E}[\mathbf{X}] + \mathbf{b}.$$

The variance of $\mathbf{X}$, denoted $\mathcal{V}[\mathbf{X}]$ is defined as an expectation:

$$\mathcal{V}[\mathbf{X}] = \mathcal{E}[(\mathbf{X} - \overline{\mathbf{X}})(\mathbf{X} - \overline{\mathbf{X}})^T]$$

where $\overline{\mathbf{X}} = \mathcal{E}[\mathbf{X}]$. It scales quadratically, as $\mathcal{V}[A\mathbf{X} + \mathbf{b}] = A\mathcal{V}[\mathbf{X}]A^T$, and is invariant to the additive constant $\mathbf{b}$, naturally enough since it is a measure of "spread" about the mean. It is also known as the "covariance matrix" of $\mathbf{X}$ and must be symmetric and "positive semi-definite" (all eigenvalues positive or zero).

Bayes' rule   Suppose a density $p(\mathbf{X})$ is given, based on prior knowledge of the state $\mathbf{X}$ of some system and its likely values. Then suppose that observations $\mathbf{Z}$ are made from an imperfect sensing device which is characterised by its **observation density** $p(\mathbf{Z}|\mathbf{X})$, specifying the likely range of observations *given* a particular system state $\mathbf{X}$. Then Bayes' rule gives the **posterior density** $p(\mathbf{X}|\mathbf{Z})$:

$$p(\mathbf{X}|\mathbf{Z}) = kp(\mathbf{Z}|\mathbf{X})p(\mathbf{X}),$$

where k is a constant, not dependent on $\mathbf{X}$, whose value can be determined if need be by insisting that the posterior be normalised. Note that $p(\mathbf{Z}|\mathbf{X})$ is also known as a **likelihood** function for $\mathbf{X}$.

Estimation   When both the prior and observation density are available, a common way to estimate the value of $\mathbf{X}$ given observations $\mathbf{Z}$ is simply to find the $\mathbf{X}$ that maximises the posterior:

$$\hat{\mathbf{X}} = \arg\max_{\mathbf{X}} p(\mathbf{X}|\mathbf{Z}).$$

This is known as the MAP (Maximum A Posteriori) estimate. Alternatively, if no prior is available, an estimator can be defined by

$$\hat{\mathbf{X}} = \arg\max_{\mathbf{X}} p(\mathbf{Z}|\mathbf{X}),$$

the MLE (Maximum Likelihood Estimator).

Normal distribution   Much use is made in the book of multi-variate normal or Gaussian distributions. A vector variable $\mathbf{X}$ distributed as a Gaussian is denoted

$$\mathbf{X} \sim \mathcal{N}(\overline{\mathbf{X}}, P) \tag{A.5}$$

where $\overline{\mathbf{X}}$ is the mean of the distribution and P is its covariance matrix, assumed non-singular. The density function for $\mathbf{X}$ is

$$p(\mathbf{X}) = \frac{1}{\sqrt{2\pi}^{N_X}} \frac{1}{\sqrt{\det P}} \exp -\frac{1}{2}(\mathbf{X} - \overline{\mathbf{X}})^T S(\mathbf{X} - \overline{\mathbf{X}})$$

where $S = P^{-1}$, the **information matrix**.

Alternatively, given a vector $\mathbf{w}$ of N_X independent standard normal distributions, so that each $w_n \sim \mathcal{N}(0,1)$ and $\mathbf{w} \sim \mathcal{N}(\mathbf{0}, I_{N_X})$, $\mathbf{X}$ can be described as a linear transformation of $\mathbf{w}$:

$$\mathbf{X} = B\mathbf{w} + \overline{\mathbf{X}}$$

where $B = \sqrt{P}$. Circles $|\mathbf{w}| < c$ map to **confidence** ellipsoids in $\mathbf{X}$ space, regions which contain the value of $\mathbf{X}$ with probability $\chi^2_{N_X}(c)$, where χ^2_ν is the "chi-squared distribution function" for ν degrees of freedom, and can be found in statistical tables. For example, for $N_X = 2$,

$$P(|\mathbf{w}| < 2) = 86\% \quad \text{and} \quad P(|\mathbf{w}| < 3) = 99\%.$$

Appendix B

Stochastic dynamical systems

B.1 Continuous-time first-order dynamics

A first-order AR process (9.7) can be regarded as a first-order "stochastic differential equation" (SDE) in continuous time that has been sampled at regular intervals. If the sampling interval is τ so that $t_k = k\tau$, then the AR process is obtained by integrating the SDE over successive sampling intervals. The SDE is expressed as

$$\dot{\mathbf{X}} = F(\mathbf{X} - \overline{\mathbf{X}}) + G\dot{\mathbf{w}} \tag{B.1}$$

where $\mathbf{X}(t)$ is a vector in shape-space, F and G are $N_X \times N_X$ matrices and $\mathbf{w}(t)$ is a N_X-dimensional vector of independent, univariate Brownian processes in continuous time. A univariate Brownian process w has the property that the value $w(t)$ has a Gaussian distribution with $\mathcal{E}[w(t)] = 0$ and $\mathcal{V}[w(t)] = t$. The derivative $\dot{w}(t)$ is a "white noise" signal, that is one with equal power at all frequencies. The coefficients F are the deterministic parameters of the process, in the sense that its eigenvalues λ_i are the so-called "poles" of the AR process, constants with units of inverse time that represent the rates of decay of the various characteristic motions of the system. (This applies to the case that all poles are real-valued and negative. Any real positive pole will cause the process to be unstable. There is also the possibility of complex poles, representing oscillations or damped oscillations.) The matrix G represents a coupling to the multi-dimensional white noise $\dot{\mathbf{w}}$ that is driving the dynamical system. As in the discrete case, there is a mean-state and a Riccati equation for continuous time:

$$\dot{\hat{\mathbf{X}}} = F(\hat{\mathbf{X}} - \overline{\mathbf{X}}) \quad \text{and} \quad \dot{P} = FP + PF^T + Q \tag{B.2}$$

where the "covariance coefficient" $Q = GG^T$.

Conversion between continuous and discrete time

A continuous-time SDE can be converted to a discrete-time form (Gelb, 1974; Astrom and Wittenmark, 1984) by computing A and C directly from F and Q:

$$A = \exp F\tau \quad \text{and} \quad C = \int_0^\tau (\exp Ft)\, Q (\exp F^T t)\, dt. \tag{B.3}$$

It is possible to evaluate the integral for C exactly by diagonalising F but in practice the following approximation, to lowest order in τ, is convenient (and particularly so for the second-order process):

$$A = (I - F\tau)^{-1} \quad \text{and} \quad C = Q\tau \quad \text{so that} \quad B = G\sqrt{\tau}. \tag{B.4}$$

The approximation for A is known as the "backward difference" approximation and is preferable to the more obvious "forward difference" $A = F\tau$ because it preserves stability: that is, any SDE that is stable is approximated as a stable AR process, with $\|A\|_2 < 1$.

Power spectrum

In chapter 9, a form (9.12) on page 199 for the power spectrum of a first-order ARP is used. That form is derived briefly here. Restricting $\mathbf{X}(t)$ in the continuous process above to be one-dimensional $X(t)$, so that F and G are scalar coefficients, suppose that

$$X(t) \propto \exp 2\pi i f t,$$

and set its mean to zero for simplicity. Then (B.1) becomes

$$2\pi i f X = FX + G\dot{w}$$

so that

$$X = \frac{G\dot{w}}{2\pi i f - F}$$

and the power spectrum

$$S_{XX}(f) = \left| \frac{G}{2\pi i f - F} \right|^2 S_{\dot{w}\dot{w}}(f),$$

where $|\cdot|$ denotes complex modulus. Now the power spectrum $S_{\dot{w}\dot{w}}(f)$ of white noise is constant, so

$$S_{XX}(f) \propto \frac{G^2}{4\pi^2 f^2 + F^2}$$

which can be rewritten directly in the required form.

B.2   Second-order dynamics in continuous time

A second-order SDE can be written, as it was in the discrete case, as a first-order one in a suitable state-space:

$$\dot{\mathcal{X}} = F(\mathcal{X} - \overline{\mathcal{X}}) + G\dot{\mathbf{w}} \tag{B.5}$$

where now

$$F = \begin{pmatrix} 0 & I \\ F_1 & F_2 \end{pmatrix} \quad \text{and} \quad G = \begin{pmatrix} 0 \\ G_0 \end{pmatrix}. \tag{B.6}$$

This normal form is consistent with a state-space representation in terms of position and velocity:

$$\mathcal{X} = \begin{pmatrix} \mathbf{X} \\ \dot{\mathbf{X}} \end{pmatrix}.$$

The white noise $\dot{\mathbf{w}}$ can be interpreted mechanically as a (generalised) force applied to a particle whose configuration is $\mathbf{X}$. Now, from (B.3),

$$\mathcal{X}(t_k) - \overline{\mathcal{X}} = A'(\mathcal{X}(t_{k-1}) - \overline{\mathcal{X}}) + B'\mathbf{w}_k$$

where $A' = \exp F\tau$,

which can be written in terms of its submatrices as

$$A' = \begin{pmatrix} A'_{11} & A'_{12} \\ A'_{21} & A'_{22} \end{pmatrix}.$$

The matrix A' does not yet conform to the normal form (9.18) on page 204 for discrete coefficients A, which would require $A'_{11} = 0$ and $A'_{12} = I$. To reach the normal form, a coordinate transformation $\mathcal{X} \to M\mathcal{X}$ with

$$M = \begin{pmatrix} I & 0 \\ A'_{11} & A'_{12} \end{pmatrix}$$

must be applied. The transformed process has $A = M^{-1}A'M$ which is in normal form, and $B = M^{-1}B'$ which is only approximately (for small τ) in the normal form for B in (9.18) on page 204.

The inverse transformation, obtaining continuous parameters F and G from discrete ones A and B requires a matrix logarithm

$$F' = \frac{1}{\tau} \log A$$

followed by a coordinate change, similar to the one above, to reach the normal form.

Power spectrum

The expression (9.13) on page 200 for a second-order power spectrum is obtained from (B.5) using a frequency analysis similar to the one in the first-order case.

B.3   Accuracy of learning

The claims concerning accuracy of learning, stated in chapter 11 on page 241, are justified here. First the proportional error in the discrete dynamical parameters a_1, a_2 and b_0 is obtained. Then this is used to derive the proportional error of the underlying continuous parameters f, β and $\bar{p}$.

Discrete analysis

The error of estimators $\hat{a}_1$, $\hat{a}_2$ and $\hat{b}_0$ is derived from the Fisher information measure for a maximum likelihood estimator (Kendall and Stuart, 1979). Asymptotically, for large M,

$$\mathcal{V}[b_0|\hat{b}_0]^{-1} = -\mathcal{E}\left[\partial^2 L/\partial b_0^2\right],$$

where L is the log-likelihood function (11.1) on page 237. Using (11.2) on page 237, this gives

$$\mathcal{V}[b_0|\hat{b}_0] = \frac{\hat{b}_0^2}{2(M-2)}$$

so that the proportional error in $\hat{b}_0$, denoted $\Delta\hat{b}_0$, is

$$\Delta\hat{b}_0 = \frac{\sqrt{\mathcal{V}[b_0|\hat{b}_0]}}{\hat{b}_0} = \frac{1}{\sqrt{2(M-2)}}. \tag{B.7}$$

Applying a similar analysis to the vector $\mathbf{a} = (a_1, a_2)^T$ obtains error variances and covariances for a_1 and a_2. This gives

$$
\begin{aligned}
\mathcal{V}[\mathbf{a}|\hat{\mathbf{a}}]^{-1} &= -\mathcal{E}\left[\partial^2 L/\partial \mathbf{a}^2\right] \\
&= \mathcal{E}\left[\frac{1}{\hat{b}_0^2}\begin{pmatrix} r_{11} & r_{12} \\ r_{21} & r_{22} \end{pmatrix}\right] \\
&= (M-2)\mathcal{E}[\frac{1}{\hat{b}_0^2}\mathcal{X}(t)\mathcal{X}(t)^T] \\
&= (M-2)\frac{1}{\hat{b}_0^2}\mathcal{P}_\infty
\end{aligned}
$$

where r_{ij} are the auto-correlation coefficients from the learning algorithm of figure 11.2 on page 238. Covariance $\mathcal{P}_\infty$ is obtained as the steady-state solution for $\mathcal{P}$ in the state equation (9.19) on page 204 for the process. After some manipulation, this gives

$$\mathcal{V}[\mathbf{a}|\hat{\mathbf{a}}] = \frac{1+\hat{a}_2}{M-2}\begin{pmatrix} 1-\hat{a}_2 & -\hat{a}_1 \\ -\hat{a}_1 & 1-\hat{a}_2 \end{pmatrix}. \tag{B.8}$$

For all 3 parameters a_1, a_2 and b_0 it seems that estimator error $\propto 1/\sqrt{M-2}$ so that a typical training sequence of 1000 video fields should lead to error of just a few percent. This is a little misleading however because small changes in a_1 and a_2 can have a substantial effect on the ARP model. Looking at continuous parameters β, f and $\bar{p}$ gives a clearer picture.

Continuous analysis

Making the assumption that $\beta\tau \ll 1$ then, from (9.25) on page 206,

$$\beta\tau \approx \frac{1}{2}(1 + a_2)$$

so that, using (B.8),

$$\mathcal{V}[\beta\tau] \approx \frac{1}{4}\mathcal{V}[a_2] \approx \frac{1}{M-2}\hat{\beta}\tau$$

and finally

$$\Delta\hat{\beta} = \frac{\sqrt{\mathcal{V}[\beta]}}{\hat{\beta}} = \frac{1}{\sqrt{\hat{\beta}\tau}}\frac{1}{\sqrt{M-2}} \approx \frac{1}{\sqrt{\hat{\beta}T}},$$

as claimed in chapter 11. A similar analysis for $\hat{f}$ shows that

$$\Delta\hat{f} \equiv \frac{\sqrt{\mathcal{V}[f]}}{\hat{\beta}} = \frac{1}{2\pi}\frac{1}{\sqrt{\hat{\beta}T}}$$

Note that the error $\Delta\hat{f}$ in the estimated frequency is defined here relative to $\hat{\beta}$. Finally, it remains to establish an error bound for the estimated value of $\bar{p}$. From (9.27) on page 206, and given that $\Delta\hat{b}_0$ (see above) can be neglected,

$$\Delta\hat{\bar{p}} = \frac{1}{2}\Delta\hat{\beta}\left(1 + \frac{\hat{\beta}}{\pi\hat{f}\sin 2\pi\hat{f}\tau}\right).$$

Appendix C

Further shape-space models

C.1 Recursive synthesis of shape-spaces

Chapter 4 ended with a discussion about shape-spaces for articulated motion and a summary comparing, for various hinged objects, the number of degrees of freedom of the object and the dimension of linear shape-space needed to represent its motion. There is a powerful general rule, presented here, for building up the dimension of a shape-space as articulated components are tacked onto a body. The dimension is built up recursively, tacking one component on at a time. It will be convenient to write the equation of the curve in *homogeneous coordinates*, a standard geometric tool in graphics and computer vision (Faugeras, 1993; Foley et al., 1990), giving

$$
\mathbf{r}_h(s) = \begin{pmatrix} x(s) \\ y(s) \\ 1 \end{pmatrix}
$$

so that the curve ranges over the shape-space swept out by

$$
\mathbf{r}_h(s) = T\bar{\mathbf{r}}_h(s)
$$

where the template $\bar{\mathbf{r}}_h(s)$ (also in homogeneous coordinates) may be either two-dimensional or three-dimensional as appropriate and T is a linear transformation. For example, the Euclidean similarities discussed earlier are represented by the following transformation T:

$$T\mathbf{r} = \begin{pmatrix} X_1 \\ X_2 \end{pmatrix} + \begin{pmatrix} X_3 & -X_4 \\ X_4 & X_3 \end{pmatrix} \mathbf{r}$$

which can conveniently be written as a 3×3 matrix in homogeneous coordinates:

$$T = \begin{pmatrix} X_3 & -X_4 & X_1 \\ X_4 & X_3 & X_2 \\ 0 & 0 & 1 \end{pmatrix}. \tag{C.1}$$

This is appropriate, of course, for a body moving rigidly in the plane. More generally, it applies to the end link of a series of hinged links attached to a base that moves rigidly in the plane.

The new component is attached so that it is free to be acted on by transformations T' relative to the end body. For example, a simple planar hinge is represented in homogeneous coordinates by

$$T'(\theta) = \begin{pmatrix} \cos\theta & -\sin\theta & 0 \\ \sin\theta & \cos\theta & 0 \\ 0 & 0 & 1 \end{pmatrix}.$$

Incremental rule: now the general rule can be stated, that when the hinged component is added, the dimension of the shape-space is increased by:

$$\dim\{T, TT'\}^+ - \dim\{T\}^+, \tag{C.2}$$

where $\{T\}^+$ is the vector space of transformations T and $\{T_1, T_2\}^+$ denotes the vector space spanned by the two transformations taken jointly (simply concatenating the elements of T_1 and T_2 into one vector). To make this clear, we will work through the rule using the examples for T and T' given above of a planar base element moving rigidly in the plane with a single hinged component.

Planar rigid body with hinged appendage

First of all, $\dim\{T\}^+ = 4$, clearly, since T has 4 independent linear parameters $X_1, \ldots, X_4$. Next we compute TT':

$$TT' = \begin{pmatrix} \gamma & -\delta & X_1 \\ \delta & \gamma & X_2 \\ 0 & 0 & 1 \end{pmatrix} \tag{C.3}$$

where $\gamma = X_3 \cos\theta - x_4 \sin\theta$ and $\delta = X_4 \cos\theta + X_3 \sin\theta$ so now

$$\dim\{T, TT'\}^+ = \dim\{X_1, X_2, X_3, X_4, \gamma, \delta\}^+ = 6$$

so that adding the hinge increases the dimension of shape-space by

$$\dim\{T, TT'\}^+ - \dim\{T\}^+ = 6 - 4 = 2$$

and the dimension of the new space is increased from 4 to 6.

Further hinged appendages

Now suppose we want to add a further hinged appendage. If it is added to the main body (figure C.1a), the argument above is unchanged (it is not in the least affected by the existence of the previous appendage) and the increase in dimension is still 2. Now the total dimension of the shape-space increases to 8. If instead the new appendage is tacked onto the end of the previous appendage (figure C.1b) we can apply the general method as follows. The matrix TT' in (C.3) above becomes the new T, and now

$$T' \equiv T'(\phi) = \begin{pmatrix} \cos\phi & -\sin\phi & 0 \\ -\sin\phi & \cos\phi & 0 \\ 0 & 0 & 1 \end{pmatrix},$$

where ϕ is the angle of the latest appendage. Now the argument proceeds exactly as for the addition of the first appendage except that we have γ, δ and ϕ where before we had X_3, X_4 and θ, so again the subspace dimension is increased by 2. Alternatively, a quicker way to get to the same conclusion, is simply to note that the shape-space of the

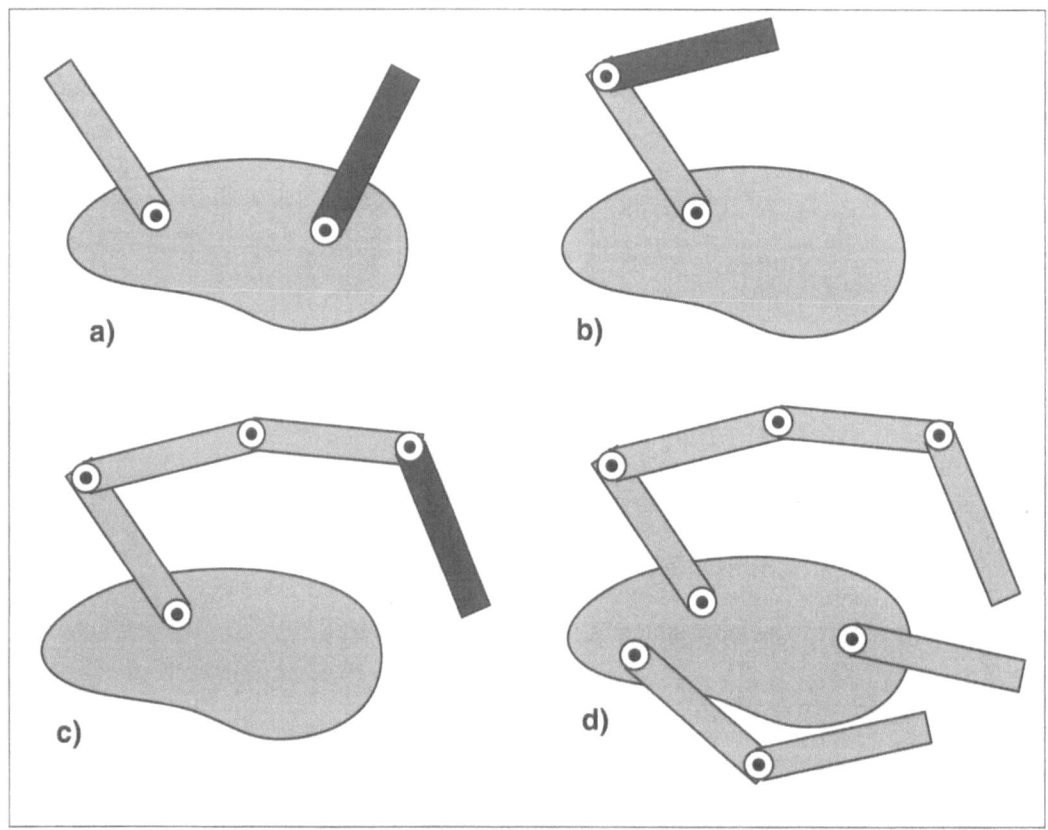

Figure C.1: Hinged appendages *Hinging an appendage onto: a) the main body; b) the end of another appendage; c) the end of a chain of appendages. d) A general planar articulated body consisting of a base in rigid motion with n hinged bodies.*

first appendage, considered in isolation, is the space of Euclidean similarities. (It can execute any rigid motion given that it is hinged to a base that can execute any rigid motion.) Therefore we are simply solving again the problem of computing the increase in shape-space dimension when an appendage is hinged to a base whose shape-space is the Euclidean similarities. Clearly we could continue to add appendages to a chain (figure C.1c), adding 2 to the shape-space dimension each time.

Now a simple inductive argument shows the following rule. A rigid planar body with n hinged appendages (figure C.1d) has a shape-space with dimension $4 + 2n$.

This is true regardless of how the hinges are arranged, provided there is no *closed* kinematic chain (sequence of hinged bodies forming a loop).

Adding telescopic appendages

How is the dimension of shape-space affected if an appendage is added with a "prismatic" or telescopic joint? In that case

$$
T' \equiv T'(d) = \begin{pmatrix} 1 & 0 & d \\ 0 & 1 & 0 \\ 0 & 0 & 1 \end{pmatrix},
$$

where d is the variable length by which the joint is extended and taking T for Euclidean similarities of the base object as in (C.1), gives

$$
TT' = \begin{pmatrix} X_3 & -X_4 & X_3 d + X_1 \\ X_4 & X_3 & X_4 d + X_2 \\ 0 & 0 & 1 \end{pmatrix},
$$

so that

$$
\dim(\{T, TT'\}^+) = \dim(\{X_1, X_2, X_3, X_4, X_3 d, X_4 d\}^+) = 6
$$

(since $X_3 d, X_4 d$ extend the basis by two elements — even though there is only one new degree of freedom d, it appears non-linearly and requires two degrees of freedom to represent linearly). Just as in the hinged case therefore, each telescopic appendage to a rigid body also raises the dimension of shape-space by two.

Planar body with co-planar appendage, in three dimensions

From earlier discussion in chapter 4, we know that images of a planar body in three dimensions form an affine space, so that $\{T\}^+$ is the usual 6-dimensional planar affine space. Unfortunately, unlike rigid planar bodies in which hinged appendages cost only 2 degrees of freedom each, in the planar affine case they come relatively expensively. Each requires 4 degrees of freedom so that insisting on linear parameterisation is relatively costly.

The argument, applying the incremental rule (C.2) is as follows. Transformations
are

$$T = \begin{pmatrix} X_3 & X_6 & X_1 \\ X_5 & X_4 & X_2 \\ 0 & 0 & 1 \end{pmatrix} \quad \text{and} \quad T' = \begin{pmatrix} \cos\phi & -\sin\phi & 0 \\ -\sin\phi & \cos\phi & 0 \\ 0 & 0 & 1 \end{pmatrix}$$

and after a little calculation

$$\dim(\{T, TT'\}^+) = \dim(\{X_1, X_2, ((X_n \cos\phi, X_n \sin\phi), n = 3, \ldots, 6)\}^+) = 10,$$

an increase of 4 over the 6 affine degrees of freedom.

Proof of incremental rule

For completeness, a proof of the rule (C.2) for incrementing the dimension of shape-
space is included here. Consider a base shape γ which we would normally represent
by a template either as a parameterised curve $\bar{\mathbf{r}}(s)$ in two dimensions or $\overline{\mathbf{R}}(s)$ in three
dimensions, or as a control point vector $\overline{\mathbf{Q}}$ in two or three dimensions. It is subject to
linear transformations T onto the image plane, parameterised (not necessarily linearly)
by a parameter set λ, giving a set of image shapes $\{T(\lambda)\gamma, \ \lambda \in \Lambda\}$. This set spans
a vector space denoted by the closure $\{T(\lambda)\gamma, \ \lambda \in \Lambda\}^+$ which generally (for non-
degenerate γ) is isomorphic to the closure of the space of transformations $\{T(\lambda), \lambda \in$
$\Lambda\}^+$, regardless of the particular shape γ.

Next, it is assumed that the body to which an appendage is about to be added
is already articulated so that, already attached to the base, are a set of components
transformed relative to the base by a set of linear transformations $T_n(\lambda_n)$, $\lambda_n \in$
Λ_n for $n = 1 \ldots N$. The nth component is thus transformed into the image plane
by the transformation $TT_n(\lambda_n)$. Finally, the appended component is attached via
$T'(\lambda')$, $\lambda' \in \Lambda'$. It is assumed that the parameters $\lambda_1, \ldots, \lambda_N, \lambda'$ are all independent
— the hinging/telescopic actions of the individual components are not coupled and this
is where closed kinematic chains are excluded. We also assume that all components
are non-degenerate so that we can continue to consider the transformations only and
drop any reference to the component shapes themselves.

The independence of parameters for components means that

$$\dim\{T, TT'\}^+ + \dim\{T, TT_1, \ldots, TT_N\}^+ = \dim\{T, TT_1, \ldots, TT_N, TT'\}^+ + \dim\{T\}^+$$

(omitting for simplicity explicit reference to parameters λ, $\lambda_1 \ldots \lambda_n$ and λ') and this is simply rearranged into a formula for the increase in shape-space dimension:

$$\dim\{T, TT_1, \ldots, TT_N, TT'\}^+ - \dim\{T, TT_1, \ldots, TT_N\}^+ = \dim\{T, TT'\}^+ - \dim\{T\}^+,$$

the right hand side of which is the required formula (C.2), simplified in that it involves only the base and the new component.

Silhouettes

For smooth silhouette curves, it can be shown that a shape-space of dimension 11 is appropriate. This shape-space representation of the curve is an approximation, valid for sufficiently small changes of viewpoint. The proof of this result follows from results in the computer vision literature about the projection of silhouettes into images (Giblin and Weiss, 1987; Blake and Cipolla, 1990; Vaillant, 1990; Koenderink, 1990; Cipolla and Blake, 1992b).

Glossary of notation

SYMBOL	MEANING	see page
$\hat{x}$	an *estimate* for the quantity x	
$\tilde{x}$	a *prediction* of the quantity x	
$\dot{x}$	*temporal derivative* of x	
$\mathbf{x} \cdot \mathbf{y}$	scalar product of vectors $\mathbf{x}, \mathbf{y}$	
$\|\cdot\|$	norm for functions/curves	47,58
$\langle \cdot, \cdot \rangle$	inner product for functions/curves	47,58
$A \otimes B$	Kronecker product of matrices A, B	282
T	superscript denoting vector/matrix transpose	

$\mathbf{0}$	vector of zeros, of length N_B	
$\mathbf{1}$	vector of ones, of length N_B	
A	deterministic coefficient matrix in discrete dynamical model	204
A_1, A_2	components of A	204
$\mathbf{B}(s)$	vector of B-spline blending functions	44
B_0	stochastic coefficients in second-order discrete dynamical model	204
B	stochastic coefficient matrix in discrete dynamical model	193, 204
$\mathcal{B}$	metric matrix for B-spline functions	50
$B_m(s)$	mth B-spline blending function	43
C	covariance coefficient $C = BB^T$	242
d	order of spline polynomial	
$\mathcal{E}[Y]$	expectation of a random variable Y	294
H	observation matrix for Kalman filter	216

SYMBOL	MEANING	see page
$\mathcal{H}$	metric matrix for curves in shape-space $\mathcal{S}$	79
$\mathbf{h}(s)$	image measurement matrix for recursive curve-fitting	124
I_r	$r \times r$ identity matrix	
$K(t), \mathcal{K}(t)$	Kalman gain	216
k	index for discrete time $t_k = k\tau$	
κ	length factor for search lines	174
L	no. of spans on B-spline curve	45
M	no. of frames in a training sequence $\mathbf{X}_1, \ldots \mathbf{X}_M$	175
N	no. of sampled image features along a B-spline curve	124
N_B	no. of control points on B-spline curve	43
N_Q	dimension of spline space	65
N_X	dimension of configuration space (shape-space) $\mathcal{S}$	69
$\mathcal{N}(\overline{\mathbf{X}}, P)$	multi-variate Gaussian (normal) distribution	295
$\mathbf{n}(s, t)$	image-curve normal	122
ν_i	innovation due to ith image feature	124
P, P', P''	covariance components for shape $\hat{\mathbf{X}}(t)$	218
$\mathcal{P}(t)$	covariance of state $\hat{\mathcal{X}}(t)$	204, 213
$\tilde{\mathcal{P}}(t_k)$	covariance of predicted state-vector $\tilde{\mathcal{X}}(t_k)$	216
$\mathbf{q}_n$	control points for spline curve	53
$\mathbf{Q}$	vector of control points	58
$\mathbf{Q}^x$	vector of x-coordinates of control points	44, 58
$\mathbf{Q}^y$	vector of y-coordinates of control points	58

SYMBOL	MEANING	see page
$\mathbf{Q}_0$	control points of template curve	74
R	rotation matrix	283
$\mathbf{R}(s)$	space-curve (3D)	83
R_{ij}, R'_{ij}	auto-correlation coefficients of training sequence	244
R_i	training-sequence sums	244
$\mathbf{r}(s)$	image curve (2D)	53
$\rho_0(s)$	root-mean-square displacement at s on a curve	161
$\overline{\rho}_0$	root-mean-square displacement of curve	161
S, S_i	statistical information matrix	127
$\mathcal{S}_Q$	space of spline curves $\mathbf{Q}$	58
$\mathcal{S}$	shape-space	69
s	spatial parameter for curve	
σ	σ^2 is variance of image measurement process	127,169
t	time parameter	
T	duration $T = M\tau$ of an image sequence	
τ	sample interval for image capture	
$U(s)$	matrix mapping control point vector $\mathbf{Q}$ to image curve $\mathbf{r}(s)$	58
$\mathbf{u}$	translation vector	76
$\mathcal{U}$	metric matrix for B-spline parametric curves	59
$\mathcal{V}[Y]$	variance of a random variable Y	294
W	shape-matrix mapping from configuration $\mathbf{X}$ to control point vector $\mathbf{Q}$	74
$W^\dagger$	pseudo-inverse of mapping W	74
$\mathbf{w}_k$	discrete noise: vector of independent standard normal variables	193

SYMBOL	MEANING	see page
$\mathbf{X}$	curve shape-vector	74
$\overline{\mathbf{X}}$	mean value of curve configuration	160
$\mathcal{X}$	state-vector for image-based dynamics at discrete time k	204
$\hat{\mathcal{X}}(t_k)$	estimated state-vector from a tracker	214
$\tilde{\mathcal{X}}(t_k)$	predicted state-vector in a tracker	216
$\overline{\mathcal{X}}$	mean value of state $\mathcal{X}$	204
$\underline{\mathcal{X}}(t_k)$	history of states $\mathcal{X}(t_1), \ldots, \mathcal{X}(t_k)$	260
$\mathbf{Z}(t_k)$	aggregated vector of visual measurements at time $t = k\tau$	127
$\underline{\mathbf{Z}}(t_k)$	measurement history $\{\mathbf{Z}(t_1), \ldots, \mathbf{Z}(t_k)\}$ up to time $t = k\tau$	213

Bibliography

Adjoudani, A. and Benoit, C. (1995). On the integration of auditory and visual parameters in an HMM-based ASR. In *Proceedings NATO ASI Conference on Speechreading by Man and Machine: Models, Systems and Applications*, 461–472. NATO Scientific Affairs Division.

Allen, P., Yoshimi, B., and Timcenko, A. (1991). Real–time visual servoing. In *Proc. IEEE Int. Conf. Robotics and Automation*, 1, 851–856.

Aloimonos, J. (1990). Perspective approximations. *J. Image and Vision Computing*, 8, 177–192.

Aloimonos, J. (1993). *Active Perception*. Erlbaum.

Aloimonos, J., Weiss, I., and Bandyopadhyay, A. (1987). Active vision. In *Proc. 1st Int. Conf. on Computer Vision*, 35–54.

Amini, A., Owen, R., Staib, L., Anandan, P., and Duncan, J. (1991). *Non-rigid motion models for tracking the left ventricular wall*. Lecture notes in computer science: Information processing in medical images. Springer-Verlag.

Amini, A., Tehrani, S., and Weymouth, T. (1988). Using dynamic programming for minimizing the energy of active contours in the presence of hard constraints. In *Proc. 2nd Int. Conf. on Computer Vision*, 95–99.

Anderson, B. and Moore, J. (1979). *Optimal filtering*. Prentice Hall.

Arbogast, E. and Mohr, R. (1990). 3D structure inference from images sequences. In Baird, H., editor, *Proceedings of the Syntactical and Structural Pattern Recognition Workshop*, 21–37, Murray-Hill, NJ.

Astrom, K. (1970). *Introduction to stochastic control theory.* Academic Press.

Astrom, K. and Wittenmark, B. (1984). *Computer Controlled Systems.* Addison Wesley.

Ayache, N., Cohen, I., and Herlin, I. (1992). Medical image tracking. In Blake, A. and Yuille, A., editors, *Active Vision*, 285–302. MIT.

Ayache, N. and Faugeras, O. (1987). Building, registration and fusing noisy visual maps. In *Proc. 1st Int. Conf. on Computer Vision*, London.

Ayache, N. and Faverjon, B. (1987). Efficient registration of stereo images by matching graph descriptions of edge segments. *Int. J. Computer Vision*, 107–131.

Azarbayejani, A., Starner, T., Horowitz, B., and Pentland, A. (1993). Visually controlled graphics. *IEEE Trans. on Pattern Analysis and Machine Intelligence*, 15, 6, 602–605.

Bajcsy, R. (1988). Active perception. In *Proc. IEEE*, 76, 996–1005.

Baker, H. and Binford, T. (1981). Depth from edge and intensity based stereo. In *Proc. Int. Joint Conf. Artificial Intelligence*, 631–636.

Ballard, D. (1981). Generalising the Hough transform to detect arbitrary shapes. *Pattern Recognition*, 12, 2, 111–122.

Ballard, D. and Brown, C. (1982). *Computer Vision.* Prentice– Hall, New Jersey.

Bar-Shalom, Y. and Fortmann, T. (1988). *Tracking and Data Association.* Academic Press.

Barnett, S. (1990). *Matrices: Methods and Applications.* Oxford University Press.

Bartels, R., Beatty, J., and Barsky, B. (1987). *An Introduction to Splines for use in Computer Graphics and Geometric Modeling.* Morgan Kaufmann.

Bascle, B. and Deriche, R. (1995). Region tracking through image sequences. In *Proc. 5th Int. Conf. on Computer Vision*, 302–307, Boston.

Baumberg, A. and Hogg, D. (1994). Learning flexible models from image sequences. In Eklundh, J.-O., editor, *Proc. 3rd European Conf. Computer Vision*, 299–308. Springer–Verlag.

Baumberg, A. and Hogg, D. (1995a). An adaptive eigenshape model. *Proc. British Machine Vision Conf.*, 87–96.

Baumberg, A. and Hogg, D. (1995b). Generating spatiotemporal models from examples. In *Proc. British Machine Vision Conf.*, 2, 413–422.

Belhumeur, P. (1993). A binocular stereo algorithm for reconstructing sloping, creased, and broken surfaces in the preesence of half-occlusion. In *Proc. 4th Int. Conf. on Computer Vision*, 431–438.

Belhumeur, P., Hespanha, J., and Kriegman, D. (1996). Eigenfaces vs. Fisherfaces: recognition using class specific linear projection. In *Proc. 4th European Conf. Computer Vision*, number 800 in Lecture notes in computer science, 45–58. Springer-Verlag.

Bellman, R. and Dreyfus, S. (1962). *Applied dynamic programming*. Princeton University Press, Princeton, U.S.

Bennett, A. and Craw, I. (1991). Finding image features for deformable templates and detailed prior statistical knowledge. In *Proc. British Machine Vision Conf.*, 233–239.

Bergen, J., Anandan, P., Hanna, K., and Hingorani, R. (1992a). Hierarchical model-based motion estimation. In *Proc. 2nd European Conf. Computer Vision*, 237–252.

Bergen, J., Burt, P., Hingorani, R., and Peleg, S. (1992b). A three-frame algorithm for estimating two-component image motion. *IEEE Trans. on Pattern Analysis and Machine Intelligence*, 14, 9, 886–896.

Beymer, D. and Poggio, T. (1995). Face recognition from one example view. In *Proc. 5th Int. Conf. on Computer Vision*, 500–507.

Bishop, C. (1995). *Neural networks for pattern recognition*. Oxford.

Black, M. and Jepson, A. (1996). Eigentracking: robust matching and tracking of articulated objects using a view-based representation. In *Proc. 4th European Conf. Computer Vision*, 329–342.

Black, M. and Yacoob, Y. (1995). Tracking and recognizing rigid and non-rigid facial motions using local parametric models of image motion. In *Proc. 5th Int. Conf. on Computer Vision*, 374–381.

Blake, A. (1992). Computational modelling of hand-eye coordination. *Phil. Trans. R. Soc*, 337, 351–360.

Blake, A. and Cipolla, R. (1990). Robust estimation of surface curvature from deformation of apparent contours. In Faugeras, O., editor, *Proc. 1st European Conf. Computer Vision*, 465–474. Springer–Verlag.

Blake, A., Curwen, R., and Zisserman, A. (1993). A framework for spatio-temporal control in the tracking of visual contours. *Int. J. Computer Vision*, 11, 2, 127–145.

Blake, A. and Isard, M. (1994). 3D position, attitude and shape input using video tracking of hands and lips. In *Proc. Siggraph*, 185–192. ACM.

Blake, A., Isard, M., and Reynard, D. (1995). Learning to track the visual motion of contours. *J. Artificial Intelligence*, 78, 101–134.

Blake, A. and Marinos, C. (1990). Shape from texture: estimation, isotropy and moments. *J. Artificial Intelligence*, 45, 323–380.

Blake, A. and Yuille, A., editors (1992). *Active Vision*. MIT.

Bobick, A. and Wilson, A. (1995). A state-based technique for the summarisation and recognition of gesture. In *Proc. 5th Int. Conf. on Computer Vision*, 382–388.

Bolles, R., Baker, H., and Marimont, D. (1987). Epipolar-plane image analysis: An approach to determining structure. *Int. J. Computer Vision*, 1, 7–55.

Bookstein, F. (1988). Thin-plate splines and the decomposition of deformations. *IEEE Trans. on Pattern Analysis and Machine Intelligence*.

Bracewell, R. (1978). *The Fourier transform and its applications*. McGraw-Hill.

Bray, A. (1990). Tracking objects using image disparities. *J. Image and Vision Computing*, 8, 1, 4–9.

Bregler, C. and Konig, Y. (1994). Eigenlips for robust speech recognition. In *Proc. Int. Conf. Acoustics, Speech, Signal Processing*, 669–672, Adelaide.

Bregler, C. and Omohundro, S. (1994). Surface learning with applications to lipreading. In Cowan, J., Tesauro, G., and Alspector, J., editors, *Advances in Neural Information Processing Systems 6*, 6. Morgan Kaufmann Publishers.

Bregler, C. and Omohundro, S. (1995). Nonlinear manifold learning for visual speech recognition. In *Proc. 5th Int. Conf. on Computer Vision*, 494–499, Boston.

Broida, T. and Chellappa, R. (1986). Estimation of object motion parameters from noisy images. *IEEE Trans. on Pattern Analysis and Machine Intelligence*, 8, 90–99.

Brown, C., Coombs, D., and Soong, J. (1992). Real-time smooth pursuit tracking. In Blake, A. and Yuille, A., editors, *Active Vision*, 123–136. MIT.

Brown, C. and Terzopoulos, D., editors (1994). *Real-time computer vision*. Cambridge.

Bruce, J. and Giblin, P. (1984). *Curves and Singularities*. Cambridge.

Bucy, R. (1969). Bayes theorem and digital realizations for non-linear filters. *J. Astronautical Sciences*, 17, 2, 80–94.

Bulthoff, H., Little, J., and Poggio, T. (1989). A parallel algorithm for real-time computation of optical flow. *Nature*, 337, 9, 549–553.

Burr, D. (1981). Elastic matching of line drawings. *IEEE Trans. on Pattern Analysis and Machine Intelligence*, 3, 6, 708–713.

Burt, P. (1983). Fast algorithms for estimating local image properties. *Computer Vision, Graphics and Image Processing*, 21, 368–382.

Buxton, B. and Buxton, H. (1983). Monocular depth perception from optical flow by space time signal processing. *Proc. Royal Society of London*, B 218, 27–47.

Buxton, B. and Buxton, H. (1984). Computation of optic flow from the motion of edge features in image sequences. *J. Image and Vision Computing*, 2, 59–75.

Campbell, L. and Bobick, A. (1995). Recognition of human body motion using phase space constraints. In *Proc. 5th Int. Conf. on Computer Vision*, 624–630.

Canny, J. (1986). A computational approach to edge detection. *IEEE Trans. on Pattern Analysis and Machine Intelligence*, 8, 6, 679–698.

Cipolla, R. and Blake, A. (1990). The dynamic analysis of apparent contours. In *Proc. 3rd Int. Conf. on Computer Vision*, 616–625.

Cipolla, R. and Blake, A. (1992a). Motion planning using image divergence and deformation. In Blake, A. and Yuille, A., editors, *Active Vision*, 39–58. MIT.

Cipolla, R. and Blake, A. (1992b). Surface shape and the deformation of apparent contours. *Int. J. Computer Vision*, 9, 2, 83–112.

Cohen, L. (1991). On active contour models and balloons. *CVGIP: Image Understanding*, 53, 2, 211–218.

Cootes, T., Hill, A., Taylor, C., and Haslam, J. (1994). The use of active shape models for locating structures in medical images. *J. Image and Vision Computing*, 12, 6, 355–366.

Cootes, T. and Taylor, C. (1992). Active shape models. In *Proc. British Machine Vision Conf.*, 265–275.

Cootes, T., Taylor, C., Cooper, D., and Graham, J. (1995). Active shape models—their training and application. *Computer Vision and Image Understanding*, 61, 1, 38–59.

Cootes, T., Taylor, C., Lanitis, A., Cooper, D., and Graham, J. (1993). Building and using flexible models incorporating grey-level information. In *Proc. 4th Int. Conf. on Computer Vision*, 242–246.

Cox, I. (1993). A review of statistical data association techniques for motion correspondence. *Int. J. Computer Vision*, 10, 1, 53–66.

Craig, J. (1986). *Introduction to robotics: mechanics and control*. Addison-Wesley.

Crisman, J. D. (1992). Color region tracking for vehicle guidance. In Blake, A. and Yuille, A., editors, *Active Vision*, chapter 7. The MIT Press.

Curwen, R. (1993). *Dynamic and Adaptive Contours*. PhD thesis, University of Oxford.

Curwen, R. and Blake, A. (1992). Dynamic contours: real-time active splines. In Blake, A. and Yuille, A., editors, *Active Vision*, 39–58. MIT.

Curwen, R., Blake, A., and Cipolla, R. (1991). Parallel implementation of Lagrangian dynamics for real-time snakes. In *Proc. British Machine Vision Conf.*, 29–36.

Davidson, C. and Blake, A. (1998). Error-tolerant visual planning of planar grasp. In *Proc. 6th Int. Conf. on Computer Vision.*

de Boor, C. (1978). *A practical guide to splines.* Springer-Verlag, New York.

Demey, S., Zisserman, A., and Beardsley, P. (1992). Affine and projective structure from motion. In *Proc. British Machine Vision Conf.*

Dempster, A., Laird, M., and Rubin, D. (1977). Maximum likelihood from incomplete data via the EM algorithm. *J. R. Stat. Soc.*, B, 39, 1–38.

Deriche, R. and Faugeras, O. (1990). Tracking line segments. In Faugeras, O., editor, *Proc. 1st European Conf. Computer Vision*, 259–268. Springer–Verlag.

Dickmanns, E. (1992). Expectation-based dynamic scene understanding. In Blake, A. and Yuille, A., editors, *Active Vision*, 303–336. MIT.

Dickmanns, E. and Graefe, V. (1988a). Applications of dynamic monocular machine vision. *Machine Vision and Applications*, 1, 241–261.

Dickmanns, E. and Graefe, V. (1988b). Dynamic monocular machine vision. *Machine Vision and Applications*, 1, 223–240.

Dodd, B. and Campbell, R. (1987). *Hearing By Eye: The Psychology of Lip Reading.* Erlbaum.

Dreschler, L. and Nagel, H. (1981). Volumetric model and 3D trajectory of a moving car derived from monocular TV-frame sequence of a street scene. In *Proc. Int. Joint Conf. Artificial Intelligence*, 692–697.

Duda, R. and Hart, P. (1973). *Pattern Classification and Scene Analysis.* John Wiley and Sons.

Durrant-Whyte, H. (1988). Uncertain geometry in robotics. *IEEE J. of Robotics and Automation*, 4, 1, 23–31.

Enkelmann, W. (1986). Investigations of multigrid algorithms for the estimation of optical flow fields in image sequences. In *Proc. Conf. Computer Vision and Pattern Recognition*, 81–87.

Essa, I. and Pentland, A. (1995). Facial expression recognition using a dynamic model and motion energy. In *Proc. 5th Int. Conf. on Computer Vision*, 360–367.

Faugeras, O. (1993). *3D Computer Vision*. MIT Press.

Faugeras, O. and Hebert, M. (1986). The representation, recognition, and locating of 3D objects. *Int. J. Robotics Research*, 5, 3, 27–52.

Faux, I. and Pratt, M. (1979). *Computational Geometry for Design and Manufacture*. Ellis-Horwood.

Faverjon, B. and Ponce, J. (1991). On computing two-finger force-closure grasps of curved 2D objects. In *Proc. IEEE Int. Conf. Robotics and Automation*, 3, 424–429.

Ferrier, N., Rowe, S., and Blake, A. (1994). Real-time traffic monitoring. In *Proc. 2nd IEEE Workshop on Applications of Computer Vision*, 81–88. IEEE.

Finn, E. and Montgomery, A. (1988). Automatic optically based recognition of speech. *Pattern Recognition Letters*, 8, 3, 159–164.

Fischler, M. and Bolles, R. (1981). Random sample consensus: a paradigm for model fitting with application to image analysis and automated cartography. *Commun. Assoc. Comp. Mach.*, 24, 381–95.

Fischler, M. and Elschlager, R. (1973). The representation and matching of pictorial structures. *IEEE. Trans. Computers*, C-22, 1.

Foley, J., van Dam, A., Feiner, S., and Hughes, J. (1990). *Computer Graphics: Principles and Practice*, chapter 13, 563–604. Addison-Wesley.

Forsyth, D., Mundy, J., Zisserman, A., and Brown, C. (1990). Projectively invariant representations using implicit algebraic curves. In Faugeras, O., editor, *Proc. 1st European Conf. Computer Vision*, 427–436. Springer–Verlag.

Freeman, W. and Tenenbaum, J. (1997). Learning bilinear models for two-factor problems in vision. In *Proc. Conf. Computer Vision and Pattern Recognition*, 554–560.

Gee, A. and Cipolla, R. (1994). Determining the gaze of faces in images. *J. Image and Vision Computing*, 12, 10, 639–647.

Gee, A. and Cipolla, R. (1996). Fast visual tracking by temporal consensus. *J. Image and Vision Computing*, 14, 2, 105–114.

Gelb, A., editor (1974). *Applied Optimal Estimation*. MIT Press, Cambridge, MA.

Gelfand, A. and Smith, A. (1990). Sampling-based approaches to computing marginal densities. *J. Am. Statistical Assoc.*, 85, 410, 398–409.

Geman, S. and Geman, D. (1984). Stochastic relaxation, Gibbs distributions, and the Bayesian restoration of images. *IEEE Trans. on Pattern Analysis and Machine Intelligence*, 6, 6, 721–741.

Gennery, D. (1992). Visual tracking of known three-dimensional objects. *Int. J. Computer Vision*, 7, 3, 243–270.

Giblin, P. and Weiss, R. (1987). Reconstruction of surfaces from profiles. In *Proc. 1st Int. Conf. on Computer Vision*, 136–144, London.

Golub, G. and van Loan, C. (1989). *Matrix computations*. Johns Hopkins.

Goncalves, L., di Bernardo, E., Ursella, E., and Perona, P. (1995). Monocular tracking of the human arm in 3d. In *Proc. 5th Int. Conf. on Computer Vision*, 764–770.

Gonzales, R. and Wintz, P. (1987). *Digital Image Processing*. Addison-Wesley.

Goodwin, C. and Sin, K. (1984). *Adaptive filtering prediction and control*. Prentice-Hall.

Gordon, N., Salmond, D., and Smith, A. (1993). Novel approach to nonlinear/non-Gaussian Bayesian state estimation. *IEE Proc. F*, 140, 2, 107–113.

Grenander, U. (1976–1981). *Lectures in Pattern Theory I, II and III*. Springer.

Grenander, U., Chow, Y., and Keenan, D. (1991). *HANDS. A Pattern Theoretical Study of Biological Shapes*. Springer-Verlag. New York.

Grimson, W. (1981). *From Images to Surfaces*. MIT Press, Cambridge, USA.

Grimson, W. (1985). Computational experiments with a feature based stereo algorithm. *IEEE Trans. on Pattern Analysis and Machine Intelligence*, 7, 1, 17–34.

Grimson, W. (1990). *Object recognition by computer*. MIT Press.

Grimson, W. and Lozano-Perez, T. (1984). Model-based recognition and localization from sparse range or tactile data. *Int. J. Robotics Research*, 5, 3, 3–34.

Grimson, W., Lozano-Perez, T., Wells, W., Ettinger, G., White, S., and Kikinis, R. (1994). An automatic registration method for frameless stereotaxy, image guided surgery, and enhanced reality visualization. In *Proc. Conf. Computer Vision and Pattern Recognition*, 430–436, Seattle, WA.

Hager, G. (1990). *Sensor fusion and planning: a computational approach*. Kluwer Academic Publishers.

Hager, G. and Belhumeur, P. (1996). Real-time tracking of image regions with changes in geometry and illumination. In *Proc. IEEE Conf. Computer Vision and Pattern Recognition*, 403–410.

Hallam, J. (1983). Resolving observer motion by object tracking. In *Proc. Int. Joint Conf. Artificial Intelligence*, 2, 792–798.

Hampel, F., Ronchetti, E., Rousseeuw, P., and Stahel, W. (1995). *Robust Statistics*. John Wiley and Sons, New York.

Haralick, R. (1980). Edge and region analysis for digital image data. *Computer Graphics and Image Processing*, 12, 1, 60–73.

Harris, C. (1990). Structure from motion under orthographic projection. In Faugeras, O., editor, *Proc. 1st European Conf. Computer Vision*, 118–123. Springer–Verlag.

Harris, C. (1992a). Geometry from visual motion. In Blake, A. and Yuille, A., editors, *Active Vision*, 263–284. MIT.

Harris, C. (1992b). Tracking with rigid models. In Blake, A. and Yuille, A., editors, *Active Vision*, 59–74. MIT.

Harris, C. and Stennett, C. (1990). Rapid – a video-rate object tracker. In *Proc. British Machine Vision Conf.*, 73–78.

Heeger, D. (1987). Optical flow from spatiotemporal filters. In *Proc. 1st Int. Conf. on Computer Vision*, 181–190.

Heuring, J. and Murray, D. (1996). Visual head tracking and slaving for visual telepresence. In *Proc. IEEE Int Conf. Robotics and Automation, Minneapolis, May 1996*, Los Alamitos, CA. IEEE Computer Society Press.

Hildreth, E. (1983). *The Measurement of Visual Motion*. MIT Press.

Hinton, G., Williams, C., and Revow, M. (1992). Adaptive elastic models for hand-printed character recognition. *Advances in Neural Information Processing Systems*, 4.

Hogg, D. (1983). Model-based vision: a program to see a walking person. *J. Image and Vision Computing*, 1, 1, 5–20.

Horn, B. (1986). *Robot Vision*. McGraw-Hill, NY.

Horn, B. and Schunk, B. (1981). Determining optical flow. *J. Artificial Intelligence*, 17, 185–203.

Huang, X., Arika, Y., and Jack, M. (1990). *Hidden Markov Models for Speech Recognition*. Edinburgh University Press.

Huttenlocher, D., Noh, J., and Rucklidge, W. (1993). Tracking non-rigid objects in complex scenes. In *Proc. 4th Int. Conf. on Computer Vision*, 93–101.

Ikeuchi, K. and Horn, B. (1981). Numerical shape from shading and occluding boundaries. *J. Artificial Intelligence*, 17, 141–184.

Inoue, H. and Mizoguchi, H. (1985). A flexible multi window vision system for robots. In *Proc. 2nd Int. Symp. on Robotics Research*, 95–102.

Isard, M. and Blake, A. (1998a). Condensation — conditional density propagation for visual tracking. *Int. J. Computer Vision*.

Isard, M. and Blake, A. (1998b). A mixed-state Condensation tracker with automatic model switching. In *Proc. 6th Int. Conf. on Computer Vision*.

Ivins, J. and Porrill, J. (1995). Active region models for segmenting textures and colours. *J. Image and Vision Computing*, 13, 5, 431–438.

Jacobs, O. (1993). *Introduction to control theory*. Oxford University Press.

Jain, A. and Farrokhnia, F. (1991). Unsupervised texture segmentation using gabor filters. *Pattern Recognition*, 24, 12, 1167–1186.

Kass, M., Witkin, A., and Terzopoulos, D. (1987). Snakes: Active contour models. In *Proc. 1st Int. Conf. on Computer Vision*, 259–268.

Kaucic, R., Dalton, B., and Blake, A. (1996). Real-time liptracking for audio-visual speech recognition applications. In *Proc. 4th European Conf. Computer Vision*, 376–387, Cambridge, England.

Kendall, M. and Stuart, A. (1979). *The advanced theory of statistics, vol 2, inference and relationship*. Charles Griffing and Co Ltd, London.

Kitagawa, G. (1996). Monte Carlo filter and smoother for non-Gaussian nonlinear state space models. *Journal of Computational and Graphical Statistics*, 5, 1, 1–25.

Kitchen, L. and Rosenfeld, A. (1982). Grey-level corner detection. *Pattern Recognition Letters*, 1, 95–102.

Koenderink, J. (1990). *Solid Shape*. MIT Press.

Koenderink, J. and van Doorn, A. (1991). Affine structure from motion. *J. Optical Soc. of America A.*, 8, 2, 337–385.

Koller, D., Weber, J., and Malik, J. (1994). Robust multiple car tracking with occlusion reasoning. In *Proc. 3rd European Conf. Computer Vision*, 189 – 196. Springer-Verlag.

Kreysig, E. (1988). *Advanced Engineering Mathematics*. Wiley.

Kutulakos, K. and Valliano, J. (1996). Non-euclidean object representations for calibration-free video overlay. In Ponce, J., Zisserman, A., and Hebert, M., editors, *Object Representation in Computer Vision II*, Lecture Notes in Computer Science, 381–401. Springer-Verlag.

Landau, L. and Lifshitz, E. (1972). *Mechanics and Electrodynamics*, 1 of *A Shorter Course of Theoretical Physics*. Pergamon Press.

Lanitis, A., Taylor, C., and Cootes, T. (1995). A unified approach to coding and interpreting face images. In *Proc. 5th Int. Conf. on Computer Vision*, 368–373.

Lawn, J. and Cipolla, R. (1994). Robust egomotion estimation from affine motion parallax. In *Proc. 3rd European Conf. Computer Vision*, 205–210.

Ljung, L. (1987). *System identification: theory for the user*. Prentice-Hall.

Lowe, D. (1991). Fitting parameterised 3D models to images. *IEEE Trans. on Pattern Analysis and Machine Intelligence*, 13, 5, 441–450.

Lowe, D. (1992). Robust model-based motion tracking through the integration of search and estimation. *Int. J. Computer Vision*, 8, 2, 113–122.

Lucas, B. and Kanade, T. (1981). An iterative image registration technique with an application to stereo vision. In *Proc. Int. Joint Conf. Artificial Intelligence*, 674–679.

MacCormick, J. and Blake, A. (1998). A probabilistic contour discriminant for object localisation. In *Proc. 6th Int. Conf. on Computer Vision*.

Marais, P., Guillemaud, R., Sakuma, M., Zisserman, A., and Brady, J. (1996). Visualising cerebral asymmetry. In *Visualisation in Biomedical Computing*, Lecture notes in Computer Science, 411–416. Springer-Verlag.

Marchant, J. (1991). Intelligent machinery for agriculture. *Machine Vision and Applications*, 151, 177–186.

Mardia, K., Ghali, N., Howes, M., Hainsworth, T., and Sheehy, N. (1993). Techniques for online gesture recognition. *J. Image and Vision Computing*, 11, 5, 283–294.

Marr, D. (1982). *Vision*. Freeman, San Francisco.

Marr, D. and Hildreth, E. (1980). Theory of edge detection. *Proc. Roy. Soc. London. B.*, 207, 187–217.

Matthies, L., Kanade, T., and Szeliski, R. (1989). Kalman filter-based algorithms for estimating depth from image sequences. *Int. J. Computer Vision*, 3, 209–236.

Mayhew, J., Zheng, Y., and Cornell, S. (1992). The adaptive control of a four-degrees-of-freedom stereo camera head. *Phil. Trans. R. Soc. Lond. B*, 337, 5, 315–326.

Medioni, G. and Yasumoto, Y. (1986). Corner detection and curve representation using curve b-splines. In *Proc. Conf. Computer Vision and Pattern Recognition*, 764–769.

Menet, S., Saint-Marc, P., and Medioni, G. (1990). B-snakes: Implementation and application to stereo. In *Proceedings DARPA*, 720–726.

Montanari, U. (1971). On the optimal detection of curves in noisy pictures. *Commun. ACM*, 14, 5, 335–345.

Mumford, D. (1996). Pattern theory: a unifying perspective. In Knill, D. and Richard, W., editors, *Perception as Bayesian inference*, 25–62. Cambridge University Press.

Mundy, J. and Heller, A. (1990). The evolution and testing of a model-based object recognition system. In *Proc. 3rd Int. Conf. on Computer Vision*, 268–282.

Mundy, J. and Zisserman, A. (1992). *Geometric invariance in computer vision*. MIT Press.

Murase, H. and Nayar, S. (1995). Visual learning and recognition of 3D objects from appearance. *Int. J. Computer Vision*, 14, 1, 5–24.

Murray, D. and Basu, A. (1994). Motion tracking with an active camera. *IEEE Trans. on Pattern Analysis and Machine Intelligence*, 16, 5, 449–459.

Murray, D. and Buxton, B. (1990). *Experiments in the machine interpretation of visual motion*. MIT Press.

Murray, D., Du, F., McLauchlan, P., Reid, I., Sharkey, P., and Brady, J. (1992). Design of stereo heads. In Blake, A. and Yuille, A., editors, *Active Vision*, 303–336. MIT.

Murray, D., McLauchlan, P., Reid, I., and Sharkey, P. (1993). Reactions to peripheral image motion using a head/eye platform. In *Proc. 4th Int. Conf. on Computer Vision*, 403–411.

Nagel, H. (1983). Displacement vectors derived from second-order intensity variations in image sequences. *Computer Vision, Graphics and Image Processing*, 21, 85–117.

Nagel, H. and Enkelmann, W. (1986). An investigation of smoothness constraints for the estimation of displacement vector fields from image sequences. *IEEE Trans. on Pattern Analysis and Machine Intelligence*, 8, 5, 565–593.

Noble, J. (1988). Finding corners. *J. Image and Vision Computing*, 6, 2, 121–128.

North, B. and Blake, A. (1998). Using expectation-maximisation to learn a dynamical model for a tracker from measurement sequences. In *Proc. 6th Int. Conf. on Computer Vision*.

O'Gorman, F. (1978). Edge detection using Walsh functions. *J. Artificial Intelligence*, 10, 215–223.

Ohta, Y. and Kanade, T. (1985). Stereo by intra- and inter-scan line search using dynamic programming. *IEEE Trans. on Pattern Analysis and Machine Intelligence*, 7, 2, 139–154.

Papanikolopoulos, N., Khosla, P., and Kanade, T. (1991). Vision and control techniques for robotic visual tracking. In *Proc. IEEE Int. Conf. Robotics and Automation*, 1, 851–856.

Papoulis, A. (1990). *Probability and Statistics*. Prentice-Hall.

Papoulis, A. (1991). *Probability, random variables and stochastic processes*. McGraw-Hall.

Pardey, J., Roberts, S., and Tarassenko, L. (1995). A review of parametric modelling techniques for EEG analysis. *Medical Engineering Physics*, 18, 1, 2–11.

Pelizzari, C., Tan, K., Levin, D., Chen, G., and Balter, J. (1993). Interactive 3d patient — image registration. In *Proc. 13th Int. Conf. on Information Processing in Medical Imaging*, 132–141, Berlin, Germany.

Pentland, A. and Horowitz, B. (1991). Recovery of nonrigid motion and structure. *IEEE Trans. on Pattern Analysis and Machine Intelligence*, 7, 730–742.

Perona, P. and Malik, J. (1990). Scale-space and edge detection using anisotropic diffusion. *IEEE Trans. on Pattern Analysis and Machine Intelligence*, 12, 7, 629–639.

Petajan, E., Bischofy, B., Bodoff, D., and Brooke, N. (1988). An improved automatic lipreading system to enhance speech recognition. In Soloway, E., Frye, D., and Sheppard, S., editors, *Proc. Human Factors in Computing Systems*, 19–25. ACM.

Petitjean, S., Ponce, J., and Kriegman, D. (1992). Computing exact aspect graphs of curved objects: Algebraic surfaces. *Int. J. Computer Vision*, 9, 231–255.

Plá, F., Juste, F., Ferri, F., and Vicens, M. (1993). Colour segmentation based on a light reflection model to locate citrus fruits for robotic harvesting. *Computers and Electronics in Agriculture*, 9, 1, 53–70.

Poggio, T., Torre, V., and Koch, C. (1985). Computational vision and regularisation theory. *Nature*, 317, 314–319.

Pollard, S., Mayhew, J., and Frisby, J. (1985). PMF:a stereo correspondence algorithm using a disparity gradient. *Perception*, 14, 449–470.

Ponce, J., Burdick, J., and Rimon, E. (1995). Computing the immobilizing three-finger grasps of planar objects. In *Proc. of the 1995 Workshop on Computational Kinematics*, 281–300.

Press, W., Teukolsky, S., Vetterling, W., and Flannery, B. (1988). *Numerical Recipes in C*. Cambridge University Press.

Rabiner, L. and Bing-Hwang, J. (1993). *Fundamentals of speech recognition*. Prentice-Hall.

Ramer, E. (1975). The transformation of photographic images into stroke arrays. *IEE Trans. CAS*, 22,3, 363–374.

Rao, B., Durrant-Whyte, H., and Sheen, J. (1993). A fully decentralized multi-sensor system for tracking and surveillance. *Int. J. Robotics Research*, 12, 1, 20–44.

Rao, C. (1973). *Linear Statistical Inference and Its Applications*. John Wiley and Sons, New York.

Rehg, J. and Kanade, T. (1994). Visual tracking of high dof articulated structures: an application to human hand tracking. In Eklundh, J.-O., editor, *Proc. 3rd European Conf. Computer Vision*, 35–46. Springer–Verlag.

Reid, I. and Murray, D. (1993). Tracking foveated corner clusters using affine structure. In *Proc. 4th Int. Conf. on Computer Vision*, 76–83.

Reid, I. and Murray, D. (1996). Active tracking of foveated feature clusters using affine structure. *Int. J. Computer Vision*, 18, 1, 41–60.

Reynard, D., Wildenberg, A., Blake, A., and Marchant, J. (1996). Learning dynamics of complex motions from image sequences. In *Proc. 4th European Conf. Computer Vision*, 357–368, Cambridge, England.

Rimon, E. and Blake, A. (1996). Caging 2D bodies by one-parameter, two-fingered gripping systems. In *Proc. IEEE Int. Conf. Robotics and Automation*, 1458–1464. IEEE Press.

Rimon, E. and Burdick, J. (1995a). A configuration space analysis of bodies in contact - I. 1st order mobility. *Mech. Mach. Theory*, 30, 6, 897–912.

Rimon, E. and Burdick, J. (1995b). A configuration space analysis of bodies in contact - II. 2nd order mobility. *Mech. Mach. Theory*, 30, 6, 913–928.

Ripley, B. (1987). *Stochastic simulation*. New York: Wiley.

Ripley, B. (1996). *Pattern recognition and neural networks*. Cambridge.

Ripley, B. and Sutherland, A. (1990). Finding spiral structures in images of galaxies. *Phil. Trans. R. Soc. Lond. A.*, 332, 1627, 477–485.

Roberts, L. (1965). Machine perception of three - dimensional solids. In Tippet, J., editor, *Optical and Electro-optical Information Processing*. MIT Press.

Rowe, S. and Blake, A. (1996a). Statistical feature modelling for active contours. In *Proc. 4th European Conf. Computer Vision*, 560–569, Cambridge, England.

Rowe, S. and Blake, A. (1996b). Statistical mosaics for tracking. *J. Image and Vision Computing*, 14, 549–564.

Scott, G. (1987). The alternative snake – and other animals. In *Proc. 3rd Alvey Vision Conference*, 341–347.

Semple, J. and Kneebone, G. (1952). *Algebraic projective geometry*. Oxford University Press.

Shapiro, L., Zisserman, A., and Brady, J. (1995). 3D motion recovery via affine epipolar geometry. *Int. J. Computer Vision*, 16, 2, 147–182.

Smith, A. (1996). Blue screen matting. In *Proc. Siggraph*, 259–268. ACM.

Sorenson, H. and Alspach, D. (1971). Recursive Bayesian estimation using Gaussian sums. *Automatica*, 7, 465–479.

State, A., Hirota, G., Chen, D., Garrett, B., and Livingston, M. (1996). Superior augmented reality registration by integrating landmark tracking and magnetic tracking. In Rushmeier, H., editor, *Proc. Siggraph*, Annual Conference Series, 429–438. ACM SIGGRAPH, Addison Wesley. held in New Orleans, Louisiana, 04-09 August 1996.

Stork, D., Wolff, G., and Levine, E. (1992). Neural network lipreading system for improved speech recognition. In *Proceedings International Joint Conference on Neural Networks*, 2, 289–295.

Storvik, G. (1994). A Bayesian approach to dynamic contours through stochastic sampling and simulated annealing. *IEEE Trans. on Pattern Analysis and Machine Intelligence*, 16, 10, 976–986.

Strang, G. (1986). *Linear algebra and its applications*. Harcourt Brace.

Strang, G. and Fix, G. (1973). *An analysis of the finite element method*. Prentice-Hall, Englewood Cliffs, USA.

Sullivan, G. (1992). Visual interpretation of known objects in constrained scenes. *Phil. Trans. R. Soc. Lond. B.*, 337, 361–370.

Taylor, M., Blake, A., and Cox, A. (1994). Visually guided grasping in 3D. In *Proc. IEEE Int. Conf. Robotics and Automation*, 761–766.

Terzopoulos, D. (1986). Regularisation of inverse problems involving discontinuities. *IEEE Trans. on Pattern Analysis and Machine Intelligence*, 8, 4, 413–424.

Terzopoulos, D. and Fleischer, K. (1988). Deformable models. *The Visual Computer*, 4, 306–331.

Terzopoulos, D. and Metaxas, D. (1991). Dynamic 3D models with local and global deformations: deformable superquadrics. *IEEE Trans. on Pattern Analysis and Machine Intelligence*, 13, 7.

Terzopoulos, D. and Szeliski, R. (1992). Tracking with Kalman snakes. In Blake, A. and Yuille, A., editors, *Active Vision*, 3–20. MIT.

Terzopoulos, D. and Waters, K. (1990). Analysis of facial images using physical and anatomical models. In *Proc. 3rd Int. Conf. on Computer Vision*, 727–732.

Terzopoulos, D. and Waters, K. (1993). Analysis and synthesis of facial image sequences using physical and anatomical models. *IEEE Trans. on Pattern Analysis and Machine Intelligence*, 15, 6, 569–579.

Terzopoulos, D., Witkin, A., and Kass, M. (1988). Constraints on deformable models: recovering 3D shape and nonrigid motion. *J. Artificial Intelligence*, 36, 91–123.

Thompson, D. and Mundy, J. (1987). Three-dimensional model matching from an unconstrained viewpoint. In *Proc. IEEE Int. Conf. Robotics and Automation*.

Tomasi, C. and Kanade, T. (1991). Shape and motion from image streams: a factorization method. *Int. J. Computer Vision*, 9, 2, 137–154.

Tsai, R. (1987). A versatile camera calibration technique for high-accuracy 3D machine vision metrology using off-the-shelf TV cameras and lenses. *IEEE J. of Robotics and Automation*, 3, 4, 323–344.

Turk, M. and Pentland, A. (1991). Eigenfaces for recognition. *J. Cognitive Neuroscience*, 3, 1.

Uenohara, M. and Kanade, T. (1995). Vision-based object registration for real-time image overlay. In Ayache, N., editor, *Computer Vision, Virtual Reality and Robotics in Medicine*, Lecture Notes in Computer Science. Springer-Verlag. ISBN 3-540-59120-6.

Ullman, S. and Basri, R. (1991). Recognition by linear combinations of models. *IEEE Trans. on Pattern Analysis and Machine Intelligence*, 13, 10, 992–1006.

Vaillant, R. (1990). Using occluding contours for 3D object modelling. In Faugeras, O., editor, *Proc. 1st European Conf. Computer Vision*, 454–464. Springer–Verlag.

Vetter, T. and Poggio, T. (1996). Image synthesis from a single example image. In *Proc. 4th European Conf. Computer Vision*, 652–659, Cambridge, England.

Waite, J. and Welsh, W. (1990). Head boundary location using snakes. *British Telecom Tech. J.*, 8, 3.

Waxman, A. and Wohn, S. (1985). Contour evolution, neighbourhood deformation, and global image flow: planar surfaces in motion. *Int. J. Robotics Research*, 4, 95–108.

Wellner, P. (1993). Interacting with paper on the Digital Desk. *Communications of the ACM*, 36, 7.

Wildenberg, A. (1997). *Learning and Initialisation for Visual Tracking*. PhD thesis, University of Oxford.

Williams, L. (1990). Performance-driven facial animation. In *Proc. Siggraph*, 235–242. ACM.

Witkin, A., Terzopoulos, D., and Kass, M. (1986). Signal matching through scale space. In *5th National Conference on AI*.

Wloka, M. and Anderson, B. (1995). Resolving occlusion in augmented reality. In *Proc. Symposium on Interactive 3D graphics*, 5–12.

Yuille, A. (1990). Generalized deformable models, statistical physics, and matching problems. *Neural Computation*, 2, 1–24.

Yuille, A., Cohen, D., and Hallinan, P. (1989). Feature extraction from faces using deformable templates. *Proc. Conf. Computer Vision and Pattern Recognition*, 104–109.

Yuille, A. and Grzywacz, N. (1988). A computational theory for the perception of coherent visual motion. *Nature*, 333, 6168, 71–74.

Yuille, A. and Hallinan, P. (1992). Deformable templates. In Blake, A. and Yuille, A., editors, *Active Vision*, 20–38. MIT.

Zinkiewicz, O. and Morgan, K. (1983). *Finite elements and approximation*. Wiley, New York.

Zucker, S., Hummel, R., and Rosenfeld, A. (1977). An application of relaxation labelling to line and curve enhancement. *IEEE trans. comp.*, 26, 4, 394–403.

Zuniga, O. and Haralick, R. (1983). Corner detection using the facet model. In *Proc. Conf. Computer Vision and Pattern Recognition*, 30–37.

Author Index

Index

Italic entries denote references in the Bibliographic notes.